ROOTED RESISTANCE

OTHER BOOKS IN THIS SERIES

*A Rich and Tantalizing Brew:
A History of How Coffee Connected the World*

*To Feast on Us as Their Prey:
Cannibalism and the Early Modern Atlantic*

*Inventing Authenticity:
How Cookbook Writers Redefine Southern Identity*

*Forging Communities: Food and Representation
in Medieval and Early Modern Southwestern Europe*

*Aunt Sammy's Radio Recipes:
The Original 1927 Cookbook and Housekeeper's Chat*

*Chop Suey and Sushi from Sea to Shining Sea:
Chinese and Japanese Restaurants in the United States*

*Mexican-Origin Foods, Foodways, and Social Movements:
Decolonial Perspectives*

Meanings of Maple: An Ethnography of Sugaring

*The Taste of Art: Food, Cooking,
and Counterculture in Contemporary Practices*

*Devouring Cultures: Perspectives on Food, Power, and Identity
from the Zombie Apocalypse to Downton Abbey*

*Latin@s' Presence in the Food Industry:
Changing How We Think about Food*

*Dethroning the Deceitful Pork Chop: Rethinking African
American Foodways From Slavery to Obama*

American Appetites: A Documentary Reader

Rooted Resistance

AGRARIAN MYTH IN MODERN AMERICA

ROSS SINGER

STEPHANIE HOUSTON GREY

JEFF MOTTER

The University of Arkansas Press
Fayetteville
2020

Copyright © 2020 by The University of Arkansas Press

All rights reserved
Manufactured in the United States of America

ISBN: 978-1-68226-137-8
eISBN: 978-1-61075-725-6

24 23 22 21 20   5 4 3 2 1

Library of Congress Cataloging-in-Publication Data

Names: Singer, Ross, 1980– author. | Grey, Stephanie Houston, 1965– author. | Motter, Jeff, 1977– author.
Title: Rooted resistance : agrarian myth in modern America / Ross Singer, Stephanie Houston Grey, Jeff Motter.
Other titles: Food and foodways (Fayetteville, Ark.)
Description: Fayetteville : The University of Arkansas Press, 2020. | Series: Food and foodways | Includes bibliographical references and index. | Summary: "This monograph uses a series of case studies to explore the agrarian myth in modern America and its intersection with movements for food system change, including the organic and fair trade movements in the United States. The authors suggest that agrarian myth continues to provide a rhetorical pathway for positive systemic change in the food system"— Provided by publisher.
Identifiers: LCCN 2019053168 (print) | LCCN 2019053169 (ebook) | ISBN 9781682261378 (cloth) | ISBN 9781610757256 (ebook)
Subjects: LCSH: Organic farming"—United States"—Case studies. | Agriculture"—Social aspects"—United States"—Case studies. | Food"—Social aspects"—United States"—Case studies. | Anti-globalization movement"—United States"—Case studies. | Agricultural systems"—United States"—Case studies.
Classification: LCC S605.5 .S565 2020 (print) | LCC S605.5 (ebook) | DDC 631.5/84"—dc23
LC record available at https://lccn.loc.gov/2019053168
LC ebook record available at https://lccn.loc.gov/2019053169

♾ The paper used in this publication meets the minimum requirements of the American National Standard for Permanence of Paper for Printed Library Materials Z39.48–1984.

CONTENTS

Series Editors' Preface   vii

Introduction:
Mythic Rhetoric Imagines the American Garden   3

I ▪ Seeds of Resistance

1. The Home Front Plants Agrarian Munitions   37
2. Country Life Defends Yeoman Democracy   65
3. The Southern Agrarians Take Their Stand   89
4. Rodale's Jeremiad Inspires the Organic Movement   109

II ▪ Threatened Harvests

5. The South Central Farmers Cultivate a Precarious Community   137
6. Chipotle Brands Agrarian Innocence   163
7. RAM Mechanizes God's Farmer   189

Conclusion:
Agrarians Greet the Apocalypse   221

Notes   239

Bibliography   263

Index   287

SERIES EDITORS' PREFACE

The University of Arkansas Press series on Food and Foodways series explores historical and contemporary issues in global food studies. We are committed to representing a diverse set of voices that tell lesser known food stories and to provoking new avenues of interdisciplinary research. Our strengths are works in the humanities and social sciences that use food as a critical lens to examine broader cultural, environmental, and ethical issues.

Feeding ourselves has long entangled human beings within complicated moral puzzles of social injustice and environmental destruction. When we eat, we consume not only food on the plate, but also the lives and labors of innumerable plants, animals, and people. This process distributes its costs unevenly across race, class, gender, and other social categories. The production and distribution of food often obscures these material and cultural connections, impeding honest assessments of our impact on the world around us. By taking these relationships seriously, Food and Foodways provides a new series of critical studies that analyze the cultural and environmental relationships that have sustained human societies.

Rooted Resistance offers a sophisticated analysis of the "agrarian myth" in modern American food and foodways. By drawing on a variety examples that range from the wartime politics of home-front gardening to the rhetoric of pickup truck advertisements, Ross Singer, Stephanie Houston Grey, and Jeff Motter reveal how diverse communities of Americans in the twentieth and twenty-first centuries have engaged, transformed, and deployed presumptions of the virtuousness of farms and farmers in their heterogenous efforts to enact agrarian visions of the future through nostalgic allusions to the past. In doing so, *Rooted Resistance* demonstrates how agrarian rhetoric emerged as one of the lowest common denominators of normative discourse in American culture, even, at times, spanning divisions of race, class, and gender as an appealing mode of communication used to challenge notions of selfhood, value, and belonging in modern American society.

At once critical and sympathetic, this book provides perspective on significant touchstones of American agrarianism that have too often been glossed. The authors' discussion of the history and historical memory of the famed South Central Farmers community garden of Los Angeles, for instance, underscores the importance of agrarianism as a rhetoric of class and race-based resistance to capitalism as well as a potent legal vision undergirding private property. Likewise, the authors provide fresh insights into the stakes held by agrarian myth in American culture and politics over the last century through uncompromising yet compassionate treatments of broadly recognized figures of modern American agrarianism like Wendell Berry, Jerome I. Rodale, and the self-styled Southern Agrarians.

By incorporating the analytical sensibilities of rhetorical analysis along with cultural geography, intellectual history and historiography, communication, and critical theory, *Rooted Resistance* contributes far more than just another empirical set of case studies. It offers a model of interdisciplinary work that will thrill popular and scholarly readers alike.

JENNIFER JENSEN WALLACH and MICHAEL WISE, *Series Editors*

ROOTED RESISTANCE

INTRODUCTION

Mythic Rhetoric Imagines the American Garden

Americans are increasingly conscious of the food and agriculture system's multiple, impactful effects on their lives, as well as the world around them. As evidenced by the proliferation of farmers' markets, urban community gardens, grocery cooperatives, community-supported agriculture programs, and farm-to-table restaurants, we are experiencing an "agrarian turn" focused on the ethics of food production and consumption.[1] Today's growing public conversation about food ethics is rife with anxiety over the decline of "real food" as a consequence of industrial agriculture. At unprecedented speed and scale, industrial agriculture supplies a predominantly urban consumer society with foods designed foremost for convenience and low customer cost. While the noted anxiety about highly processed industrial food often centers on the diminished quality or nutritional value of food, it is also due to other issues: food safety, animal abuse, labor exploitation, rural depopulation and decay, and a host of environmental problems to which industrial food is a leading contributor—most notably, climate change.[2] Reframing individual consumer taste and food choice as intricately intertwined with collective, societal issues such as social justice, public wellness, and environmental sustainability exposes the realization that without deep change in the food and agricultural system, society itself is at stake. A new agrarianism rooted in local production on manageable scales may, therefore, be crucial to staving off disaster.

This book responds to this anxiety and hope with a historical and rhetorical perspective on agrarianism, specifically the endurance and evolution of American agrarian myth. As scholars of rhetoric working in the academic discipline of communication in the United States, we approach the new agrarianism and today's food movement as a network of discourses directed toward social change through altered relations between food, people, and nature. However, the agrarianism that we explicate remains deeply tied to a past vision of connectedness to community and land. This vision predates industrial agriculture, the petroleum economy, and consumerist culture. Despite the violence and exploitation that has sometimes characterized rural life, the ethos of transformation offered by living closer to the land remains an evocative presence in the American cultural imagination. As we will explain with historical detail, agrarianism's social identity is linked to a democratic vision, is imbued with spiritual meanings, and is flexibly adaptable to a multitude of circumstances. At root, this identity is discursive and imaginary, emanating from a foundational myth of exceptional American character. This myth suggests that Americans derive defining moral and political virtues from farming as a way of life.

We examine this American agrarian myth in its "modern" manifestations, with emphasis on its forms, functions, opportunities, and constraints as a rhetoric of social resistance. No longer a dominant rhetorical form, agrarian myth takes on new qualities as a latent vision of bedrock Americanism that retains a cultural resonance and power to be tapped by a variety of social actors, for diverse purposes. By the twentieth century, agrarian myth had become a richly ambiguous form that lent its commonplaces, drawn from the nation's formation, to narratives of food, family, community, and democracy. In our time, agrarian myth remains a discourse of authenticity used to legitimate diverse phenomena, ranging from political campaign platforms to consumer products. While the cases in this book trace some meaningful "moments" defined by the practice of agrarian mythmaking in modern America, their unique insights all support a general conclusion: agrarian myth is powerful, even "bankable," and has retained a remarkably widespread appeal across today's urban majority. Resistant agrarianism may entail reclaiming meanings that seem buried or forgotten. Yet, as we demonstrate in this book, agrarian discourse's versatility, perpetual resonance, and mythic qualities allow it to continue to impact contemporary culture and politics. In these times of profound divisions in society—perhaps especially in such times— agrarian myth and memory provide a renewed sense of stability and meaning, a conjuring of a shared destiny emanating from the land and our relation to it. Simultaneously, agrarian myth's potential for empowerment may be stifled as some social actors appropriate it superficially, as a shield of public defense for their own self-interested, unethical agendas.

As other scholars have already argued, fostering "agricultural literacy"—a capacity to critically engage with and act upon messages about food and agriculture—is crucial for informed citizenship and the public interest.[3] This is one of the primary goals of this book.

With the intended functions of knowledge building, critique, and advocacy, the chapters that follow closely adhere to the premise that resistance is never free of elements of domination and is often assimilated and absorbed by hegemonic systems. Critics of new agrarianism have rightfully worried that it may be another manifestation of exclusionary privilege that benefits an affluent and White suburban population. This population has the disposable income to purchase pricey organic produce or to eat at a trendy farm-to-table restaurant. This charge has validity but also overlooks new agrarianism's myriad connections to allied movements such as those for food justice and food sovereignty. These movements include rural as well as urban farmers, activists, and nonprofit organizations, as well as people from diverse racial, ethnic, and economic groups. Internationally, these movements promote fair-trade standards and practices for agriculture. Organizations and advocates have fought for higher pay and better conditions for migrant farmworkers. Local activists have sought to redress limited access to affordable, nutritious, and fresh foods in urban places.[4] As has often been said, the viability of just food as an idea depends not only on the commitment of farmers or the fervor of advocates, but on changes in governmental policies, as well as shifts in patterns of consumption. However, unleashing the potential of just and sustainable agriculture to thrive in diverse communities also depends upon the retelling of agrarian narratives.

In this book, we return to new agrarianism's unfulfilled democratic promise as a food- and agriculture-based politics of respectful, sustainable, and just connections with all human and more-than-human life. While most people will not become farmers, reviving the idea of Americans and humans as agricultural beings is a worthy and workable goal. To put it bluntly, we will not arrest climate change or end social, economic, or environmental injustice without transforming our practices of living and our definitions of the human. Further, we suggest that the transformative potential of agrarianism is that it may help to bridge rifts and bring a growing body of Americans, and others, into its fold. This may seem surprising: after all, amid widespread attention to agriculture and food, there remains a large body of consumers who may care little about where food comes from or how it is produced, or who may consider the fruits of new food movements a performance of affluent taste that is impractical, idealistic, and expensive. Even a public increasingly concerned with healthy eating does not always recognize the degree to which the industrial food system inflicts various costs, financial and otherwise, on their bodies, society, and the planet. Additionally, following the discourse of powerful industry interest groups, skeptics warn against limiting consumer choices and call instead for better health and nutrition education, and informed parenting to instill personal responsibility.[5]

Despite ambivalence and countervailing arguments, people with widely divergent ideas on other subjects often hold values that align with agrarianism. Skepticism about new food movements does not preclude "traditionalists" from valuing the connections among food, community, and place. As the cases in this book suggest time and again, agrarian ideas that connect working the land to the practice of democracy endure. These ideas retain resonant rhetorical power even for populations that may seem distant from the farm. In a modern world that many experience as artificial, fragmented, and unfulfilling, agrarianism exudes authenticity as a life deeply bound to land, water, crops, and community. Even though agrarianism is no longer the basis for American life, and has not been for many decades, it returns not only through traditional practices but in the resonant power of myth and memory.

New-agrarianism scholar William H. Major notes that despite growing anxiety about what we are putting in our bodies and doing to the planet, a culture of convenience continues to obscure the realities of food production, "relegating the general embeddedness of agriculture to the realm of myth—when it is thought of at all."[6] As we suggest in this book, however, the realm of American agrarian myth provides a potent repository of cultural imagery drawn on by many social actors. Past scholarship has shown that, ironically, those at the forefront of the food movement and those unaware or uninterested in it share an enduring idealistic belief in the virtue of the mythic figure of the farmer and farming as a way of life.[7] *Rooted Resistance* examines this reservoir of myth as a source of trusted social truth and persuasive commonplaces. As we trace it here, the new agrarianism is an ethic of resistance intertwined with this myth, that also contains the seeds of its own undoing. Indeed, the flexibility and ambiguity of rhetorical appeals to agrarianism too often allow for its disingenuous appropriation by industrial agriculture and other anti-agrarian interests. Still, the social-change potential of agrarian myth as a rhetorical form stands out. Capable of more than generating nostalgia, images of farms, family, and community life stir a collective soul. Agrarian myth summons a cultural identity that may not have been realized in fact but unfolds outside of time, where memory is destiny.

This introduction frames the chapters that follow in several contexts. First, we continue the discussion begun above by placing agrarian myth within a conversation about food and its connections to social identities. We explicate our rhetorical perspective as a way to engage this conversation while situating modern agrarian myth as a discourse of resistance to dominant social identities and agricultural practices. We also place the present study in the context of the new agrarianism and how twentieth and twenty-first century writers have conceptualized it. In addition, to further explore the shift between old and new agrarianisms, we trace the meaning and history of American agrarian myth to early ideas of American democracy. By exploring American agrarian mythmaking from Thomas Jefferson onward, we prepare to examine the modern manifestations of this precarious but resilient, and still surprisingly relevant, form.

Expanding a National Conversation

In this book, we examine American agrarian myth as a discursive and ideological formation with the potential to raise critical consciousness at the nature-culture nexus from which our food derives. We again follow Major, who asks, "Is it not possible to tap this resource, this influential fantasy, as a countercultural force in the twenty-first century? . . . If farming and rural life still have an ideological toehold in this country, they might very well be put to good use—not in the form of a lament for bygone times, though, but as a material friction to an ecologically destructive machine."[8] Evidence abounds that this agrarian myth is contributing to a grassroots movement of the sort Major suggests. Conversely, there are plenty of signs of powerful interests co-opting the localized, place-based discourse of the new-agrarian movement to refuel the industrial food machine amidst growing public disfavor. Agrarianism has become a hotly contested discursive terrain, with parties drawing upon the common mythos and its reservoir of celebrated cultural values for purposes that often conflict with one another.

New agrarianism's co-option in corporate food marketing and advertising clouds the transformative promise of new food movements. Popular food writer Michael Pollan observes the contemporary proliferation of romanticized agrarian imagery on food labels, such as appeals to farm-raised meat, cage-free chicken and eggs, and the image of a happy cow on dairy products. He refers to this symbolic menagerie as an aesthetic, the "supermarket pastoral."[9] Pollan notes that patterns of particular words and phrases impart narratives about food and farming in the public imagination. "'Organic' on the label conjures a whole story, even if it is the consumer who fills in most of the details, supplying the hero (American Family Farmer), the villain (Agri-businessman) and the literary genre."[10] Food scholars and activists agree, adding that the agrarian imagery that Pollan identifies in the form of misleading advertising also influences food and farm lobbying and policymaking.[11] As Pollan rightly concludes, closer inspection reveals cracks in the narratives. The rebranding of a product as "organic" may be prompted by corporate buyouts or minor changes in ingredient sourcing practices. While not all agrarian marketing appeals are mere window-dressing or romance, dishonest food-marketing "greenwash" is a threat to the ideological vision for which a new-agrarian turn stands.[12]

Scholars of food discourse have a key role in holding accountable those social actors who co-opt agrarian imagery. We share in the responsibility to critique and call public attention to observable patterns in the form and content of this influential imagery. In particular, scholars must deconstruct the constant and widespread appropriation of the romanticized symbolism of the humble farmer and family farm working in harmony with nature. Purity symbols such as these are widely appealing, yet also ambiguous, ideological, and contestable modes of greenwashing. Such images and the broader set of cultural discourses that give rise to them invite further questioning about the degree to which they signal significant change in the food system or everyday foodways. Is the ethos and aesthetic of a new agrarianism a passing trend that is "hot" for the season? Is there an emerging "agrarian" market to be exploited by capitalist interests and inevitably incorporated into a system where global and local mix for profit? While agrarian discourses may invite an ethical culture of food-literate citizens who critically engage and take informed action in relation to the places, processes, and people from which their food originates, this group also forms a coveted consumer demographic. While the new agrarianism, at times, seems a revolution, it may instead, or additionally, be a wave in the consumer culture that will result in profits for those who can properly scale it.

This fraught dynamic of agrarianism as an imaginary caught between change and the status quo, resistance and control, is a theme that we expand upon through the cases examined in this book. These cases constitute key moments in modern agrarian mythmaking in the United States during the twentieth and early twenty-first centuries. Following social theorists Ernesto Laclau and Chantal Mouffe, we view these cases as "nodes" within a co-articulated web of enduring but shifting discursive practices and meanings.[13] Each node demonstrates a variation of new-agrarian mythmaking while retelling the myth in a distinctive manner fit for a rhetorical circumstance or occasion. These moments reveal the allure and resilience of agrarian myth, while also showing how this malleable myth may be adapted to comment upon or critique cultural conditions structured by forces including but not limited to capitalism, government, technology, and urbanism. As these cases show, the new agrarianism mediates uneasily between a mythic past with resonant meanings and the present food system driven by corporate profits and commodified products.

One may ask, why rely on rhetoric as our guide to this conflicted site? *Rhetoric*, viewed as public and persuasive discourse in its various verbal and nonverbal forms, shapes and is shaped by culture and politics. As a vehicle for navigating conflict and generating cultural meanings, rhetoric defines major debates, comprises and leverages existing identifications with societal values, projects moral character, and shapes pathways for thinking, feeling, and acting. Rhetoric is essential for sounding cultural resonance, as it allows us to see discourse in its multiplicity and heterogeneity while also illuminating common themes and potential gathering points.

Through rhetoric, we may also change scales of value by emphasizing social meanings. We cannot measure the relative success of a movement such as the new agrarianism by economic trends alone. Rather, we must gauge success or failure by its relative contribution to the broader values that new-agrarianism scholar Norman Wirzba describes as "the health and vitality of a region's entire human and nonhuman neighborhood."[14] We are not suggesting that economic measures do not apply to the new agrarianism; in a capitalist culture, the viability of an idea is typically recognized by its monetization in profitable ventures. However, when it comes to addressing injustice, sustainability, and the decline of the democratic public sphere, even the most conscientious consumerism is unlikely to provide the necessary depth or breadth of critique, much less guide systemic transformations. In the food system today, food industry lobbying influences the policies of "captive" government agencies to further entrench the industrial model.[15] Concurrently, food insecurity in the United States and abroad rises from issues of access and exploitive labor relations across the food production chain, while increases in mortality and morbidity follow the spread of the "American" high-fat diet to other parts of the world. Energy-intensive industrial agriculture and the transportation network that supports it are central to a hydrocarbon economy that destabilizes culture and radically alters climate. Given the host of food-system problems tied to structural and policy issues, the limits of consumer-citizens as change agents are evident. While we may adjust our consumptive practices, support fair trade or grow gardens in our back yards, individual actions only exhibit transformative potential when they are joined to broader, collective rhetorics that open pathways for structural change. A collective and transformative new-agrarian rhetoric of resistance must address systemic conditions by envisioning new, just, and sustainable forms of food and agricultural enterprise while demanding changes at the policymaking level.

The case studies that we present demonstrate the generative nature of new agrarianism for resisting industrial agriculture and reconstituting social meanings based, in part, on food. While the discourses of the new agrarianism are likely to remain richly varied, this book provides a nexus by viewing American agrarian myth as constituting a distinctive social identity. These engrained narratives and images exude morality while affirming an ideal for living tied to the land, the seasons, family, community, and agricultural practices. As the next section shows, conceiving agrarian myth as a generative rhetoric of resistance is part of the broader scholarly project that has reengaged and reappraised agrarianism, not as an archaic type of agriculture but as a multifaceted social form regaining currency.

American Agrarianisms, New and Old

Scholars have noted that although the term *agrarianism* receives little attention in current public discussions of food, it provides important philosophical insights for productively reorganizing food, agriculture, and community.[16] When these authors use the terms *agrarianism* and *new agrarianism*, they refer to sets of attitudes and practices guiding a way of living. In this view, agrarianisms old and new provide normative visions that constitute a culture rooted in the values, practices, and structures of local food systems. *Rooted Resistance* focuses episodically on the revival of a resistant agrarianism as an evolving engagement with the mythic narratives at the core of agrarian social theory. In our view, agrarian theory and myth continue to be reconstructed and revised to fit modern occasions and purposes. We trace the pattern of succession that Janice Hocker Rushing calls "mythic evolution," while attending to the hybrid combinations of ideological discourses that Tarla Rai Peterson describes as "mythic permutations." Cultural myths pervading food and agricultural discourse in twentieth and twenty-first century America shift with the scene, often taking forms that fit, alter, or contradict earlier manifestations of agrarian myth and its values.[17] For this reason, we believe that attending to agrarian discourse in a case-by-case, contextualized fashion is essential for witnessing new agrarianism as a rising form of social action. By examining a range of examples, we begin to recognize the variegated but persistent presence of agrarianism in our cultural discourse. Rhetorical representations of food and farming are omnipresent, if often hidden, in American culture. Narratives and images of food and farming are part of the commonplace foundations of our rhetorical lives, to the point that they often pass unobserved.

Although there is no precise timeline on which an old or traditional agrarianism ends and new agrarianism begins, most writers identify traditional agrarianism as a dominant social form practiced by a rural-agricultural majority until around the turn of the twentieth century. In contrast, writers associate new agrarianism with twentieth and twenty-first century voices that speak against the corporatist, mechanistic, and fossil-fuel-intensive paradigm of industrial agriculture. This new agrarianism is "always-already," anti-consensual, insurgent, and empowerment-seeking.[18] Wendell Berry, who has long been the most prominent champion of new agrarianism, helped to define the concept by noting that agriculture is fundamental to the rise and fall of community, health, and the environment. For Berry, the common perceptual decoupling of food from agriculture is one of the root causes of many problems in modern society. Berry's agrarian corrective to is to reorganize society around the principle that "eating is an agricultural act."[19] Berry describes agrarianism as beginning with the practice of "farming as the proper use of an immeasurable gift" and a cultural identity centered on a "willingness to receive gratefully, use responsibly, and hand down intact an inheritance, both natural and cultural, from the past."[20] Berry and other leading new-agrarian voices such as Wes Jackson build upon previous generations of agrarian thinkers, including early twentieth-century horticulturalist, environmentalist, and Country Life Commission leader Liberty Hyde Bailey. One of Bailey's key arguments adopted by today's agrarians is that nature's permaculture of perennials and its wondrous cycles of soil renewal are the ultimate measure of agricultural integrity.[21] Other new agrarians—such as Jerome I. Rodale, the progenitor of organic agriculture and critic of the rising paradigm he termed "chemical farming"—further developed this principle of integrity.

Even as the new agrarianism looks to a sustained relationship to the land as a source of wisdom, it does not call for turning back time to the nineteenth century, when most citizens farmed and lived in rural areas. By most measures, American agrarianism as a rural way of life and cultural attitude has been in decline since the mid-nineteenth century. In response, new agrarianism focuses on adapting to an urban and global society, including mediating the relationships between urban and rural, and local and global. New agrarianism projects cultural meanings in relation to forms of agriculture and the farm rather than focusing only on agricultural practices. As an ethic of resistance, it critiques the industrial food system and proposes an evolving agrarianism. This evolving agrarianism seeks sustainable and just alternatives to the forms of community, consumption, and production authorized by modernity, consumerism, and their misguided sense of progress.[22]

As Norman Wirzba contends in the introduction to *The Essential Agrarian Reader*, the agrarianism of today is a critically reflexive project: "Agrarianism is this compelling and coherent alternative to the industrial/technological/economic paradigm. It is not a throwback to a never-realized pastoral arcadia, nor is it a caricatured, Luddite-inspired refusal to face the future. It is, rather, a deliberate and intentional way of living that takes seriously the failures and successes of the past."[23] Given this breadth, it is important to recognize that new-agrarian critiques have emerged from both conservative and liberal-progressive quarters, as well as the more ambiguous positions between them. Most new-agrarian writers agree, however, on the need to overcome the racism, sexism, and rural insularity found in some past agrarian discourses.[24] Progressive agrarian writers such as William Major and Kimberly Smith theorize agrarianism's existing and potential convergence with, among other camps, ecofeminism.[25] These scholars offer fruitful ideas on which a twenty-first century new agrarianism of equality, justice, and care may thrive at the nexus of rural and urban, and local and global, culture. Enacting this vision will require not only a myriad of diverse projects, but systemic and sustained changes involving economies and policies. Here the advocates of a new agrarianism frequently seek to recover aspects of the agrarian past, including a place-based localism attuned to the environment that offers a marked counterpoint to, and immanent critique of, the superficiality of consumerism.

In some respects, then, new agrarianism nurtures rhetorical roots that took hold over a century ago. Studies in the new agrarianism tend to focus on late twentieth and early twenty-first century perspectives, extending the work of Wendell Berry in particular.[26] Essays assembled in Eric T. Freyfogle's *The New Agrarianism: Land, Culture, and the Community of Life*, a key collection for scholars of new agrarianism, were originally published no earlier than 1986. The advent of the new agrarianism, however, goes back much further in time. For instance, the new agrarianism echoes anti-corporate grievances of "the people" from the late nineteenth-century agrarian Populist movement against corporate monopolies and trusts.[27] Voices of Populist dissent in the second part of the nineteenth century protested that urbanization and industrialization had resulted in the decline of agrarian self-sufficiency. As paid labor shifted toward narrow specializations, the household became increasingly dependent on goods and services created outside the home. Further, while political power had been decentralized in an agrarian economy, modernization and urbanization were tightly coupled with corporate-state consolidation of political power. In *The New Agrarian Mind*, Allan Carlson sees a "decentralist" new-agrarian movement emerging at the turn of the twentieth century in reaction to industrialization and urbanization.[28]

The present book attends to the modern history of agrarianism while adding to the multidisciplinary project of food studies. Along with scholars from sociology, geography, history, and English, the fields of communication and rhetorical studies from which we write have contributed to the growing scholarship on food as cultural and political artifact. A thorough review of the literature and history of food studies is beyond the scope of this project. Constance Gordon and Kathleen Hunt provide an excellent account of past research at the nexus of communication studies and food studies, and argue that further work in this area is crucial for informing social change on several fronts.[29] The intent of our book-length study is to add to this growing research with a fresh perspective on American agrarian myth. Through agrarian myth, we engage what it means to talk about food in a pluralistic society that projects a shared identity in the imaginary of an agrarian past. Positioned as a rhetoric of resistance, agrarian myth draws the advocates of social change, but it may also be used to legitimate the status quo or propel developments that are hardly agrarian in intent or form. Agrarian myth is a contested site and thus a suitable subject for the multidisciplinary perspectives of food studies and social theory enacted through historical and rhetorical analysis.

We hope that this work will draw further interest in agrarianism as a subject for additional scholarly studies. Observing that agrarianism continues to receive relatively limited scholarly attention, some scholars have argued that agrarianism has suffered from a distinctly cosmopolitan bias, including an affirmation of rootlessness, that characterizes contemporary academic culture.[30] Whether for this or other reasons, the burgeoning and wide-ranging food studies literature tends to marginalize agrarianism as a topic of study. The food-studies literature gives limited attention to how historical myths and public memories of rural culture inflect new food movements and discourses of place-making, identity, and authenticity. We believe, however, that engagement with agrarian myth as a conceptual heuristic is crucial for unearthing meaningful connections between historical and contemporary narratives that intertwine food, culture, and nature.

Additionally, the study of agrarian myth reveals how ideological discourses of consumption, production, and civic identity condition one another. Agrarian myth emerges through verbal and visual appeals to family farming, farmers, local production, honesty, fair trade, and unadulterated food derived from nature. Agrarian rhetoric conjoins criticism with historiography to cast the bounty of the land and the virtue of its inhabitants as the symbolic emanations of the mythic past.[31] We contend that new-agrarian mythmaking does not simply invoke the imagery of a pastoral place of tranquility for the urbanite's Arcadian weekend retreat. To limit agrarian mythmaking in this way underestimates its efficacy as a discourse of democratic resistance utilized and enacted by the new agrarianism. Some of today's agrarian mythmakers mediate the historical divide between country and city, while reanimating fading memories of rural life through political calls for change in hotly contested arguments over the future of food. In its visionary moments, this evolving myth prepares us for a world beyond divides, where agriculture is engrained in urban experience and the public engages in food production wherever they are, contributing to a collective sense of the local re-placed within the global.

Throughout this book, we view American agrarianism in its new and old forms as profoundly and complexly linked to agrarian myth's political, economic, and cultural functions. The cases that we draw on illustrate enduring and evolving qualities of agrarian myth in modern America, from World War I onward. These cases exhibit perennial tensions between the American agrarianism of the yeoman of the colonial era and the new agrarianism growing within the modern the age of industrial agriculture. In this ambiguous space, agrarian myth becomes a rhetorical vehicle for supporting ideologies that may not be readily visible. Here, the notion of agrarian myth as a ready reserve takes hold, with rhetorical analysis serving as an instructive method for laying bare motives and effects.

We have organized this book into two parts; the first historical, the second contemporary. Part 1, "Seeds of Resistance," examines key figures and movements in the history of modern agrarianism, including the war-garden movement on the World War I home front; the postwar Country Life movement for the vindication of farmers' rights; the Southern Agrarians' critique of the social costs of industrialism; and the practical and spiritual prophecy of organic farming offered by Jerome I. Rodale. Together, these sources decry a headlong rush into a bleak modernity stripped of the traditional practices through which society was moored to shared meanings and values. Agrarianism in each case becomes a discourse of resistance that points the way "back," to a future salvaged from the tyrannies of modern life, whether these are seen in the form of a rootless industrialism that would strip the soil of its sustenance and the human world of its value, or the unparalleled terror of modern war. Each case evokes a social practice that while nearly lost, has an intrinsic power that can rise again.

These seeds of new agrarianism sprout in part 2, "Threatened Harvests," in cases that track and critically analyze agrarian myth across a spectrum of places, ideologies, and media. The cases examined include a community garden developed by Latinx residents of Los Angeles; an idealized agrarian branding campaign by the convenience-food chain Chipotle; and a Super Bowl ad connecting the resilience of the farmer with RAM trucks. Together, these chapters track rhetorical, political, and cultural patterns characterizing modern American agrarian mythmaking, while also demonstrating the complex relations between agrarian myth and the discourses of industrialism and consumerism. While these cases reveal the adaptive qualities of an agrarianism that is open to diverse parties who wish to transform food systems, practices or imaginaries at local or national levels, they also reveal the ambiguities implicit in a myth with broad appeal that may be coopted by and made to serve dominant social forms or moneyed interests.

We hope readers will take from this book an appreciation for both the unities and complexities of agrarian myth in its many manifestations, including differences in its ideological articulation across contexts. We invite other scholars to add to the knowledge construction, critique, and advocacy that we perform here, namely by tracking this mythos as it reappears and shape-shifts across time and space. Toward this end, we turn to the concept of American agrarian myth driving this book—a concept of heuristic value for scholars interested in the critical study of food and agricultural discourses. Let us begin by examining earlier work establishing and explicating this mythic form. The section below offers a historical and theoretical synthesis aimed at delineating what we mean by the term *American agrarian myth*.

The Meaning of Myth in American Agrarianism

In spite of its ubiquity, the legacy of agrarian myth in American culture has received limited scholarly attention. From the grand narratives of the nation's mission as a global actor to practices as basic as the food we eat, agrarian ideas are hidden in rhetorical commonplaces and daily habits. To explicate this omnipresent agrarianism, we turn to a group of thinkers who have remained at the boundaries of critical attention but are central to this study, as they speak to agrarianism as a matter of social identity. Not all the voices examined in this book are self-identified agrarians, and some were likely unaware of their contributions to the rhetorical tradition of agrarian mythmaking. Indeed, this is how mythmaking often works; as part of the cultural fabric, it allows a social unconscious to speak from the annals of everyday life.

Studying mythic narratives entails unearthing the values, ideologies, and ends these narratives promote, even if unintentionally. Most studies of cultural myths are not focused on deciphering the factual validity of a story or argument; instead, scholars of myth identify narratives commonly believed to be true among a group of people—social truths—and examine how they function toward particular ends. As Richard Hughes suggests, "Contrary to colloquial usage, a myth is not a story that is patently untrue. Rather, a myth is a story that speaks of meaning and purpose, and for that reason it speaks truth to those who take it seriously."[32] While we are not the first scholars to examine the mythic qualities of American agrarianism, the subject remains rather obscure. In framing our study, we draw on the original conceptualizations of American agrarian myth developed by scholars such as Henry Nash Smith, Richard Hofstadter, and Richard Slotkin.[33] After their work, we approach American agrarian myth as a malleable formation of ideas regarding the citizen-farmer's exceptional virtue and agriculture's vital role in the nation's democratic prosperity. Following Hughes as well as Hofstadter, we recognize myth as an idea that so embodies a set of values that it significantly impacts how those who hold on to it view reality and act upon it.[34] Like Hofstadter, we set out in search of agrarian myth as a "complex of ideas" with "component themes that form a clear pattern," equipped with an understanding of the myth as "homage that Americans have paid to the fancied innocence of their origins."[35]

In myths, the characteristics and mission of a people are symbolically enacted and revised to accommodate contingencies. Perhaps it would be simpler to regard myth as an over-determined view of cultural identity that remains static; however, this is not how myths function rhetorically in the dynamism of social relations. Individuals in groups act as mythmakers by recasting narrative to reorder their world and mediate its conflicts and changes. In rhetoric, this process is described as "invention," which Sharon Crowley describes as the art of "how to find and use arguments made available by the cultural contexts that give rise to disagreement."[36] Individuals and groups adapt by recalibrating mythic narratives that align core values to meet present circumstances. This remaking is an act of resilience that allows the culture to change while retaining a recognizable form.

This process of strategic adaption and reconfiguration is clearly seen in the enduring, evolving forms of American agrarian myth. As a performance of the roots of national identity, we look to this myth for iterations of "Americanism" that fit the present occasion while aligning with the past. American agrarian myth is a persuasive political form to which a diverse cadre of voices across the nation have contributed. It is not limited to farmers or other agricultural and food practitioners and has no essential owner. Scholars of agrarianism, too, have contributed to agrarian mythmaking and inform today's food movements.[37] The Southern Agrarians, whose ideological discourse we examine in this book, were a group of university professors. This group drew upon the mythic antecedents of previous generations and eras. Victor Davis Hanson traces mythic agrarian inheritances to the rhetorical vision of citizenship articulated by the *mesoi*, the agrarian, egalitarian middle class of smallholding farmers foundational to ancient Greek democracy.[38] In the early days of the United States, the myth of the mesoi was restated in the writings of Thomas Jefferson, who remains the nation's most impactful agrarian voice. For most scholars, American agrarianism begins with Jefferson's vision linking land and farming to citizenship, which endures as a distinctive element of American ideology. In the next section, we review the significance of what is often called the Jeffersonian myth for the study of the new agrarianism.

Jefferson, Turner, and American Agrarian Myth

Thomas Jefferson's agrarian ideal celebrated the fertility of the land and the virtuosity of its people as the source of material abundance, moral fortitude, and national resilience and stability. Jefferson identified the availability of fertile land and the virtuous nature of farmers as the foundations upon which an exceptional, democratic nation could be built. He believed that as a young nation on a large continent, America had "lands enough to employ an infinite number of people in their cultivation."[39] Agrarians cared for the land and embodied the virtues necessary for a free people; they were, in short, "the most valuable citizens."[40] Jefferson's ideal relied on a citizenry of agrarians who cultivated the earth and manifested moral conduct. He saw this agrarian ethic as the safeguard against the risks that beset the nation, including those implicit in competing ideologies, as well as the specter of authoritarian rule. An engrained agrarianism would otherwise be a safeguard against centralization of the economy or the unbridled growth of cities. In the Louisiana Purchase, he forestalled indefinitely the threat posed by the end of open land. The United States, as Jefferson envisioned it, would be an expansive nation with a citizenry staked to its success as they built homes and communities upon the land.

Jefferson articulated his position in debates with Alexander Hamilton and others. Jefferson's ideas about the nation's proper commercial and civic bases help explain why the farmer remains a resonating cultural and political symbol today. In American agrarianism scholar Paul B. Thompson's estimation, "Jefferson believed that tying a person's economic interest to land also cultivated the virtues of patriotism and citizenship."[41] Jefferson's ideal described agrarian practices as the means by which citizens become rooted to a place and the life force from which national identity and virtues grow. The yeoman-farming middle class that Jefferson envisioned would serve as the young nation's most vigorous citizens, the core of an agrarian democracy that would protect the nation from moral decay by providing a vital counterbalance to the disturbances of urban industrialism.[42] As Drew R. McCoy explains, the yeoman agrarian democracy Jefferson wrote about was based on a republican notion of practicing civic virtue to contain corruption and defend against threats such as luxury and wealth—especially when obtained dishonestly. Agrarianism would also guard against idleness and lethargy, as well as the displacement of identity arising from the division of labor characteristic of the manufacturing economies emerging in Europe. Agrarian republicanism sought to preserve a vision of national identity based on classical civic virtues such as simplicity, frugality, prudence, diligent and dignified labor, tranquility, and physical and spiritual wellbeing.[43] Further, it strongly correlated the sanctity of the spirit of citizens and the community with practices of husbandry and cultivation. Labor on the land and farm was the great teacher of virtue.

Jefferson's emphasis on virtue may constitute a reason to balk at the notion of an agrarian ideal that would later be codified in myth. After all, claims to virtue have frequently been used to rationalize suppressing those who do not conform to a social norm. But to dismiss the importance of virtue to agrarianism would be to miss why it possesses a resilient rhetorical force. In agrarianism, virtue constitutes a sense of solidarity with one's place and promotes the communal good above narrow self-interest. Russell L. Hanson notes that, for Jefferson, "agricultural pursuits instilled a particularly valuable kind of republican virtue" that inculcated "an active and vigilant citizenry."[44] With this agrarian foundation in place, the whole of the culture could draw on its durability and strength. This vision of small freehold farmers continues to inspire quintessentially American visions of citizenship for urban as well as rural persons; in American myth, citizens are property-owning, self-sufficient, and independent. The figure of Jefferson's citizen-farmer dominated nineteenth-century conceptions of rural life and continues to animate nostalgic notions of the struggles and joys of rural life as a basis for national identity. For Jefferson, working the land seeded loyalty to an agrarian nation. In the consumer urbanity of the twentieth and twenty-first centuries, this position has been codified even as it is ironically reversed: we show a deep sense of loyalty to the nation by identifying with romantic images of rural experience. It is as if we are all farmers at heart, as we recognize that the virtues of rural life are integral to American identity.

While Jefferson's role in forming an agrarian myth for a new nation is undeniable, scholars including Richard Hofstadter contend that popular literary discourse developing and affirming the virtuous image of a farming majority fueled the myth's growth and popularity. Hofstadter explains that this literature casts farmers as countering the profit motive, emphasizing their self-sufficiency and the practice of honest labor performed close to nature. This way of living allowed farming families and communities to "produce and enjoy simple abundance."[45] By the early nineteenth century, he notes, agrarian myth had moved beyond literary imagination and "become a mass creed, a part of the country's political folklore and its nationalist ideology."[46] On the rhetorical qualities of agrarian myth, Hofstadter writes:

> Its hero was the yeoman farmer, its central conception that he is the ideal man and ideal citizen. Unstinted praise of the special virtues of the farmer and the special values of rural life was coupled with the assertion that agriculture, as a calling uniquely productive and uniquely important to society, had a special right to the concern and protection of government. The yeoman, who owned a small farm and worked it with the aid of his family, was the incarnation of the simple, honest, independent, healthy, happy human being. Because he lived in close communion with the beneficent nature, his life was believed to have a wholesomeness and integrity impossible for depraved populations of cities. His well-being was not merely physical; it was moral; it was not merely personal, it was the central source of civic virtue; it was not merely secular but religious, for God had made the land and called man to cultivate it.[47]

This vision identifies the farmer as landholder—with a direct, propertied stake in the nation—as "the best and most reliable sort of citizen."[48] As Hofstadter points out, while agrarian myth's manifestations are literary, it could not have become popular without the existence of a rural farming majority that held common social and political ideals. The myth therefore functioned ideologically by articulating a moral righteousness with which the public identified. He notes, "The American mind was raised upon a sentimental attachment to rural living and upon a series of notions about rural people and rural life."[49] Aided by its popularity, this nascent ideological narrative became a social truth beyond reproach.

Following Jefferson and his agrarian-mythmaking descendants, today's new agrarians tend to advance the belief, rendered with varying degrees of explicitness, that farming has a special place among other forms of living and working. New agrarians hold onto the conviction that a particular form of rural life provides a model for politics, economics, and social relations, even when this life is viewed as an exception rather than the norm. This dualistic sense of farming versus other types of occupation or social organization, too, arises from the early days of American agrarian myth. Jefferson and agrarian contemporaries such as John Taylor and the French visitor to America, J. Hector St. John de Crèvecoeur, employ dualistic rhetorics that not only describe and locate virtue in rural experience, but pinpoint the evil among the enemies of agrarian ideology. Through this moral frame, agrarian mythmakers have long identified the centralized power of industry, government, and elites, concentrated in cities and characterized by abstractions, corner-cutting, and greed, as chief threats to agrarian ideals. As the American agrarian mythos has developed over time, it has recast this portrayal of threats, contesting not only the many manifestations of urbanization and ceaseless material consumption, but also related consequences of industrial agriculture's master narrative of techno-scientific progress. Against these threats, new agrarianism and its myth rise as an ethics of resistance that seeks to reconnect the growing of food to democratic practice.

To track the contemporary forms of this moral agrarian frame, we must examine shifts in American myths during the nineteenth century as the nation expanded westward. According to Richard Slotkin, the Jeffersonian myth captured the imagination of settlers who envisioned a garden of peace and prosperity built on open frontier land. However, a second narrative, centered upon the White-Indian conflict, also became part of popular lore. Emerging between 1823 and 1850, novelist James Fenimore Cooper's frontier-romance genre of dime novels circulated a sense-making narrative that promoted social and racial stereotypes, and justified violence. Cooper codified and systematized the imagery of White-Indian conflict as key to interpreting other fundamental oppositions within a narrative plot, such as divides between social classes, men and women, and masters and slaves. Cooper's acolytes carried on this genre for generations.[50]

These myths promoted the virtue of westward expansion in sharply divergent ways. Slotkin writes that, in contrast to the Cooperian frontier myth, "the ideology of agrarianism only rarely makes mention of the Indian wars." It focuses instead on "the real action of history, which involves the clearing and cultivating of the soil by the diligent democratic husbandmen." Agrarian myth displaces violent colonization with "the interaction of man with pure and inanimate nature."[51] Moreover, agrarianism appeals to mythic transcendence as it "asserts that progress can proceed harmlessly, and that the bases of conflict are essentially immaterial: the abundant resources of land are sufficient to make all conflicts of class and interest unnecessary. Agrarian myth proceeds "by treating the Indian presence as if it were insignificant or marginal."[52]

Near the end of the nineteenth century and into the early twentieth century, the rise of anti-corporate Populist movements in which small-scale farmers played a major role, as well as the institutionalization of Progressivism, explicitly symbolized a sharp ideological split.[53] This split had become apparent between an "agrarian reading of the past" and the "Indian war" reading.[54] The major influence that Henry Nash Smith suggests agrarian myth had in the formation of nation-building policies such as the Homestead Act was now being challenged.[55] New voices were adapting this myth to times of urbanization and industrialization. As Slotkin notes, these regenerated mythic forms were most famously set forth in the very different versions of American history told by Frederick Jackson Turner and Theodore Roosevelt. The former posited agrarianism as antidote to class antipathies of industrialization, whereas the latter claimed the Indian Wars as a model for "the rationalization of class subordination at home and imperialism abroad."[56]

While giving some credit to Roosevelt, Richard White describes the two dominant narratives of the American West as encapsulated by Turner and Buffalo Bill Cody. These figures' themes and iconography synthesized elements of the public imagination, while adding unique stylizations. Buffalo Bill made himself into a performer embodying, loosely but dramatically, his life as a scout, cowboy, rancher, hunter of buffalo, and killer of Indians. For Buffalo Bill, the frontier was the Wild West, centered on conflict with savages; the West needed to be civilized, but interaction with wilderness was tangential to this project. By contrast, Turner, a prominent historian, evoked an Edenic vision of open and "free land" in which a "race of savages" appear incidentally at the periphery of the pioneer's settlement. For Turner, the treacherous wilderness, not the Indians, was the primary obstacle that settlers faced in taming the frontier.[57] The iconography of these competing stories juxtaposes the carnivalesque imagery of cowboys and Indians with romantic images of covered wagon trains, log cabins, and emerging farms.[58] The mythic settings likewise varied from an uncivilized frontier of lawlessness and conquest, to a virgin garden at the outer edge of civilization, where traders and backwoodsmen such as Daniel Boone made contact with the natives.[59]

Turner's widely influential "frontier hypothesis" integrated, updated, verified, and promoted a mythic American agrarian conception of national character that had circulated since the nation's founding. As White explains, in Turner's rhetoric, the land and place-making served as a reservoir of virtue. For Turner, "American progress began with a regenerative retreat to the primitive, followed by a recapitulation of the stages of civilization."[60] The Turnerian narrative drew upon the existing public belief in Americans as egalitarian and representing "great achievements from primitive beginnings."[61] Slotkin writes that Turner's narrative of successive stages of development and sentimental attachment to the life of the yeoman closely match those in J. Hector St. John de Crèvecoeur's *Letters from an American Farmer*, written over a century earlier. The difference is that Turner's narrative affirms the urban, industrial, and westward expansion discourses of its time.[62]

Also engaging Turner's thesis, Henry Nash Smith contends that Turner assured the survival of the Jeffersonian myth into the next century. Examining agrarian visions in early American political culture, American literature, European travel writings about the American frontier, and in Turner's thesis, Smith articulates a "Garden of the West" myth.[63] Smith succinctly summarizes this myth as follows: "The image of this . . . constantly growing agricultural society in the interior of the continent became one of the dominant symbols . . . a collective representation, a poetic idea . . . that defined the promise of American life. The master symbol of the garden embraced a cluster of metaphors expressing fecundity, growth, increase, and blissful labor in the earth, all centering about the heroic figure of the idealized frontier farmer."[64] Smith further suggests that Turner's explanation of the distinctly American character embodied by pioneers and permanent agricultural settlement would, in the twentieth century, connect agriculture with nostalgic notions of purity, innocence, and simplicity.[65]

On the impact of the agrarian myth after the nineteenth century, Smith notes, "It is true that with the passage of time this symbol, like that of the Wild West, became . . . a less and less accurate description of society transformed by commerce and industry."[66] Nevertheless, "embodying group memories of an earlier, a simpler and, it was believed, a happier state of society," this "image of an agricultural paradise in the West" stood the test of time as a powerful and vivid image of the authentic, unspoiled America.[67]

Agrarian myth had become central to the imagery of the West as a land of opportunity, functioning as a key part of a romantic moral creed of American nation-building. Later in the nineteenth century, the agrarian myth also served as a political defense of democracy. As traditional agrarian society declined in the second half of the nineteenth century, agrarianism became the rhetorical device of a waning rural majority's moral economic grievance against what early agrarian writers had called "corruptions" manifested in the onslaught of urban industrialism and the rise of corporate power.[68] Hofstadter observes that the more commercialized the country became and "the more farming as a self-sufficient life was abandoned in favor of farming as a business, the more merit men [women/they] found in what was being left behind."[69] Moreover, while the agrarian myth had often been propagated by politicians and urban merchants throughout the nineteenth century "with a kind of genial candor," as time passed it increasingly attained "overtones of insincerity" and became a matter of paying "ritual obeisance" to the idea of the farming as fundamentally more important and virtuous than other occupations. Hofstadter cites the example of President Calvin Coolidge posing for a series of photographs which show him harvesting hay on a farm wearing a fresh pair of overalls over his formal attire. In one photograph, his car and driver appear in the background, waiting to remove him from the scene. Further, Coolidge showed "monumental indifference" to the farmer's plight in legislation and policy decisions.[70]

As the motif of agrarianism has appeared in public discourse throughout American history and appealed to ordinary people and enduring values, some scholars have questioned the sincerity and motives behind its circulation. They have argued that politicians and big business interests have ambiguously drawn on agrarian myth for the sole purpose of moral legitimacy. Since the Progressive era, some of the parties and institutions contributing to the demise of agrarian society have ironically drawn on the power of agrarian myth to advance industrial projects. As James Montmarquet observes, starting in the twentieth century, advocates of industrialized agriculture began to "march in some fashion or other under the banner of agrarianism."[71] Other scholars, including William J. Browne, Jerry R. Skees, Louis E. Swanson, Paul B. Thompson, and Laurian J. Unnevehr, similarly argue that powerful agricultural interest groups who shape federal agricultural policymaking rely significantly on co-option of agrarian myth.[72] To legitimize proposals that may not actually help farmers sustain agrarian values and practices, these parties regularly invoke vague appeals to the moral symbolism of the family farmer and evoke "family farming as a way of life worthy of public support."[73] They wager that "cynical uses of agrarian myth to promote private interests, whether they be by agribusinesses or farm interest groups, will further erode the cultural foundations of agrarian ideals."[74] Browne and his colleagues conclude that the focus must be on the public good and the discourse used must actually foster traditional agrarian ideals and values still relevant to farming and food today.[75]

The present study, too, joins in the explication of the co-option of agrarian myth to further corporate interests. However, as we will suggest, the resilience and lure of American myth may finally resist exploitation. The figures of the farmer and the farm continue to evoke a meaning-laden way of living rendered all the more compelling by being nearly forgotten.

A Precarious Form

This book traces inventive permutations of agrarian myth in modern America, from the era of World War I to the first two decades of the twenty-first century. While this myth shapeshifts to fit changing purposes, occasions, audiences, and social structures, emanations of Jefferson remain. Modern agrarian mythmakers retain different forms of his insistence that "true" or "original" American virtue is rustic, rural, and cultivated from the soil. If the nation's moralistic meta-narrative is that Americans possess innately exceptional character based foremost on economic and political self-determination, then agrarian myth reminds us that the farmer is our most revered archetype of civic virtue. Within a free-market economy, however, the image of the farmer morphs to that of the resilient entrepreneur whose hard work and self-reliance leads to deserved rewards, and whose story inspires others to find the independent, resilient, hard worker—the farmer—within. Even as we journey through city life, we continue to embody the same set of values that made this country, fed the world, won wars, and governed the peace.

While the present scene of agrarianism in the United States is radically different from those of Jefferson's republic or Turner's garden frontier, agrarianism continues to be what Michael McGuire calls a "working myth" about farmers in agricultural communities: "a story of a hero superior in kind to us who nonetheless is a real person acting in a real social setting."[76] As a working myth, agrarianism uses renderings of the past to filter current ideas about food, farming, citizenship, and land. In spite of its demonstrated durability, American agrarian myth is precarious and nomadic; it is not essentially or exclusively limited to use by specific social actors or ideologies. Instead, virtually any person, or interest, can attempt to weave the myth into its narrative discourse to gain an advantage. The myth's ambiguous nature allows it to be appropriated widely yet remain rooted in cultural memory formulated for new circumstances. Agrarian myth resides uneasily as the nexus of past and future, in a contested present where its rhetorical force remains unguarded and available.

In this manner, agrarianism functions *heterotopically* to demarcate and complicate geographies of discourse. Michel Foucault explores the concept of heterotopia as a distinctly mixed space that allows secondary purposes or ideologies to be etched onto real-time experience. Unlike an idealized utopia that is an unattainable product of the imagination, heterotopia is a resistant space that coincides with real space—providing a place of refuge or a platform for critique.[77] Interestingly, among the spaces Foucault initially delineates as heterotopic is the garden, a space which illustrates one of the key principles of heterotopia: "the power to juxtapose in a single real space, several emplacements that are in themselves incompatible."[78] Foucault recalls that the garden, a space with ancient roots, initially had deep meanings beyond the production of food; the garden of the Persians was "a sacred space that was supposed to bring together inside its rectangle four parts representing the four parts of the world."[79] At its center was the most sacred space of all "like an umbilicus, the naval of the world."[80] Persian carpets represent this garden, and perfection, in mobile form. By taking in the whole world into a symbolism realized by the production of plants, the garden becomes a universalizing heterotopia where all things are shown in their connection to the sacred.

It has long been recognized that space functions rhetorically, allowing communities to carve out discursive terrains that provide a sense of place, while, in some cases, also providing a bridge to the dominant public.[81] In cases when dominant spaces cannot be challenged or altered through other means, these counter-spaces emerge in corners and on peripheries, exercising rhetorical influence by providing a geography in which to perform non-sanctioned identities. In this manner, we associate heterotopias with the formation of counterpublics and satellite publics in patterns of resistance that may also be open for further conquest. Agrarian myth is re-formed as an imagistic repository for cultural critique and generative alternatives. It continues the symbolic and spatial projection of the sacred in discourses that readily associate farming, rural life, and the symbolic figure of the farmer with philosophic, political or moral concepts, including citizenship, civic virtue, and redemption—whether of individuals, society or the land itself. However, in an irony perhaps implicit in the concept of heterotopia, this energy may also be co-opted, the myth tapped to support the next round of consolidation. It as if the heterotopic disruption was prefigured to expand a dominant culture that absorbs resistant energies while repurposing emergent social structures or forms.

The significant symbolic force of agrarian myth manifests in its unheralded persistence in time and presence in everyday interactions. Agrarian myth most often appears in fragments and incomplete forms such as fleeting images, brief sound bites, and simple written or visual appeals, such as those on food labels. Something as simple as an image of a farmer, a town meeting in a television or print advertisement, or an image of someone who is doing the right thing when it is not in their best interests are all possible fragments of agrarian myth. These fragments appear all around us and illustrate the precariousness of myth. These fragments alone do not adapt myth to a new context; rather, mythmakers create fragments and adapt them to the present, projecting an experience of simultaneity in which time collapses. These precarious adaptations of myth mark what Kenneth Burke refers to as their seemingly "pre-political" status. Such myths appear as self-evident social truths that mythmakers rarely present in fully developed narrative forms.[82] As Thomas R. Burkholder writes in his study of the agrarian myth of the Kansas Populists, mythic discourse displays a unifying, enthymematic function, inviting an audience "to supply their own vision of the mythic ideal." Myth does this by way of missing premises, preexisting beliefs in the social truth of mythic appeals, and aligning the myth to fit neatly within their ideologies.[83] Hence, while the open nature of agrarian myth functions through participation to enact provisional forms of social solidarity, this depends upon the parties agreeing to the myth's qualities and parameters.

The contention that any myth, agrarian or otherwise, is at once ideologically enabling and constraining originates with cultural myth scholarship, including studies of ideological appropriations of agrarian myth.[84] This earlier work provides an important conceptual foundation for continued study, as it highlights the precariousness of agrarian myth as a complex process of meaning-making traceable in public discourse. This line of scholarship differs significantly from the literature in rural sociology that engages agrarianism as a quantified measure of urban and rural perspectives toward farming, farmers, rural life, and so-called traditional American values.[85] Instead, American agrarianism, when viewed as a mythos, constitutes a powerful rhetoric of critique and corrective. It takes the form of the loyal opposition as it avows the values and practices from which the nation arose and from which it has strayed. The myth presents a bedrock of Americanism as a foundation for resistance and restoration. It summons its subjects from their technologies and commodities to return to the real and right.

While the cases in this book each reveal the call of agrarian myth in a unique set of circumstances and a key moment in history, they also function together to tell a unified story, of the roots of agrarian myth in its modern formulations and the manifestations of the myth in contemporary discourses as agrarianism shows up in unusual places. Through the instances foreshadowed below, we develop the story of modern American agrarian myth while also attending to its ideological functions. As the cases show, we are particularly interested in examining how Agrarian myth is used to legitimate both social transformations and consolidations of capitalized power. In a vexing manner, agrarian myth becomes both an ultimate tool toward meaningful change and a way to channel this change into places where it may be contained or repurposed.

From Seed to Harvest

This book casts agrarianism as a defining and fundamental symbolic link across cultural counter-narratives of food, agriculture, and culture, and intertwined issues ranging from environmental protection to social justice. We will illustrate how agrarian myth functions rhetorically as a proposed corrective to forces of modernization, urbanism, and social inequities seen as threatening citizenship, democracy, and community. The discourses we study seek to restore a sense of American values in the throes of a threat formed not by a foreign enemy, but by the entrenchment of a social order that places chemicals over nutrients, space over place, consumerism over richness of experience, or profits over people. As we shall see, however, these discourses are also shadowed by a lack of efficacy, the consolidation of corporatism, and even forms of violence hidden in agrarian legacies. While agrarian discourses often seek to resist the dominant social order, the danger that they will instead replicate or advance this order is ever present.

As indicated above, the book proceeds in two parts: the first extends the historical analysis opened in this introduction. We explicate manifestations of agrarian myth that begin to link Jefferson's ideals to a national era defined by rural depopulation, industrial revolution, the Great Depression, and world wars. This part is titled "Seeds of Resistance," as it explicates rhetorics of critique and invention that will be echoed or recast in the new agrarianism of the twenty-first century, where issues of food and farming are increasingly seen in their myriad connections to economics, politics, social injustice and environmental catastrophe, on both global and local scales.

Chapter 1 opens this section with an examination of American food propaganda during World War I as an original rhetoric of food politics that drew upon agrarian myth to mobilize food production as a key weapon for both an unparalleled war and a peace that waited undefined. Led by figures such as Herbert Hoover, the engineer who headed the federal food administration; the ingenious propagandist George Creel; and Charles Lathrop Pack, the architect of a movement for war gardening, the American home front pulled together to enact the vision of a society of abundance based on virtuous action on plentiful lands. This mobilized agriculture on the mass scale of modern war drew on the imagery and values of agrarian myth, setting in place contradictions that modern agrarianism has not escaped. The vision of a nation of gardeners, projected by the propagandists and enacted by Americans on the home front and largely abandoned in the retrenchments of peace, remains a road not taken, but never entirely closed off, in the American imagination.

In chapter 2, we examine the aftermath of World War I for American farmers in a close analysis of myth and resistance in the civic rhetorical practices of the American Country Life Association. In the years after World War I, this organization upheld the status of rural citizens as mythmakers who granted legitimacy to American political and social life. While farmers had been heroes feeding Americans and their allies during the war, afterward they were viewed as poor businesspeople who had rashly over-expanded. Farmers sensed an existential threat in the neglect of farming by federal policies and brought public attention to what they saw as the descent of the citizen-farmer into the landless condition of peasantry—a social structure more reminiscent of the failed powers of Europe than the ascendant America projected by the propagandists and war culture of World War I. Following Jefferson, figures in the Country Life movement saw the farmer engaged with land, family, and community as the foundation of the republic. They argued that neglecting this legacy placed not only farming, but democracy itself, at risk.

We further explore the collective struggle for rural-agrarian voice amidst urbanization and industrialization in chapter 3, through rhetorical analysis of the Southern Agrarians' defiant and searching discourse. This group of intellectuals led a movement driven by the sentiment that Southern identity was under threat from the encroachments of technologically advanced industrialism. They advocated a return to idealized, spiritual, Edenic concepts of farming practices associated with the South. However, they were also harried by the specter of the South's own legacy of murderous violence and social and economic exploitation, as well as their awareness that any "old order" of yeomen distinct from plantation culture was fast retreating. While the movement did not manage to buffer an agrarian mindset from change, it introduced the powerful idea of agrarianism as a critical engagement with, and resistance to, the apparently irresistible advances of modernity, with its implicit displacements of human scale and values. Our analysis traces the spectrum of this group's progression, from their original status as a conservative retrenchment against urbanization and modernization, to their current alignment with the leading edge of food movements in the new agrarianism.

Chapter 4 explores the rhetorical leadership and noted influence of Jerome I. ("J. I.") Rodale. Rodale was the first prominent United States organic agriculture advocate and a leading critic of industrial agriculture for three decades during the mid-twentieth century. Striving to explain the impact of Rodale's landmark 1945 book *Pay Dirt*, we critically examine how Rodale combines the rhetorical genre of the American ecological jeremiad with an evolving mythology of "natural" agrarian living as the basis of a virtuous citizenry. Positioning himself as a lay-scientific prophet of the future of food, Rodale sanctifies organic methods and calls for reclamation of the ancient practices of caring for the land. *Pay Dirt* condemns the evils of "chemical agriculture" and urges farmers to atone for their sins by restoring nature's covenant through organic farming. While Rodale sought to extend the success of the World War II Victory Garden movement toward a long-term postwar project, he also warned of the consequences for soil, farmers, and communities if organic methods were not urgently adopted.

The seeds of resistance open in the second part of the book, "Threatened Harvests," which examines contemporary manifestations of agrarianism through three compelling cases involving community capacity-building, agrarian branding, and the nostalgic appeal of farming and rural life in the American imagination. As William Major writes, key tasks for a new agrarianism include adapting to imperatives regarding cultural diversity and better situating itself in relation to forms of community beyond the local and the rural.[86] While these cases operate in diverse and consumerist settings, they also show the continuing appeal of agrarian myth and the durable practice of farming as a center of moral virtue, whether reconceived through vernacular discourses in urban settings, projected as an ideal for fulfilling consumer longing, or remembered as a familiar American story.

In chapter 5 we examine the case of the SCFarmers of urban Los Angeles, where the traditional mythic images of agrarianism are contested and recast toward a new agrarianism of place-making resistance and food sovereignty spearheaded by marginalized communities. The SCFarmers, an association of hundreds of mostly Latinx community gardeners from the poorest part of Los Angeles, first gained a national spotlight in the 1990s. Now part of the folklore that mobilizes food and environmental activists, the pro-SCFarmers narrative has become synonymous with the Academy Award–nominated documentary *The Garden*. Through a close analysis, we argue that *The Garden*'s cinematic storytelling draws upon agrarian myth, while it also contributes to the evolution of a rising cultural myth of urban agrarianism. We contend that *The Garden* articulates a mythic, vernacular, Latinx narrative of new-agrarian resistance; it adapts agrarian myth to new communities and contexts. While highlighting pathways for decolonizing food rhetorics, the experience of SCFarmers also depicts the constraints that new-agrarian counter-myths face as they attempt to rewrite Western colonialist narratives based on narrow conceptions of the farmer's identity.

One cannot engage new agrarianism in the twenty-first century without examining how the mythology of the farmer has been distorted and utilized by corporate interests. Chapter 6 focuses on Chipotle Mexican Grill's *Food with Integrity* social media campaign. As a rising fast food chain with a brand linked to social responsibility, Chipotle grafted the anti-corporate agrarian movement into a sophisticated branding strategy centered upon pathbreaking videos that have been widely viewed and critiqued. Within these new media platforms reside nostalgic narratives of agrarianism told through a distinctive imagistic style of disarming simplicity that identifies Chipotle as a modern manifestation of the methods and values of the simpler times abandoned by industrialized food. Analyzing this campaign's imaginative agrarianism that initially took social media by storm while fueling the expansion of the brand, we consider the implications of Chipotle's attempts to overcome what Jean Baudrillard terms the "crisis of hyperreality" by linking fast food to a natural food revolution.[87] While Chipotle's projected agrarianism was compelling in its combination of convenience with traditional values of food as a social good, the food-borne illness outbreak that befell the brand attests to the difficulties of upholding the rhetorical boundary between purity and pollution.

Chapter 7 further explores the expansion of the agrarian myth across the postmodern mediascape and shows how agents beyond those working directly in food and agriculture deploy this cultural formation. Consolidating the mythic permutations of American agrarianism theorized through the book in the form of rhetorical topoi, we examine RAM's "So God Made a Farmer" television commercial, aired during the 2013 Super Bowl. In it, a distinctly nostalgic agrarian myth focused on the enduring social roles of the farmer is adapted toward the commercial marketing of pickup trucks. We argue that RAM romanticizes the American farmer's virtue and lived experience while obscuring the significant differences in the ecological, political, economic, and cultural outcomes of industrial and non-industrial agricultural forms. RAM, we contend, naturalizes the corporate colonization of food and farming, while framing "the farmer" as an enduring monument beyond politics, ideology, or changing times. In the ad, the farmer, appointed by God, is less a living presence than a specter of virtue that can be aligned with any product that has the right look and feel: for instance, the large, powerful, gas-guzzling pickup. The ambiguities of modern agrarianism are realized in this ad as it memorializes farming and the endurance of rural experience while supporting the destructive status quo of the industrial food and agriculture system.

We conclude with a discussion of new-agrarian futures in difficult times by resuscitating the original meaning of *apocalypse* as bringing to light or summoning forth. Our conclusion weighs the prophetic capacity of emerging developments to generate new possibilities and form an inclusive new-agrarian consciousness, rather than give in to dystopian visions of displacement or doom. The development of "intelligent" technologies, including autonomous farm equipment that communicates with the land and crops through responsive sensors, promises "farming without farmers" as farm operators become managers with high-tech expertise. Meanwhile, emergent agrarianism proceeds in the opposite direction, with a "hands-on" approach to agriculture cropping up, heterotopically, in unexpected places. With urban agriculture and sustainable and fair-trade practices on the rise, we appear to be experiencing an agrarian awakening with surprising vibrancy and breadth. However, with rising populations, patterns of food insecurity, and climate change, this agrarianism is met with challenges beyond even those posed by the usual enemies of industrial farming or processed foods. We consider the prospect of a new-agrarian identity informed by myth and spreading across the world, capable of dismantling the industrial food system and culture of death at its root. Although we cannot predict the future of food and agriculture, it is possible that the industrial and agrarian paradigms will be further stretched and melded, through technologies and social forms now emerging or still over the horizon.

PART I ▪ Seeds of Resistance

CHAPTER 1

The Home Front Plants Agrarian Munitions

Around 11 a.m. on February 20, 1917, officials and workers peering out the windows of city hall in New York City faced a disconcerting prospect: 400 mothers were outside the building, calling for food. The protest had begun that morning in Rutgers Square, where a crowd of at least 1,000 gathered, desperate to halt the perilous spikes in prices besetting the nation. The crowd moved through the Bowery, toward city hall. There, the *New York Times* reported a desperate scene: "Tears were streaming down the faces of many. The babies in arms increased the uproar with their wails, and for a time no one who attempted to address the throng could be heard."[1] This "food riot," as it came to be known, was not an isolated incident. At several places across the city that day, mothers overturned the carts of vendors.[2] On a Saturday evening, a crowd of approximately 4,000 appeared at New York's Waldorf Astoria, refuge of the powerful and rich, demanding bread.[3] In Philadelphia, police opened fire on a picket line of workers and protestors, killing one and wounding nine.[4] The International News Service quoted a commodity broker from the "wheat pit" in Chicago about the supply of food hiding in plain sight: "Thousands of [railroad] cars loaded with provisions are standing in the yards of Eastern cities waiting to be unloaded. That is a fact. The food is there. And if it were destined for domestic owners it would be unloaded. But it is not. It is all for the export trade."[5] These events occurred a few weeks before the declaration of war, marking the official entrance of the United States into the titanic conflict of World War I. The nation was already deeply engaged, however, in the struggle to feed the hungry of Europe, as the modernized, international system for trade began to disintegrate. World War I is noteworthy for fearsome technological "advances" in killing— machine guns, long-range heavy artillery, aircraft, tanks, as well as the poison gases since considered too hideous even for war. However, as a war of attrition that pitched not only massive armies, but whole societies, against one another, World War I is also the story of food, in which shortages led to protracted hunger and starvation, and victory was assured in large part by an unprecedented effort to produce and distribute food aid on a prodigious scale.

This chapter examines the American home-front effort to conserve food and bolster production across society as a key episode in the modernization and mobilization of American agrarian myth. When war planners faced cultural anxieties manifest in the food riots, as well as increasing demands for help from allies in Europe, they turned to the resonance of agrarian myth at the bottom of American memory. No longer emblematic of farmers alone, this myth would now form the core of a common identity for an increasingly urban society that was polyglot to an extent that the world had seldom known. In the crucible of a war in which food became a leading weapon, Americans recast agrarian myth not as a vestige of the past but as the ultimate sign of American power to end the war and dispel the ancient enemy of starvation. A modern nation of farmers, democratic and virtuous, would supplant the old order, the moral bankruptcy of which was laid bare in the trenches. Through mobilization of land and people, approximately 28 million metric tons of food was shipped from the United States to allies in Europe during the war, including flour, grain, pork, beef, and sugar.[6] Further, as Helen Zoe Veit notes, the experience of World War I "clarified that world power in the new century would hinge on the ability to marshal and coordinate resources, both within and without national borders."[7] Participation in programs at home to increase international aid to Europe "was the most direct and meaningful way that most Americans experienced their country's rise to power."[8]

While we have discussed agrarian myth as the wellspring of democratic virtue and the structural basis for an enduring republic, this chapter reveals another aspect of the myth: as a protector against hunger and the fear of starvation. Jefferson envisioned the nation of yeoman farmers as a stratagem for warding off an aristocratic or authoritarian state not only by spreading political control over a wide geography but also by giving the population the direct ability to produce food. As scholars such as Piero Camporesi have observed, the fear of starvation for centuries stalked European consciousness, generating "dreams of bread" and visions of mythical lands of plenty.[9] A nation rooted in farming on bountiful land placed this age-old wish for material and spiritual sustenance on a practical basis. In World War I, however, the myth was subjected to a stern test. An increasingly urban population was not only at risk of privation but detached from the land and the demand to feed populations in dire need overseas. As the riots of 1917 indicate, at its entry to the war, the United States was a nation at risk. While leaders spoke of the need to unite in the cause of democracy and civilization, it seemed possible that the nation would instead devolve into violent conflicts and social disarray. Food security was at the center of this unrest. The specters of hunger and starvation were at loose in America, challenging the hopeful vision of a new economics and a culture of abundance.

President Woodrow Wilson and leaders in his administration realized that food was not only central to the American war effort but essential to the continued ability of allies to sustain the fight. In response, America mobilized the home front, through new structures of control that sought to unite government, private interests, and the public, including the United States Food Administration under Herbert Hoover and the Committee on Public Information under George Creel. These agencies disseminated the values and practices of agrarian myth across the segments of American society and formed a powerful message of American know-how and productive capacity for the rest of the world. This chapter approaches these agencies as, in part, advances in social discourse, as the new rhetorical form of state-sponsored propaganda was deployed to create a permeating reality that called Americans into a common effort and destiny.

Within this assemblage of propaganda, we examine retellings of agrarian myth that etched the emerging identity of a pluralistic nation of immigrants to the older story of resilient democracy born from connection to a fertile, expansive land. In reviewing the combined efforts to conserve food and ramp up production, the chapter focuses on the overlooked but vital work of the National War Garden Commission in spreading agrarian practices and values to urban Americans. The commission's leader, Charles Lathrop Pack, came to see the spread of gardening across society not only as an efficient method to solve the immediate crisis provoked by war, but as a transformative power now unleashed on a nation hungry for change. Americans engaged in the production of food, no matter their location or other occupations, were an irresistible force, capable not only of ending a war, but securing a peace freed from hunger for allies and former enemies.

The home-front experience in World War I marks a transition, and growing ambiguity, in agrarian myth, as Americans were mobilized into a unified, mighty enterprise for raising and conserving food. In the new discursive form of state-sponsored propaganda, familiar images and ideologies were re-spun to remind Americans that the democracy and liberty that sprung from the land was now in peril. While affective motifs of fear spurred diverse Americans to join the common cause, home-front propaganda also projected messages of hope and aspiration, promising that the world, especially the fallen powers of Europe would gratefully recognize the rightful place of the young nation at the head of a new order.

While war planners sought to mobilize agriculture as part of an industrialized war effort, the agrarian myth that they summoned could not be entirely contained by war, weaponization or mass production. As we will see, the war-garden movement, closely aligned with agrarian myth, heralded another type of transformation and the promise of mobilization in reverse. In times of unparalleled modern war, Americans discovered older ways of work, with rich social meanings. As Pack found so rapturous, gardening repurposed people and reordered groups with a strangely compelling power. The gardening society he envisioned and sought to enact might have rivaled or even unwound the rise of mass society based in consumerism. For a moment, an expansive agrarianism appeared as a quintessentially American social order. While the agrarian reordering did not come to pass and the war-garden effort waned in peace times, the image of the urban gardener working with peers on a plot of land returned to the public imagination, first in the victory-garden movement of World War II, and recently in the community-garden movement that has brought agriculture into urban spaces. Here, we examine how it arose from the new discursive form of state-sponsored propaganda, directed by a cadre of leaders who saw in war the opportunity to recast the world.

A Weaponized Garden

World War I was an epoch-making modern experience that left the world so changed that the prewar era was seen as a time of innocence that could never be recaptured. This rift in history was shown on the battlefield, as the tactics of earlier times, such as massed attacks by infantry and cavalry, became mass murder when repulsed by modern artillery and machine guns. The weaponization of food, too, reached a new threshold as the integrated nature of the world trade system that had arisen before the war led to its exploitation and near-destruction by the combatant nations. As Frank Trentmann observes, the period beginning in the 1870s leading up to World War I marked a "historical breakthrough" when an international food system took hold.[10] The nations of Europe formed symbiotic relationships while increasingly relying on trading partners to provide staple items: for instance, the English obtained sugar from the Germans, while Italy imported Russian wheat. As Pack noted, these nations depended on shipments of food as American urbanites did a daily delivery of milk. This international system, as he put it, functioned in sync, like the parts of watch. While reliable under the conditions of prewar life, the food system was vulnerable to the impacts of violence.[11]

The war struck the fragilities of agriculture and trade like a hammer. Farms were deprived of farmers to work them, and agricultural lands turned into battlefields. Food exports within Europe fell off to nothing, while strict rationing programs were adopted by the early combatants. Production was further hampered by a shortage of fertilizer due to the use of nitrate in the manufacture of munitions.[12] Widespread crop failures, particularly acute in 1916, left all of Europe more vulnerable.[13] While allies and non-combatant nations were desperate for American aid, shipments were subject to attack by German U-boats. Food, as well as munitions and raw materials, were sent to the bottom of the ocean with regularity. The Germans were subject to a British blockade and mining of their ports that led to shortages, malnutrition, and starvation.[14] The crisis came in the "Turnip Winter" of 1916 and 1917, with harsh cold, shortages of fuel oil, and an absence of food, save for the hearty turnip.[15] While riots at home and mutinies among the troops led Germany to cease combat operations in 1918, the British blockade continued until the Germans accepted the draconian Treaty of Versailles.[16] All told, approximately 420,000 Germans starved or perished from causes related to severe malnutrition.[17]

Revolution in Russia began with the fear of starvation and was ignited by riots led by women who had spent long hours in lines or slept outside bakeshops to get bread while the daily supply lasted. The riots spread to striking factory workers and then to elements of the army, who placed the Tsar and his family under house arrest.[18] Colonialism and conquest also led to the weaponization of food, as subjugated populations were deprived of their own resources and any aid. Lebanon was occupied by the Ottoman Turks and subjected to the confiscation of harvests and a British blockade; 100,000 are thought to have perished.[19] The case of Persia, where the colonial British appropriated grain to feed their troops and civilian population, was particularly grave. Millions of people—some estimates run as high as 10 million or 40 percent of the population—died of starvation or causes related to malnutrition.[20]

For a time, the United States remained detached from this unfolding mass tragedy. America was an ocean apart from Europe, self-sufficient in the production of food, and consumed with internal affairs. As Robert H. Wiebe notes, America had been a nation of "island communities," with weak communication between regions, dispersed political power, and local autonomy.[21] He attributes the development of railroads spanning the continent to the formation of a nation in the public imagination: "A new United States, stretched from ocean to ocean, filled out and bound together, had miraculously appeared."[22] While the United States was linked, symbolically and practically, its population was increasingly diverse. In the decade before the war, immigration reached record levels.[23] By 1917, out of a national population of 100 million, 17 million were foreign born.[24] They had been lured by the mythos of prosperity, pitched in contrast to cultural memories of hunger and want. The United States, conversely, looked to the future. Simon Patten, an influential economist in the early years of the twentieth century, declared that a new economy of abundance, rooted in the production and consumption of consumer goods and embodied in the United States, would soon supplant the economics of scarcity that had for centuries reigned in Europe and elsewhere.[25] Potentially, the war would accelerate this transition, but it also called the new economy into question. Did the United States have the capacity to reshape a world when its own identity was still uncertain?

Shocked by the conflagration of war abroad as well as by the riots at home, American leaders were also galvanized by what they saw as a historic opportunity. By coming to the aid of allies in Europe, the United States would seize the mantle of civilization and summon a new order into being. To grasp the opportunity, they would weaponize food in a different manner. While other nations starved their enemies, the Americans would deploy material profusion. As Woodrow Wilson said, "Hunger does not breed reform; it breeds madness and all the ugly distemper that makes an ordered life impossible. . . . The future belongs to those who prove themselves the true friend of mankind."[26] The United States would be this friend. It would sustain its allies with provisions, and when it entered the battlefield, it would do so with a well-fed army of massive strength. It would send the hungry enemy the immediate message that an agrarian people would reign victorious in a war that had drained other nations. Americans would construct the peace through the ascendancy of abundance. If we were to fight in this cataclysmic war, we would do so to recast the world in our image.

As we discuss below, this war for the future was waged by mobilizing the past. The architects of war fused the original American identity of a nation of farmers with the emerging imaginary of consumerist bounty to unify Americans and unleash their productive spirit. The war effort involved not only unprecedented structures for social action, but the projection of collective memories and identities. In a polyglot society harried by the specter of disarray, war planners returned to agrarian myth as American bedrock. From here, they told the collective story of a rising nation ready to reset the world. By enacting agrarian values and practices in countless locales, Americans would mobilize their founding myth. An unprecedented reign of mass death would end with a bounty on a similar scale, offered by the industry and thrift of an emergent nation that had recalled its roots in the remarkable figure of the noble farmer.

The Field of America

A year before the declaration of war, President Wilson addressed the Common Counsel Club on Jefferson's birthday. Jefferson, Wilson said, was the personification of the spirit of humanity, as exemplified in "the field of America."[27] Jefferson was an intimate comrade of the thinkers of Europe as well as the frontiersmen of America; he set "an example of organization and concerted action for the rights of men, first in America, and, then, by America's example, everywhere in the world."[28] Through Jefferson, Wilson explicated the position of the nation in judging whether to enter the war: "God forbid that we should ever become directly or indirectly embroiled in quarrels not of our own choosing"; however, if the country were to be drawn in, he asked the audience, "Are you ready to go in only where the interests of America are coincident with the interests of mankind?" Those who are ready, he said, "have inherited the spirit of Jefferson."[29] Wilson's retooled Jefferson is not only the visionary of an agrarian republic but a transcendent figure who embodies a greater humanity, from a distinctive, American perspective.

The war effort to come followed this speech as if it was not an homage but a framework for action. Under Wilson, these aspects of Jefferson—the Jefferson of the farm and of the world—were fused into a singular vision of Americanized civilization called into conflict and ready to establish its presence. Leading figures in the administration followed Wilson's suggestion of an American form of organization and collaborative action by activating and building upon social networks to produce omnipresent persuasions directed toward "coercive voluntarism."[30] This strategic choice sent a moral message to Americans and others. As Veit explains, in a war that the American government sought to cast as a battle for the future of civilization between democracy and autocracy, "Americans described the internal self-control of individuals in a democracy as vastly superior to a dictatorship's external demands."[31] Americans would model another way of organizing the world and demonstrate its results in battle and at home.

With the declaration of war in April 1917, the anxiety of hunger afflicted even more Americans. As Elaine F. Weiss puts it, "America seemed more worried about bread and beans than bullets."[32] While the federal government was confident in ramping up industrial capacity to produce munitions, acquiring farm commodities in sufficient quantities was another matter. American agriculture was stressed by the European relief effort already underway, and soon it would have a huge armed services population to feed. Further, the effort to raise a massive army would draw farmers, the sons of farmers, and other agricultural workers. The Wilson administration deliberated the possibility of making those in the agricultural sector exempt from the draft.[33] While Wilson eventually decided against this exemption, the issues remained. J. Ogden Armour, founder of the meat packer and distributor Armour and Company, made headlines in the press around the country with a stern warning for government and an implicit charge to Americans: "Unless the government takes immediate steps to conserve the food supply of this country and have two or three meatless days for each family a week, in less than six months this country will be short of foods as many European nations."[34] The government's response was an unprecedented management and messaging effort that sought to enlist home-front Americans in securing supplies not only to feed Americans in the armed services and at home but also to provide the aid allies abroad desperately needed.

The United States Food Administration, headed by Herbert Hoover, was perhaps the most impactful of governmental agencies on the lives of Americans on the home front. Starting before the nation's entry into the war, Hoover led a campaign of relief for occupied Belgium that eventually fed 9 million people a day in Belgium and France.[35] When charged with leading the broader efforts of the Food Administration, Hoover knew the urgency of the task. After the war, he wrote, "At our normal rate of export, the Allies would have starved.... The organization and expenditures of our food resources contributed no less than the Army and Navy to winning the war."[36] Already wealthy, Hoover refused a salary for leading the Food Administration; instead he was its first volunteer. Under his direction, the Food Administration became a vast coalition, involving other governmental agencies at all levels, charitable organizations, churches, social clubs, businesses, farms, the press, families, and other institutions of American life.

To gain and symbolize widespread "buy-in," the Food Administration developed a pledge for Americans to sign and return, indicating their support. The pledge was widely distributed, including by armies of children scouring neighborhoods and towns;[37] 70 percent of American families eventually signed it.[38] They prepared their daily diet in accordance with published schedules for meatless or wheatless meals and substituted widely available items for those needed for the troops or allies. Homemakers performed their duties clad in the uniform of the Food Administration: a blue dress, with white collar and cuffs and an insignia in red, displaying the national colors.[39] Hoover himself was synonymous with this effort. Unsolicited, Americans sent him inventories of their pantries, in case he needed to know what was at his disposal. They kept portraits of him on their dining tables to remind them of their commitments to the war effort; children were told that Hoover would "get them" if they wasted any of their portions.[40]

According to Christopher Capozzola, the tactics and impacts of the American home front blurred the lines between the public and private. The war effort "marked an unprecedented mobilization of social institutions, human labor, and popular will."[41] Obligations to employers, families, schools, and churches once considered outside of the powers of the state, were now mobilized, "prompting state intervention into American bedrooms, kitchens, and congregations."[42] For Hoover, this captive interiority enabled a continual focus on food, often targeted for specific audiences. For instance, younger women, as future homemakers, were shown that diet was central to a modern perspective on the health and fitness of citizens and society. "All our questions now center in food: its production, its distribution, its use, its conservation," Hoover wrote in an introduction to a textbook directed to women students. "The more you know about these things, the more valuable you will be, and the greater will be your service to humanity."[43]

Working again with systems at his disposal, Hoover adapted cartel capitalism and international finance to acquire and distribute food aid. Empowered by the Lever Act to set commodity prices, Hoover appealed to the business sense of farmers, setting prices on the high end of reasonable to ensure profits while encouraging extended production; farm income rose by 200 percent from 1914 to 1919.[44] He created a licensing system to control purchasing that eliminated the ability of distributers or farmers to reach overseas markets. Instead, crops were purchased by a government-controlled corporation that Hoover created. The financier J. P. Morgan was contracted to purchase commodities, armaments, and other materials to sell to allies in Europe. He became the world's largest purchaser of goods.[45] As Robert H. Zeiger states, "the scope of these arrangements was staggering. Between January 1915 and April 1917, Morgan bought more than $3 billion worth of goods for the Allies."[46] American financiers loaned the allied nations capital to buy foodstuffs and other goods. Nations including England and France were thus beholden to the United States for money as well as food. The English economist John Maynard Keynes complained that with Britain borrowing enormous sums, it was only a matter of time before "the American executive and American public will be in a position to dictate to this country on matters that affect us more dearly than [they do] them."[47]

Hoover marshaled resources to quell anxieties at home and further American interests abroad through a strategic combination of benevolence and leverage. The success of this mission, however, relied on the continual engagement of the American people with the messages and purposes of the war effort. In large measure, this job fell to the head of another expansive wartime network: George Creel, who led the Committee on Public Information (CPI). In his memoir of the CPI, aptly titled *How We Advertised America*, Creel gauged the success of the CPI in the eyes of the world: "From being the most misunderstood nation, America became the most popular. A world that was either inimical, contemptuous, or indifferent was changed into a world of friends and well-wishers."[48] Creel sought to project an image of the capable, generous American, who was not only a friend to the world but the embodiment and champion of a new civilization. A determined Progressive, and a muckraking journalist earlier in his career, Creel saw prejudice, injustice, and social violence as emblematic of deep rifts in American culture. The war provided an opportunity for transformation. Surface unity was insufficient; the American people, regardless of social status, politics or origins, had to draw from a common passion that would weld them into "one white-hot mass instinct" with "fraternity, devotion, courage, and deathless determination."[49]

Creel's CPI provided persistent, persuasive, multichannel communications aimed at building this unity. Discussing World War I as the rise of a modern order that collapsed barriers of distance and time, Wiebe points to the rapid spread of information and ease of communication that "astonished even patriotic optimists. Information on allocations and production, draft procedures and bond sales, war propaganda and peace proposals, all depended fundamentally on an extensive private network which crisscrossed the nation. Alert to the government's directives and eager to act in a quasi-official capacity, newspapers, chambers of commerce, schools, citizens' committees, and many more marshalled the home front."[50] As the font of this information and head of networks, the CPI delivered its messages across the nation and into the war-torn world. When advising Wilson on a censorship law that would control the flow of information of in wartime, Creel recommended the opposite path; he saw "expression" rather than "suppression" as the proper exigency for a democracy ready to assume leadership internationally. As Alan Axelrod puts it, "the problem was not to stop potentially damaging communication but to flood every possible media outlet with positive communication useful to the allied war effort."[51]

Creel drew members of his staff from the emerging professional fields of advertising and public relations; Edward Bernays, the nephew of Sigmund Freud and a founding figure in the field of public relations was a member of the CPI.[52] This new expertise helped the CPI craft discourse at once public and personal to cultivate collective actions. Creel's greatest innovation was the Four-Minute Men, an army of volunteers who delivered short, heart-felt speeches, often while the reels where changed at movie theaters, reminding American of the reasons for the war and explaining why supporting it through actions was vital. Creel also reached out to immigrants; the hundreds of foreign language newspapers operating during the war received daily doses of CPI copy in the language of publication. To further their messages, the CPI formed pro-war committees within immigrant communities, including groups for Swedes, Danes, Norwegians, Finns, Italians, Czechs, Slovaks, Poles, and Serbs. The German committee was carefully charged to designate German-Americans as Americans and to recognize Germans as a democracy-loving people who had temporarily fallen prey to a militaristic and autocratic regime. The group was known as the "American Friends of German Democracy."⁵³

While CPI content covered various home-front themes, the most common were interwoven messages concerning the conservation and production of food. Americans were told to stick with voluntary meal schedules and to substitute items as Hoover suggested. They were told, in myriad ways and across languages, to conserve and not waste a scrap. The work of Americans to increase production was signaled out as meritorious, with the expanding role of women in agricultural production pointed out again and again. At the bottom of this discourse lay agrarian myth as the foundation of virtuous action, based in the production and value of food. Americans were originally people of the farms and they would be known by this identity again. The foundational agrarian myth, however, was also recrafted to reach the diverse body of the nation.

Prosthetic Farmers

The iconic image of World War I for Americans is Uncle Sam looking directly at the viewer and pointing, with the famed caption, "I want you." This image was created by one of the era's most successful illustrators, James Montgomery Flagg, originally for an article in an illustrated weekly on the need to expand the armed services before America's entry to the war.⁵⁴ The image became the basis for an army recruiting poster but was soon reproduced wherever informal authority was needed, to push Americans to buy Liberty bonds, conserve food, or watch for espionage among their neighbors.⁵⁵ As Capozzola notes, the character inspires a sense of duty and obligation, with his pointing figure singling out viewers to do their part.⁵⁶ Flagg based the image on a British recruiting poster that featured Lord Kitchener, the secretary of war, pointing to the audience with the message "Your country needs YOU."⁵⁷ While Kitchener was an actual person known to British viewers, Uncle Sam was a fictional character, who seemed drawn from an American family with its share of oddballs. As Capozzola suggests, Uncle Sam is an ambiguous figure, authoritative but personal, individual but institutional. He reads as an uncle summoned to urgent family business who wants to impress on an erstwhile youngster that this is indeed serious. While the heavy brows and bags under his eyes remind us that he is no stranger to worry, the slight upturn of his lips, even on this occasion, suggest that he is prone to mirth, perhaps known for it in the family. As Capozzola puts it, Uncle

Sam "reassures viewers that war is not in America's lifeblood; the nation, like its uncle, would rather be doing something else."[58]

By making the awesome power of government in wartime into a personal mini-drama, Uncle Sam embodied a peace-loving nation at war while illustrating the genre of the propaganda poster as developed in World War I. As Pearl James observes, the genre of the poster was positioned in the seam of history, as a modern form of persuasion designed to appeal to urban viewers that bears images and forms from the past.[59] Uncle Sam restates the figure of a common American rousing to war, reminiscent of the drummer with the billowing white hair leading rough-hewn farmers to battle in the Archibald Willard painting *The Spirit of '76*. While Sam has come from a familiar and cultural past, his dramatic gesture, the isolating finger-point from which there is no escape, evokes a modernized war of mobilization where every person is enlisted.

Susan Sontag has termed the propaganda poster an art of "quotation" that bears an image of drama already known.[60] But these posters are also works of cultural imagination, based not on artifacts of the past, but projections or memories of what might have been. Alison Landsberg sees "prosthetic memory" as a transference that "emerges at the interface between a person and a historical narrative about the past."[61] She notes, "In this moment of contact, an experience occurs through which the person sutures himself or herself into a larger history."[62] Prosthetic memory blurs distinctions between reality and myth, lived experience and cultural narratives that render collective identities. Such memories allow people who may be on the margins of a culture or considered outside of it by many to form "a more personal, deeply felt memory of a past event through which he or she did not live."[63] While prosthetic memories "do not erase differences or construct common origins," those who experience them are drawn to a collective past while also being made aware of their position in relation to it.[64] Prosthetic memory is, therefore, an affective structure through which feelings of connection or belonging and distance or removal may mix or continue to mutually deepen.

For Landsberg, the larger-than-life qualities of cinema make it a medium of choice of prosthetic memories. However, the powerful imagistic propaganda posters of war also convey prosthetic memories designed to reinforce social unity while shaming those who do not do their part; imagine, for instance, the guilt of disobeying Uncle Sam in the raw period before he became a cliché. Creel took note of the power of images in public display to persuade even those reluctant to participate in the war effort. While newspaper articles often went unread, and hearing a four-minute speech might require a ticket to the theater, an image on a billboard or prominently displayed on a poster was hard to miss, especially when such images were omnipresent.[65] Pearl James notes, "It was in part by looking at posters that citizens learned see themselves as members of the home front."[66] Such posters were widely distributed, appearing as billboards; covering fences; posted in shop windows, churches, libraries, banks, schools, town halls, factories, and even in private homes; or mounted on carts or wagons for public circulation. As James says, "they redefined the boundaries of public space, bringing national imperatives into private or parochial settings."[67] Hanging a poster indicated not only that one was part of the war effort but was encouraging others to be as well. Displays were often customized or commented on to draw further attention to a poster's imperatives. Through posters, the themes and messages of the home front became a dense, imagistic web.

While propaganda posters spoke with governmental authority, they did so with methods tested in the consumer economy. Charles Dana Gibson, head of the cadre of illustrators in Creel's CPI, was the creator of the "Gibson Girl," an athletic ideal of American femininity that had been used to sell a wide range of merchandise before the war and would remain popular well beyond it. He brought this commercial sensibility to the CPI. Working with agencies to support war messages, he wanted his illustrators to get beyond "the merely material view of things."[68] There was no need to picture wheat, coal, or ammunition directly; instead he charged them with dramatizing the content, showing, for instance, a child in danger of starvation in Belgium or a serviceman slain when he ran out of ammunition. Such images reveal the stakes of actions on the home front in vivid human terms. Gibson was drawn to the grander purposes set out by President Wilson, whom he saw as an idealist and "the great Moses of America" who "points out the promised land, the milk and honey." Gibson aspired to depict, through illustration, "the spiritual side of the conflict . . . the great aims of the country in fighting this war."[69]

Like Uncle Sam, another famous image was created by James Montgomery Flagg, this time for the National War Garden Commission. Here we see the figure of a woman clad in clothes like the American flag, but with sandals climbing her ankles like those of a Roman, sowing seeds on rolling, furrowed land. "Sow the seeds of Victory," the message at top left reads, "plant & raise your own vegetables." Whereas posters of Uncle Sam used direct address as an imperative to the viewer, this poster offers a lyrical, but jarring slogan: "'Every Garden a Munition Plant.'" If, as James theorizes, posters are poised at the juncture of history where modern experience meets traditional forms, the accordion effect in this poster is pronounced, as it transports the embodiment of a mythical past of an American garden to the extreme modern experience of mass mobilization in war. Jeffery T. Schnapp observes that media predominant in World War I, including telegraphed "breaking" news from the front, stories illustrated with photographs in magazines and the new mode of documentary newsreels in film, all sought to recreate the feeling of being "live" at the scene.[70] Posters, however, operated differently by attempting to capture enduring or poignant meaning in a deepened sense of time. Posters provided "elaboration of idealized images," including "allegories of the nation and stereotypes of the soldier, citizen or collectivity or of monstrous doubles—the enemy combatant, the foe as faceless horde."[71] Where other characteristic media of modern war project immediacy, posters conveyed mythic meanings.

As William Leach suggests, America has traditionally been portrayed as a garden in which "paradisiacal longings" are satisfied.[72] For Protestant settlers this utopian impulse formed a millennial vision of a New Jerusalem that would provide not only spiritual salvation but also an end to bodily hunger and privation. Before World War I, this myth was "transformed, urbanized, and commercialized," and reframed as a narrative of personal desires that visualized potential satisfaction in the "pleasure palaces" of department stores, restaurants and amusement parks."[73] In creating the "seeds of Victory" poster, Flagg bore down to the heart of this culture as well as backward in time. His design asks Americans to demonstrate their allegiance to this consumerist paradise by partaking in its primordial action, the planting of seeds on American lands. This prosthetic memory of agrarian myth promises victory and belonging through the actions and virtues of the farm.

While Uncle Sam appeared as a relative from a shrouded familial past, "seeds" shows the mythic figure of Liberty/Columbia, who was immediately recognizable by Americans in World War I, with meanings that were richly ambiguous. Especially for new arrivals, the figure of Liberty evoked the promise of American plenty and their desire to belong. In the Statue of Liberty, the public memorial, a memory device preferred in the nineteenth century, was recast on a colossal scale for an expanding nation. Dedicated in 1886, the statue had come to symbolize the United States as a beacon welcoming the world. The poem by Emma Lazarus in which Lady Liberty asks to be sent the world's "huddled masses, yearning to breathe free," was written in recognition of the statue's prominence in the imaginations of those coming to America.[74] As a figure representing a new America, tied to ancient ideals through the Roman goddess Libertas, Liberty became a prosthetic memory of America open to new arrivals and a beacon to those contemplating the journey. Commercial interests capitalized on the statue's appeal. By the time of World War I, Liberty was a leading brand name and symbol in the consumer economy. As Eric Foner puts it, "Consumption was a central element of freedom, an entitlement of citizenship."[75] When the government issued war bonds, they strategically chose to call them Liberty Bonds to appeal to all Americans.

In Flagg's poster, Liberty appears as a portal to an agrarian past and an earlier mythic American imaginary. In the first years of European presence on the American continent, this "new world" was represented by an older presence, a native woman warrior often pictured with exotic animals or severed heads. As the vision of American plenty took hold as a spur to further colonization, the representational figure changed, to a full-bodied Native queen surrounded by a bounty of food.[76] By the nineteenth century, the figure had morphed into Columbia, a White woman who embodied ideals of beauty, purity, and vitality and was a familiar symbol of America. Like the British Britannia, Columbia drew from representations of Classical goddesses, including the "liberty cap" of Roman Libertas.[77] By World War I, the cap was often replaced by the crown of the Statue of Liberty.[78] Flagg's figure also draws on the tradition from the ancient Greeks onward to represent military victory with a goddess. While the Roman Victoria was borne through the air with wings, however, Flagg's figure is bound to earth. Victory in this war would be assured by the ancient practice of sowing seeds on fertile land. Now, the myth would be mobilized and rescaled to join the mobilization of an industrial society.

As Schnapp notes, World War I posters often included "composite figures" from a common stock customized to capture modern identities. The figure of the medieval knight, for example, was used by German propagandists to symbolize the virtues of composure and obedience and by the British to convey observance to a moral calling.[79] Flagg produced a composite image of the goddess as a modernized American, tied to a deep sense of the past but recast for the demands of war. Americans from farming backgrounds were reminded of the peaceful qualities of their work while being shown that it was the foundation of the war effort.

The jarring line turning gardens to munition plants also suggested that something was slightly amiss. Was agriculture now a war machine? For immigrants the poster was equally ambiguous. Liberty reminded them that they were in America, where prosperity is earned and identities can be remade. While the poster evokes a latent promise of cultural belonging through participation in the war effort, an undertow of fear is also present. If one were judged to have not done one's part, or seen as in league with the enemy, would the mythos of a new beginning still hold?

The CPI produced another poster in multiple languages that showed immigrants arriving by ship in New York harbor and approaching the Statue of Liberty. An old woman clutches a basket of provisions she has conserved, while a young man, perhaps her son, takes her arm and offers a sweeping gesture as if to say that their lives would be transformed now that they were in America. The poster, however, offers a veiled warning: "You came here seeking freedom, you must now help to preserve it. Wheat is needed for the allies. Waste nothing."[80] The terms of expulsion and cause of fear are thereby established. With Liberty, the affective structure is subtler. While viewers are welcome to join her, it is also clear that liberty is to be earned, not given. All Americans, regardless of background, were charged with taking agricultural actions, but those furthest removed from the founding myth of the nation are the ones whose status is held in question.

While the poster of Liberty as the sower in an agrarian setting conveys affective structures to Americans, it also holds meanings for America's allies and enemies. As a pastiche of goddesses, the figure carries Western civilization in a compact form. However, she is set not in a world capital or among the ruins of Athens or Rome, but on the fertile farm lands of the America. In this manner, the figure provides both memories of a collective past that far predates the young nation, and a vision of a surprising future. American ascendancy seems inevitable given the figure's purposeful stride, as well as the bold message of the might of agriculture as armament. In its gestures toward a new type of civilization based on the American performance of ancient ideals, the poster fulfills the logic that has built throughout the war, captured in the propaganda of nations. Harold Laswell recounts how what started as a battle to vindicate international law quickly became a struggle to liberate occupied peoples, whether Belgian, Irish, or Indian. When the British proclaimed that civilization was at issue, "the defense and nurture of *Kultur* became a duty and privilege of all good Germans."[81] Liberty now claims the mantle through the sowing of seeds that will come to fruition in ravaged Europe. Agrarian liberty as a symbol of civilization is at once peaceful and fearsome. While sowing seeds seems to have no part in war, Europeans knew hunger as war's primary condition.

This vision of a militarized plenty echoes both the concept of an economy of abundance and the war strategy for a weaponization of food based on production for the nation and its allies. At times, American propagandists portrayed the Germans as barbarians, for instance, in an image of a simian Hun as potential rapist appearing as a precursor to King Kong in the American imagination.[82] The British, however, often focused demonization on actions related to diet and eating. While British propaganda showed a disciplined home front following rationing programs, the Germans were portrayed as undisciplined, eating all they can of whatever is available, even non-edibles like nettles.[83] The most infamous British propaganda cut deeper, exposing taboos. In the most infamous incident, British sources rigged a government report to claim to have evidence of a near-cannibalistic practice. It was falsely claimed that the Germans with no feed for their livestock had taken to feeding the ground-up remains of their human dead to hogs held at a secret factory. Once fattened on this human feed, the hogs were themselves slaughtered and consumed by hungry, subhuman Germans.[84]

In pursuing their vision of an ascendant nation, American propagandists took a different approach. Their nation was not part of the failing, and flailing, order that would allow even its most disciplined subjects to starve. America was, instead, a place apart, in effect another world, where a civilization at a slight remove had taken root in rich and well-tended soil. Now Americans would come to Europe to claim their rightful inheritance. A German-language leaflet produced by the CPI, distributed widely across enemy lines, illustrated this point simply by itemizing the substantial rations of American GIs, aptly known in this war as "Doughboys." The leaflet outlined what the Germans sensed, that their American adversaries would finish the war with their very presence.[85] While the Germans starved, the Doughboys ate a diet of 4,000 calories a day, embodying a future in which a surplus of food was a constant. Whether Germany would be included in this future was now the open question. For Americans, policy decisions regarding aid in the postwar world grew pressing. Was food a right for all people, a gift to bestow with benevolence, or a strategic asset to use for punishment as well as reward?

War Gardens over the Top

At the end of the war, Charles Lathrop Pack, head of the National War Garden Commission (NWGC), was poised for further mobilization. Writing just after the armistice, he recognized that international trade was ready to resume without the threat of German U-boats or British blockades. However, the world's production capacity was seriously diminished; a long period of reconstruction would be required. In the meanwhile, America's help was needed to ward off starvation. The demand would be greater in the postwar era. If providing for 120 million allies had strained our resources, "What will happen now that 180 [million] starving neutrals also come to us for food?"[86] What of "Russia's helpless 160 [million] who thrust their hands across the sea to us, even as sinking Peter appealed to Christ, saying, 'Save me or I perish?'"[87] America's former enemies, Germany and Austria, would be readmitted to the society of nations with another 100 million hungry people. "Already our former foes are begging piteously for food, and President Wilson has assured that that their appeals will be heeded."[88]

Pack identified two sources for expanded production: farms where professionals handle production or the agrarian gardens of American cities and towns. With farm land having reached its productive capacity for the present, Pack claimed that the needed increase "must come from the only remaining source, the small gardens in our urban and suburban communities."[89] Anticipating this postwar challenge, Hoover had authorized the expansion of the garden movement as essential not only to feed the world but also to consolidate the place of the United States at the head of an order where food equated to power.[90] According to "conservative estimates," by 1918 American gardeners were producing food with a cash value of $525 million from about 5.2 million gardens.[91] With a focus on the preservation of perishables, the NWGC further estimated that 1.45 billion cans of fruits and vegetables were produced by gardeners in 1918.[92] Pack believed, however, that this capacity could be expanded. While the production of the garden movement was impressive, this was certain to grow, for gardening had become a physical and spiritual occupation through which Americans had returned to the land.

The postwar-needs analysis was similar to the one Pack and others had made with America's entry to the war. As they realized that without an increase in cultivated land and farm labor, Americans could not produce the supplies required, no matter what steps were taken to conserve food or reduce consumption. The problem was complicated by a shortage of agricultural labor. Pack explained his use of the biblical analogies he favored: "The children of Israel could not make bricks for the Pharaoh without straw; and when we attempted to create food for famishing Europe we experienced similar difficulty, though our shortage was of man-power. For a decade or more there had been a tremendous exodus from our farms. Our farmers cried for help but their cry went unheeded until we found ourselves facing hunger."[93] Solutions were myriad. Volunteers from towns, cities and neighboring farms assisted with planting and harvests. Women workers in agriculture greatly expanded; a coalition known as the Woman's Land Army sent recruits, most often young adult women from cities, to work the farms. The price supports enacted by Hoover helped to stop the flight of workers from farms due to higher wages in the city. As Pack notes, however, attempting to halt the labor shortage was akin to putting Humpty Dumpty together again.[94] To meet the escalated demand, agriculture would need to be expanded and transformed; if the mountain would not come to Muhammad, Muhammad would go to the mountain.[95] Thus, "The idea of the 'city farmer' came into being."[96] He captures the success of the movement in modern terms, "Gardening came to be the thing."[97]

A businessman with a fortune made in timber, Pack became the architect of a widespread movement that saw American lands turned to agrarian cultivation. During the war, the NWGC produced and widely distributed an illustrated gardening manual, complete with how-to materials on topics we now associate with organic farming, from varietal selection to composting. Methods of preservation, including canning and drying, were discussed in another popular manual. After the war, Pack collected these manuals for posterity and added a narrative recounting the progress of war gardens. This narrative is noteworthy not only for its contents but the arc of its progression. Witnessing the impacts of gardens, Pack converted from an able businessman faced with a difficult problem to a prophet spreading the "gospel of gardening" keen to exhort its transformative power.[98] Beginning with the pressing practical matter of how to grow more crops to ease the perilous shortage, Pack explores the various manifestations of war gardens in a manner that gradually restores, then exceeds, the elements of Jeffersonian myth. Moral virtue and democratic practices for individuals and communities become a modernized vision of a future secured by gardeners at work across the land.

Recounting the use of propaganda posters in building support for the war garden movement, Pack states, "Many of the slogans sent ringing throughout the country by the Commission breathed the spirit of America and of democracy."[99] Referring to Flagg's poster, he notes how "the beautiful figure of Liberty" called on gardeners to "Sow the Seeds of Victory."[100] He explained his conception of munitions as any material that supported the war. Gardens were thus munition plants, or perhaps planted munitions, and part of the mobilization to produce materials essential to the war.[101] Another poster sponsored by the NWGC defined the role of gardening in the great war effort through a fanciful image. It shows a woman farmer-gardener climbing a hill of dirt in the company of anthropomorphic vegetables.[102] The largest, a pumpkin, raises a mighty cry; the beet looks winded and unhappy. The attack is led by the brave, perhaps foolhardy, turnip, with the tomato hustling to keep up, followed by the solid, determined potato. "War Gardens Over the Top," the banner at top reads, with the caption below extending the thought, "The Seeds of Victory Insure the Fruits of Peace." The analogy is clear; the piled dirt is a trench, the woman a soldier brandishing a hoe rather than a rifle, with a broad-brimmed hat in place of a helmet. The vegetables are an advancing squad taking fate into their hands. Like a military unit, this outfit boasts its own American flag.

NWGC leaders believed that gardening would put American production "over the top," ensuring supplies for Americans and Europe and establishing the basis for an American order that would extend into the postwar period through reconstructed trade in food. While gardeners could not produce the quantities of commodity grains useful to feed armies or for shipment overseas, they could help to feed themselves, their families and neighbors. While this reduced the commodities diverted to the home front from larger scale American farms, it also allowed for further savings: in required farm labor, rail cars or wagons to transport crops, the work of handlers and distributors, and fertilizers and fuels. By gardening, Americans would consolidate the wartime economy to reduce expenditures and maximize productivity.

To realize this vision, public persuasion loomed large. "Before the people would spring to the hoe as they instinctively sprang to the rifle," Pack writes, "they had to be shown, and shown conclusively, that the bearing of the one implement was as patriotic a duty as the carrying of the other."[103] Bringing gardening to the people required an information and social networking campaign that made use of the press, churches, social organizations and existing garden clubs around the country. New garden clubs sponsored or inspired by the Commission also cropped up. The Commissioners worried that would-be gardeners would feel the strain of the massive production they knew was needed for the war effort and believe that a garden they raised would make no difference. As Pack recounts, the NWGC had to restate the lessons of a forgotten children's verse: "Little drops of water, little grains of sand / Make the mighty ocean and the pleasant land."[104] Hoping to overcome the potential impasse caused by misperceptions of scale, the NWGC embedded the development of gardens among strategies for conservation. By feeding themselves, even in part, gardeners reduced their consumption of products from the farm, reserving these for war planners to direct to troops and allies.

Pack came to see the true impact of the garden movement, however, not in its ability to change the production and distribution picture but in its impacts on individual and collective efficacy and identities. More than a measure necessitated by the exigencies of war, gardening became an alternative occupation for millions of Americans: "No single occupation born of the war has affected a greater number of people than gardening. Starting from a mere nothing before the United States entered the war, this form of service grew into a new occupation, which numbered its followers in the millions, and, in the number of people employed, exceeded any other branch of gainful occupation with the single exception of actual farming."[105] While gardening had been ramped up as a collective mission in the war effort, it remained an intimate, familiar and neighborly practice that revitalized an agrarian practice and spirit that had been in danger of being lost. In gardening, Americans found satisfactions beyond the fragmentation, repetition, and abstraction of work in factories or offices, or modern living detached from the land. As Pack saw, war gardening was a return to first principles practiced in progressive forms.

While millions of war gardens were sited at private homes, the innovations of the NWGC involved the creation of war gardens in other public places. Fort Dix was the first military base to have its own garden; as Pack states, it was "appropriate and fitting" for war gardens to be affiliated with actual soldiers.[106] With the NWGC noting that nearly all military installations had vacant land, other bases and war hospitals established gardens. Publicizing these gardens at bases impacted the civilian war gardeners who "redoubled their efforts because of the knowledge that the men in the American army were doing similar patriotic work."[107]

While gardening was useful in preparing enlistees for the physical rigors of outdoor life, it also provided opportunities for labor for soldiers not fit for combat duty and convalescents in search of healthful measures. Conscientious objectors were assigned to work in gardens at military bases, as were some "enemy aliens" who had been in the country when it entered the war.[108]

The innovation of community gardens began with cataloging parcels of unused land. Pack describes the solution in language that juxtaposes mobilization with morality:

> Once embarked upon participation in the war it became evident that this nation would need to exert every ounce of her power in the prosecution of the conflict. In various localities anti-loafing laws were speedily enacted to put every man to work. Since food was even more necessary than man-power, it was of still greater importance to put to use every particle of "slacker land"—idle soil so located that it could be worked. In our city and towns, where the manpower was available to cultivate these areas, were thousands upon thousands of acres of idle real estate.[109]

Pack estimated that at least fifty acres were available in every town or city, while in the larger cities, where the need for locally grown food was greatest, "the aggregate area of vacant lots was astonishing": in Minneapolis, 5,000 acres; in Greater New York, 186,000 vacant lots.[110] The NWGC worked with local parties to secure sites and provided advice, nearly a template for organization, recommending that the garden be formed as a modern association, with a charter, executive director, and committee structure. This would provide both internal governance and an external face of the organization to government, landowners, or others. Mirroring private ownership, the NWGC recommended assigning plots to individuals, rather than sharing the work on a single plot or pooling harvests. Pack notes that a spirit of healthy competition spurred gardeners to produce at a higher level. However, he supports the sharing of tools to reduce expense and common access to expertise including experienced gardeners serving as advisors.

Pack also recounts a broader sense of mutuality and cooperation around the gardening movement. Governments, landowners, agricultural experts, libraries and the press all supported the gardening efforts. He describes volunteerism among the privileged in biblical terms: "Like that young man of great possession who came to Christ inquiring 'What shall I do to be saved,' hundreds of men who possessed or represented immense wealth, captains of industry or leaders of big business, came forward in this present-day struggle against pharisaism and demanded: 'What can we do to help?'"[111] Factory owners provided land and equipment for gardens at their facilities. Railroad personnel planted gardens along rights-of-way. Gas companies sponsored demonstration kitchens. Water utilities provided free water for irrigation. Gardens provided a multifaceted return on these investments. Pack notes that employers who had allowed gardening at their locations reported a more engaged and efficient workforce, with reduced turnover. The value of crops from a community garden typically ran into the millions; however, the ancillary values of gardening in beautification and the cultivation of civic pride and community spirit, Pack says, was worth far more.

"The war garden is a forge that is daily strengthening the links in our chain of democracy," Pack writes, and nothing is more potent for building a common spirit than the democratic practice of community gardening.[112] Gardening in a common space creates shared interests and unites all in the purpose of producing food. It also reduces differences in social status: "Rubbing elbows in their garden patches, lawyers and laborers, tradesmen and housewives, speedily discover that they have much in common."[113] They are united not only in growing crops but also in the advancement of their community and its garden. They are tied to one another and to the land; a provisional agrarian community, rife with democratic practices, is formed.

War gardens at schools provided training in citizenship and democracy. While gardens taught children the practical skills of cultivation, they also trained young citizens "in the intelligent application of the principles of thrift, industry, service, patriotism, and responsibility."[114] Beyond this, gardening allowed children to find an agrarian sense of connection to the land. Through gardening they came to see in nature "a generous giver who requires only to be encouraged."[115] Children recognized themselves as aspiring Americans who practiced morality through the land and deserved a bounty in return.

The transformations were perhaps most pronounced, however, at gardens sited at factories. Factory workers who gardened on company-sponsored plots not only practiced agrarian values but came to recognize their worth as individuals. Gardening thus profoundly counteracted the alienating effects of industrial society. Pack states: "A man who is a cog in a vast machine cannot put individuality into the driving of continuous pegs into a shoe; but when he gets outside the walls of a factory into the little forty by sixty vegetable plots he is cultivating under the shadow of the mill, he can put himself into this work."¹¹⁶ Work performed in the garden, Pack notes, is the worker's own. Outside the abstractions of labor in industrial capitalism, this factory farmer assumes the means of production. The labor of gardening directly benefits the worker, the worker's family, and the collective home-front effort. This worker has become an industrial yeoman, a surprisingly agrarian figure for a modern age and a mass society engaged in a hegemonic war.

Pack's narrative recounts the transformation of Americans from an anxious people entering an uncertain war to a nation of farmers on an expanded scale and in a widely dispersed form. While war planners and propagandists strove for social unity and an engineer refitted and channeled the nation's food supply, Pack evoked a deeper sense of value as Americans discovered, again, the value of farming. For Pack, working the land was not only a root of cultural identity, or a prosthetic memory connecting Americans of all background and walks of life. It was also, and most notably, a future in action by which a nation would realize the destiny made by its hard-earned plenty.

By the end of Pack's narrative of transformation, the terms of the trope in the poster of Liberty/Columbia had been reversed. Where once gardens served war, now war had been waged to produce an agrarian peace. The munition plants had been replanted as gardens. This reversal is also captured in a poster distributed for the NWGC that updates the story of "Over the Top." In the second poster, the same farmer marches over open land, a hoe over her shoulders, at the head of the company of vegetables. The war is over; peace is won. The banner at top proclaims boldly, "War Garden Victorious." The slogan at the bottom reads, "Every War Garden a Peace Plant."¹¹⁷ By incorporating agrarian myth into the fearsome culture of modern war, Pack and his gardeners summoned an alternative identity for modern America grounded in a world both new and ancient. He offers this vision of America as a place where gardens are everywhere: "In riding across the country one sees them beside the railroad right of way, in back yards, small and great, on lawns and in open fields, in every conceivable place and of every imaginable size—see these living emblems that tell, as truly as the tiny Liberty Loan button on the coat-lapel, where the owner stands and what [they stand] for, because a war garden is a service badge of living green."¹¹⁸ A century later, Pack's words seem salient and current as American agrarians and their allies struggle in the tensions between industrial agriculture and its alternatives in the local, organic or fair-trade movements. For Pack, the resolution to "the food problem" was profound: plant your own crops, in service to yourself, the nation, and the world.

Over Here

To picture the legacies of World War I for modern agrarianism, we begin with a description of another American propaganda poster, this one projecting an affective structure of foreboding among the hungry of Europe. Here, humans appear as shades in a bombed-out city under an eerie orange sky. Emaciated, in rags, the figures in the rear raise their arms skyward. They are faceless, destitute, and helpless. The central figure, however, rises in resilience. She is a mother with children in the folds of her tattered garment and a baby crying in her arms. Her face is starkly portrayed. Hope and fear mix in her eyes as she looks skyward, her mouth open as if she is about to give voice to a revelation. Her humanity is evident, her vitality sparked by the prospect of deliverance. Transfixed by the prospect of sustenance from above, she embodies the desire to survive, to be among those who merit aid and life. A simple, urgent message explains to Americans their role in this drama: "Don't waste food while others starve."[119] This image and message are reminiscent of the line from Kenneth Burke that a totalizing ideology is like a "god coming down to earth where it will inhabit a place pervaded by its presence."[120] The terror is not only the primordial beast of starvation, but the relief in the sky. The poster captures the massive system of production and consumption as it descends to the ground. While the woman is dependent on the system's benevolence, the American viewer receives the threat implicit in the directive. Wasting food would betray not only humanity but the system as well. All parties are held tensely in its thrall.

Scholars have viewed the social reorganization that occurred during World War I as the rise of the modern bureaucratic and welfare state.[121] Veit describes the American home front in World War I as a social experiment for the management of food on a national scale and in the lives of Americans, as the state extended its reach "into the home, onto dinner plates, and into kitchen tables."[122] Veit argues that this experiment has had lasting effects: "The way we think about food now has its origins in this moment."[123] Food—modified, marketed, mass-produced, wasted, weaponized, and exploited—remains the basis for modern life and the ultimate guarantor of power. A century beyond the war, it is hard for most Americans to imagine a system that does not deliver an astounding abundance through an intricate web of relations, not dissimilar from the watch that Pack once described.

Pack believed that the gardening movement he had helped to inspire and direct would continue and expand. He quotes Robinson Crusoe, "'Hunger, knows no friend, no relation, no justice, no right and therefore is remorseless and capable of no compassion.'"[124] If allowed to fester, "'hunger will endanger the peace of any community or nation.'"[125] While Americans may harbor ill will toward their former enemies, the good of the world, and American interests, demands that even hungry Germans or Austrians be fed. "Otherwise there can be no settled peace, no progress, no reconstruction."[126] Fate had placed on American shoulders the necessity of supplying sustenance for much of the world, and, Pack believed, Americans were prepared for the challenge. He cites factory owners, civic officials and railroad executives, all of whom pledged their continued support and plan to make their gardens permanent.

Hoover, too, was anxious to sell to the former adversaries in Europe. As the war ended, he controlled a surplus and, like Wilson as well as Pack, believed that instability and hunger in Germany and Austria-Hungary was no good for anyone. Eventually, after Versailles, some food aid was provided to the Germans, but the massive inflows Hoover and Pack believed necessary never came to pass. Appalled by the horror of war, the nation retrenched. Congress refused to allow the nation to join Wilson's cherished League of Nations. An influenza virus that killed 50 million people, 3 percent of the world's population, provided further reason for America's withdrawal.[127] As we will recount in the next chapter, farmers encouraged to increase production during the war were left over-extended and cut off from political influence or power. With the reduction of foreign aid and the end of price supports, commodity prices fell. Community and factory gardens waned; no such gardens dating from World War I have remained. Pack's prophecies went unheard until a second world war resurrected the messianic garden.

"Over There," the memorable American anthem of World War I, prepared the other nations fighting the war in Europe for the Doughboys: "The Yanks are coming, the Yanks are coming." It also promised commitment: "We won't come back until it's over, over there." The haunting words and familiar melody echo in time, making a distant time present. Doughboys all, we have never returned from over there. Amid lost opportunities, there was no return to an agrarian past. From the perspective of agrarian myth, we make our home on foreign ground. Before World War I, the modern food system had arrived, and in the years to come, despite the Great Depression, World War II, and other shocks, it would continue to strengthen its hold. As the chapters to come recount, agrarianism after World War I is increasingly an ethic of resistance that attempts to redirect history away from an industrialism and consumerism on a mass model toward an agriculture and way of being replete with richer meanings. At the center of this resistance, agrarian myth is again reclaimed, as a resonant and evolving story of the value of living on, and with, the land. A century later, Pack's modernized vision of a nation of farmer-gardeners changing themselves and the world seems surprisingly current. The chapters to come weigh the prospects of cultivating a new agrarianism—an agrarianism adapted to an increasingly urban, technological, and culturally diverse society.

We continue to explore the evolution of agrarian resistance in the next chapter, through the Country Life movement and the advocacy of the American Country Life Association. We move from gardeners to farmers and into the era after World War I, when agriculture interests and rural agrarian communities were losing their standing at the center of democratic life. As we show, the task of untangling the relation between the farm and the powerful coalitions of industries and government during this period must begin with close study of discourses of legitimation. Where once citizens, including rural folk, provided legitimacy to the government, key figures in the Country Life movement perceived that the dynamic had been reversed. Farmers had slid from their yeoman status toward a peasantry, while the quest for profits among concentrated, moneyed interests accelerated. The Country Life movement sought to restore the rights of farmers and we remember it as a chorus of resistance that formidably questioned the legitimacy of those in power.

CHAPTER 2

Country Life Defends Yeoman Democracy

Democracy is an attitude rather than a right or political system. It is won, not on the battlefield or at the ballot-box, but in the change of the personal attitudes of those who have toward those who have not.

—Dwight Sanderson, "The People on the Land"

For scholars familiar with agrarian traditions in America, it is no surprise that the epigraph above was written by an American farmer. Dwight Sanderson's view of democracy came from his experiences farming during the 1920s and 1930s, decades that were equally devastating for those who lived on the land.[1] His view of democracy as an attitude of equality "of those who have toward those who have not" was indicative of rural folks' belief that equality was a ground upon which democracy could grow and country life was central to that narrative.[2] Sanderson was but one among thousands of farmers in the American Country Life Association (ACLA), working "at the grass roots" of society to uphold a vison of rural life as essential to democratic practice.[3]

The ACLA was the nexus of a broad social movement. While it was not founded until after World War I, the plight and status of farmers in an increasingly urban and industrial nation drew public attention in the decade before the war. In 1908, President Theodore Roosevelt formed a Country Life Commission and appointed horticulturalist and champion of rural life Liberty Hyde Bailey as its chair. After extensive public hearings, the commission issued three recommendations. The first, for a national extension service, resulted in the founding of agricultural

65

extension offices associated with land-grant universities. The second, calling for continued fact-finding surveys of rural life, led to federal support for the emerging field of rural sociology. The third recommendation was for a campaign for rural progress with the support of a national institution or agency; this led to the eventual founding of the ACLA.[4]

The ACLA was formally organized in 1919 as a body of rural Americans and allies concerned with the declining economic, environmental, social, and political position of farmers and rural life. While most of its membership and leadership were farmers and ranchers, the association included educators, clergy, and merchants in rural and small-town America, as well as some urbanites who saw a vital need to protect rural life and interests. The ACLA was also culturally diverse. African Americans, Jewish Americans, and Mexican Americans were included among members of the association, as were women who served in its leadership. While members of the association disagreed on issues such as the degree and type of modernization warranted in farming and the practices of social life, their diverse interests were brought together by concern with the increasing marginalization of those who owed their livelihood to the land.

The analysis in this chapter follows the central strand within the Country Life movement by attending to the key voices found in the written record of ACLA publications. Through these publications, the association spoke to its membership, preparing them for engagements with other members of their communities as well as the wider public. The ACLA communicated with constituents through a monthly magazine, *Rural America,* and state and national conferences with published conference proceedings. *Rural America* covered topics ranging from methods of planting and harvesting to political problems facing farmers and their domestic and international counterparts. Articles arguing for or against federal rural policies were common, and both female and male farmers served as the magazine's reporters. ACLA conferences were held in every region of the country and proceedings, including speeches and committee minutes, were delivered to conference attendees who often numbered in the thousands. Collectively, this body of work vividly portrays the concerns of rural Americans and their allies in the years after World War I and is often notable for its stark warnings regarding not only the decline of rural citizenship but its dire implications for American democracy.

With the end of World War I, American farmers found themselves in a precarious position. They were mired in an economic depression made especially oppressive due to debts accumulated in the war effort, as well as the end of the price supports put in place by Herbert Hoover and the United States Food Administration during the war. As noted in the preceding chapter, farmers heeded the government's call to grow more food on more land as a moral duty to feed the United States and its allies. It is no surprise that farmers answered that call. After the war, however, as the market for their crops shrank, farmers found themselves caught in a financial bind. As Sarah Phillips observes, "Besieged by this postwar contraction, farmers found themselves caught between the low prices they received for farm products and the high prices they paid for nonfarm items."[5] For example, in 1919, wheat cost $2.19 per bushel but fell to $1.05 by 1929, plummeting even further after the stock market crashed in October.[6] According to R. Douglas Hurt, "by 1920, European recovery, international competition from Australia, Argentina, and Canada for the beef and wheat trade, and high American surplus production caused prices to fall precipitously."[7] Farm tenancy had dramatically increased, bankruptcies were at an all-time high, and commodity prices were at their lowest level in more than a decade. Low prices coupled with war-time debt accumulation precipitated a social insecurity and decline in status among rural Americans that appeared irreversible. In the Country Life movement, rural Americans sought to arrest this decline and regain their central position in American political life by retelling agrarian myth to meet the challenges of modern realities.

The myth of agrarianism has adapted to distinct moments in American history, particularly when economic concerns are at the forefront. As we have seen, historically, this myth focused on virtue, patriotism, and stewardship, rather than the economic elements of living in a capitalistic society. The era after World War I, however, marks a turning point in adapting the agrarian myth to present circumstances. Rather than continue to assert agrarian virtue as the primary value of farming, the ACLA asserted economic and participatory equality as central to agrarian life. By challenging the legitimacy of the state's authority, the organization began to publicly question how the state could so quickly turn its back on the very citizens it had leaned on—and praised—a few years earlier. Domestically, figures in the Country Life movement questioned democratic legitimacy by explicating the rural decline that state had authored. As the government formed alliances with urban industrialists, rural citizens had regressed from yeomen to peasants in one generation. Members of the ACLA also surveyed the international scene, recognizing and advocating for any reform or revolutionary movement seen as supporting equal opportunities for all citizens.

From a Country Life perspective, a state's legitimacy to govern depended on how well it preserved and promoted citizen equality. In this manner, the case of the ACLA demonstrates the symbolic import of the legitimation process in specific contexts. Focusing on publics as mythmakers who generate norms of democratic culture requires shifting the critical lens from critique to production. The ACLA mapped and revealed the patterns of privilege and exclusion that they saw as behind the rural decline; they also offered inventive rhetorics of resistance designed to reclaim the central role of rural life by reclaiming the link between working the land and building democracy. These rhetorical processes are linked by the question of legitimacy. In critique, members of the ACLA questioned the legitimacy of governments that seemed to have abandoned democracy as well as large and vital segments of the American population. In attempting to preserve rural culture, the leading lights of Country Life sought to reestablish the grounds of legitimacy in democratic culture as arising from connection to the land. Like the expansive gardening movement envisioned by Pack at the end of World War I, discussed in the preceding chapter, the ACLA envisioned a democratic rebirth associated with the virtues of farming, for Americans and their communities. To further explicate these movements, we turn to a consideration of counterpublics and satellite publics, and their resultant rhetorics, for describing sites of generative resistance. The ACLA offers a compelling case of how a satellite public participates in rhetorical mythmaking by challenging the legitimacy of public authority while regenerating democratic practices.

Publics, Counterpublics, and Satellite Publics

Studies of publics are often accompanied by theories of democracy that elucidate the relationship between citizens and the state. Jürgen Habermas's account of the structural transformation of the public sphere has served as the starting point for many of these studies and continues to provide a basis from which theories of publics and public spheres are affirmed and contested. Publics confront public authority and, according to Craig Calhoun, this type of authority is unique because it is both state-related and "constituted as an *impersonal* locus of authority."[8] Thus, public authority is a type of impersonal state authority. A public consists of citizens who contest public authority, while counterpublics are, according to Nancy Fraser, "parallel discursive arenas where members of subordinated social groups invent and circulate counterdiscourses, which in turn permit them to formulate oppositional interpretations of their identities, interests, and needs."[9] The relationship of publics and counterpublics to public authority is positional: publics contest from within, while counterpublics contest from without. While publics exercise their rights as bodies of citizens in public spheres, counterpublics may be unable to exercise those rights from their more contingent position. As outsiders, however, counterpublics may produce inventive hybridizations of ideology, identity, and voice.

Both publics and counterpublics challenge exclusion, yet are different due to their position vis-à-vis public authority. In some cases, however, neither of these critical categories sufficiently explains how a group challenges public authority. Catherine Squires's notion of *satellite publics* provides a third kind of public that emphasizes the inventive rhetorical work a public can perform when its members exercise their rights as citizens yet are excluded from decision-making mechanisms. Squires suggests that satellite publics "desire to be separate from other publics" and "aim to maintain a solid group identity and build independent institutions."[10] These publics blur the binary attributed to publics and counterpublics because "satellite publics can emerge from dominant or marginalized groups" in order to engage in "wider public debates when there is a clear convergence of their interests with those of other publics or when their particular institutions or practices cause friction or controversies with wider publics."[11] Neither dominant nor marginalized, satellite publics step into a discursive space of communicative engagement that invents argumentative strategies, challenges the legitimacy of public authority, and generates civic rhetorical practices capable of altering prevailing mythic formations.

Figures in the Country Life movement questioned the legitimacy of the state because of the spatial and participatory separation between the federal government and rural folk. At its most basic level, according to Habermas, "*Legitimacy means a political order's worthiness to be recognized.*"[12] Habermas's normative definition of legitimacy is citizen-centered because legitimacy is performed by public authority in front of citizens as "good arguments for a political order's claim to be recognized as right and just; a legitimate order deserves recognition."[13] Joshua Cohen contends that "the authorization to exercise state power must arise from the collective decisions of the members of a society who are governed by that power."¹⁴ Seyla Benhabib suggests that a society where legitimacy is not seen as a public good will inevitably be thrown into crisis and, more importantly, "legitimacy in complex democratic societies must be thought to result from the free and unconstrained public deliberation of all about matters of common concern . . . [and] is essential to the legitimacy of democratic institutions."¹⁵ Habermas, Cohen, and Benhabib position citizens as the exclusive grantors of democratic legitimacy. O. A. Payrow Shabani asserts, "Habermas believes . . . that in today's pluralist societies social norms are valid only if the subjects they govern see themselves also as the authors of the norms."¹⁶ In short, democratic legitimacy is granted by citizens to public authority.

Democratic legitimacy is never permanent, and public authority must continually prove its worthiness before citizens. Habermas explains that a legitimation crisis most often occurs when the actions of a political order are challenged. The political order and those challenging that order are involved in a constant back-and-forth that demands a political order continually demonstrate its legitimacy to rule.¹⁷ When the legitimation process reverses and citizens must demonstrate their worthiness to be recognized by the state, publics emerge to challenge this reversal and engage in generating civic rhetorical practices from which democratic culture can be reimagined and myths altered.

Although the ACLA arose in a long line of US agrarian publics, their actions in the 1920s and 1930s illuminate how these publics can function rhetorically to reverse a legitimation process that now required citizens to demonstrate their legitimacy before the state. In effect, this approach focuses on how publics invent strategies to challenge the legitimacy of public authority, yet also generate civic rhetorical practices that enable them to live out their commitments despite the existing formations they encounter. The case of the ACLA suggests how a focus on democracy as it exists may limit the ability to view publics as sites of rhetorical invention. Explaining how a public's exclusion occurs and continues is an important first step; however, the experience of the Country Life movement suggests that exclusion can be the context in which democratic possibilities emerge. The importance of satellite publics to democratic culture rests with how, in moments of conflict with public authority, they invent the rhetorical resources necessary to confront, contest, and contain its expanding reach.

The Fall of the Yeoman

Democracy was "rule of the people, by the people, and for the people," as Country Lifer Robert D. Baldwin recalled, but now this democratic practice was in peril.[18] In the 1920s and '30s, Country Lifers were galvanized by what they saw as the betrayal of democracy in the devolution of rural citizens from yeomen to peasants. The circulation of *Rural America* and the ACLA's annual conference proceedings spurred discussions about the precarious status of rural people. If the state was not interested in preventing the creation of a peasant class because of its continued privileging of elites, America could no longer be characterized as a democracy, but was instead an oligarchy. Country Lifers confronted unjust state policies that would prevent rural folk from participating in every aspect of public and political life. As ACLA president Frank O. Lowden asked in an address at a 1929 ACLA convention, "What kind of a farming population do we want?"[19]

This question manifested a concern that had been circulating in the pages of *Rural America* since the end of the war and arose from a framework that saw farming as grounded in democratic participation rather than solely an economic sector or means to meet aggregate demands. Farmer Clarence Poe suggested, "One of the great problems facing America is that of how to save the farmers of America from drifting into the condition of most European peasants."[20] For rural Americans, drifting towards peasantry was not merely a matter of their economic conditions but, more importantly, a question of their place and position in a democratic society. Poe went on to argue, "Not only must financial profits be increased, but the farmer must rise to the dignity of being an actual force in the government of all the conditions affecting [their life]."[21] The critical categories of *yeoman* and *peasant* illustrate the different positions that publics and counterpublics have in relation to public authority. A yeoman public consists of citizens with the capacity to challenge the current democratic regime, whereas peasants exist on the outside of political life. Certainly, a mass of peasants with limited rights was never intended by the supposedly democratic society of the United States, which had been founded on the bedrock of the yeoman citizen-farmer. Country Life rhetorics drew from an agrarian mythic narrative that positions independent freehold farmers as central to society, while at the same time altering that narrative to include economic interests. In this manner, Country Lifers remade agrarian myth to respond to the precarious position of the farmer as the twentieth century unfolded.

Country Lifers seized on the mythic narrative of American agrarians as yeomen to position themselves as firmly in political culture and necessary to the life of a democratic nation. The contrast between yeoman and peasant is pronounced and especially apparent to those recent immigrants who had fled countries where rural peasants remained among the lowest social castes. Lowden explained in his convention address: "History discloses but two types of farmers: what we may call the peasant type and the yeoman type. Which of these do we desire? Up to the present time our farmers have been composed largely of the yeoman type—independent, self-respecting, demanding education for their children and social equality with all other classes."[22] The discursive choice appeared obvious: yeomen are a free people who have political and property rights, while peasants are subjects of oppression, inequality, and dependence. Lowden's articulation of American agrarian myth locates this distinction as a choice between oligarchy and democracy. Yeomen were central to the health of United States democratic life where all participate as citizens, in contrast to European peasants, who "are mentally and physically unfit for other pursuits."[23] The distinguishing feature of yeoman farmers rests with their ability and opportunity to participate fully in political life. From this perspective, the projected "descent" of America's yeoman farmers into peasantry symbolizes the erosion of the grounds of American democracy and the repetition of the mistakes of the old order in Europe. The core of the American public, as proclaimed by Jefferson and others, would be relegated to the status of a counterpublic, on the outside of public authority, while the special quality of American democracy would be lost. The Country Life movement staked American identity, traditionally tied to agrarian myth, to the faltering social and economic standing of rural people.

While discussions of peasantry began in a quest for economic and participatory parity, Country Life rhetorics placed these concerns within the broad cultural narrative of the future of American democracy. The dark days following the war prompted farmers to question the rapidity with which they moved from prominence to poverty. During the war, farmers answered the state's call for increased production, incurring significant debt; yet after the war, farmers were viewed as poor businesspersons who had unwisely expanded.[24] From the Country

Lifers' perspective, "the farmer contracted debts on a high price basis. The farmer has suffered ruinous losses and now [the farmer] is beginning to inquire about the distribution of wealth in the country."[25] Compelled by the state, farmers purchased more land so that they could produce more food, leading key figures in Country Life to question how those living on the land could move from essential to irrelevant in a few short years. The first ACLA president, Kenyon Butterfield, pointed out that farmers had always increased production for the good of the nation. In short, the American farmer "enlarged [their] acreage of crops and sought, under great difficulties, to increase [their] yields. [The farmer] has fed [their] own armies as well as the civil population of [their] allies."[26] However, the current political situation led Butterfield to declare, "There will always be a need for a fighting farmer's organization."[27] Butterfield's declaration arose from a belief among farmers that public authority was pushing them to the margins of society and that they had to resist. Country Lifers strongly believed, in the words of M. L. Wilson, that "democracy is impossible without intelligent comprehending citizens"; however, rural folk were devolving into a peasant class that would soon be shut out of democratic participation.[28] These concerns were expressed in a mythic narrative of yeoman versus peasants with not only the status of farming, but America's democratic future, hanging in the balance. If yeoman farmers transformed into peasants, democracy would collapse as permanent economic classes were established. By calling out this imminent fate, the ACLA linked the standing of the farmer to the prospects for democracy while symbolically projecting the position of rural people at the center of national culture. America would stand or fall with its rural citizens.

Throughout the 1920s, the ACLA portrayed rural life as on the margins of American life and described an America divided by a government which provided "special privilege" to the wealthy and "special discrimination" to farmers.[29] In the early 1930s, figures in the ACLA became increasingly pessimistic about the state's interest in preventing rural folk's further descent into poverty, indebtedness, and separation. Country Lifer V. H. Culp lamented that politicians did not govern for the good of the nation, but in favor of those special interests that funded their political campaigns. He offered a question that expressed the frustrations of rural Americans: "Has the government succumbed to selfish organizations and become helpless in this great struggle for economic equality?"[30] The legitimacy of public authority was in question because the state enacted political power in a way that limited rural folk's ability to participate and, thereby, resist the insidious onset of peasantry. Culp contrasted the kind of interest-based politics practiced by politicians toward corporations with farmers' collective struggle to remain participants in democratic culture. This economic and political concern signaled the separation between rural folk and an unresponsive state. Country Life rhetoric probed for a way to redirect the state from corporatization toward democracy. Yet the voices of rural Americans and their allies in the Country Life movement remained external to the new reality, even as they sought to reclaim the central place of citizen-farmers on which democracy was based in agrarian myth.

While figures in the movement crafted their arguments to include economic interests, they were careful to define these to complement, rather than contradict, the older forms of agrarian myth. Yeomen were connected to the state through property interests, and this, in part, made them valuable citizens, distinct from peasants who did not hold a stake in an ownership economy. The ACLA did not simply argue for increased wages, regulations that would end corporate price gouging, and laws that discriminated against the business of farming, but also for a sense of justice that would vindicate rural people. In the midst of rural decline in 1928, a group of eleven Indiana farmwomen published "A Declaration of Farm Women" in *Rural America*. The Indiana farmwomen declared, "Farmers are being robbed by law. Our government is being subverted from a free democracy to an oligarchy dominated by industrial and financial leadership. The doctrine of equal rights to all and special privileges to none is an order passed away."[31] These women argued for equality, both participatory and economic, noting that by abandoning farmers, the state voided the principles on which it was founded. They argued, "The annual toll of predatory wealth reaped from the products of our lands must cease."[32] Again and again, the movement expressed the belief that diminishment and exploitation of farmers placed American democracy at risk.

Leaders in Country Life looked to the history and tradition of peasantry around the world to inform their situation. While at times buoyed by progressive movements among farming people in other nations, they also saw the long history of oppression that withheld education, economic justice, and political equality from peasant classes. Lowden remarked in his convention address: "Shall we permit our farming population gradually to descend to the lowly status of peasant, content if they but wring a meager living from the soil? And, mind you, if we shall ever come to this, we shall have a peasant population on the land inferior to that in older countries where peasant farming is the rule. For, in those older countries the peasant population has inherited for centuries their place in the social scheme."³³ This assessment demonstrates the sophistication with which rural Americans approached the problem of farm life as multifaceted, with economic, social, and political aspects. Lowden suggested economic and political status defined peasants as people living off the land by earning meager wages. Unable to participate as citizens apart from the production of agricultural commodities, peasants' social status created a class of citizens incapable of democratic participation because, according to Lowden, "neither they nor their forebears were given an education and therefore a larger outlook upon life."³⁴ Education, in this formulation, was not merely a tool of technical training for farm youth; rather it served as their first point of contact with the world outside their local context. It was a tool for democratic participation that looked to a greater public good. Education had been a defining feature that separated American farmers from European peasants and gave Americans the capacity to participate fully in American public life.

Figures in Country Life further saw peasantry for rural people as an inevitable consequence of an autocratic state. L. J. Taber, master of the National Grange and a member of ACLA, contrasted an autocratic form of government characterized by militarism and economic decline with a democracy that "cannot endure unless it gives to the average [person] a reasonable chance to secure a fair start toward success in the race of life."³⁵ If the state did not provide a fair start for all citizens, it no longer possessed democratic legitimacy. The opportunity for success, defined not only economically, but as democratic participation, should be built into the structure of society for all people. This is equality not by chance or circumstance, but as a natural right supported by conscious design and cultivated by democratic participation.

Farmers were essential to the economic life of the nation yet were not treated as political or economic participants. In times of plenty farmers were neglected, but in times of crisis they were cast as society's saviors. Farmers were tired of moving between these extremes. John J. Tigert explained how present modes of group organization promote cooperation, yet this cooperation could not function properly when the government privileges one group over another, as "government by majorities becomes impossible under such a regime and democracy disappears from the land."³⁶

In short, members of Country Life thought it unconscionable that farmers were becoming relegated to the production of food and tangential to the political life of the nation. The specter of peasantry transformed economic injustice into an issue of common concern that linked the future of democracy with the capacity of citizens to participate. Further, the widening gulf between public authority and farmers served as the impetus for dissociation of democracy from the state. This critique opened a discursive space between publics and counterpublics in which the rhetorics of Country Life could rearticulate democracy as made and maintained by the practices of citizens. Democracy was a common field to be tending and it increasingly required resistance to a captive corporate state that no longer respected the interests of citizens.

The economic plight of rural Americans remained significant as the members of the ACLA questioned the legitimacy of the state through the dissociation of democracy from public authority. The movement sought a transformation in democratic culture that would make these exclusions, symbolized by marginalization to the status of peasantry, unthinkable. The specter of peasantry provided the discursive imaginary for articulating the economic condition of farmers as a symptom of authoritarian regimes. From here, figures in Country Life sought to generate possibilities from which a participatory culture could be reconstituted. Country Life discourse remade American agrarian myth by simultaneously challenging public authority and explicating and attempting to actualize rural experience as the root of American democracy.

A Legitimacy-Challenging Public

Reinventing the legitimation process began when Kenyon Butterfield of the ACLA challenged President Wilson's claim, justifying the entrance of the United States into World War I, that, "the world must be made safe for democracy." Instead, Butterfield argued that "Democracy must be made safe for the world."[37] United States democracy had become increasingly separated from citizens, leading to the belief that if exported from America, democracy would be little better than other governmental forms where power resided with the elite. Butterfield concluded, "The war is over, but the great struggle of the new day has but just begun."[38] Following Butterfield, the ACLA took the problem of peasantry as the driving necessity to challenge the state's legitimacy because the future of democracy and rural life were in question. As early as the first national ACLA conference in 1919, Country Lifers concluded in their official discourses that rural life should be a vital part of attaining world democracy because, "If democracy throughout the world is our goal, we must make rural life democratic."³⁹ Local and global democracy relied on the equality and participation of citizens, regardless of place. If a government failed to enact the principles of democracy, it could not legitimately claim to be a democracy. Due to its shoddy treatment of its rural citizens, even the United States was not exempt from having its legitimacy questioned.

As a satellite public, the ACLA challenged the legitimacy of public authority by dissociating democracy from the state and repositioning democracy with citizens. Dissociation functions as a "technique of separation," according to Chaim Perelman and Lucie Olbrechts-Tyteca, that assists in the disassembly of concepts within a reified system of thought.⁴⁰ Dissociation pulls apart concepts that are presumed to fit together in order to invent new rhetorical possibilities for understanding and action. The governance experienced by Country Lifers could not be democratic because they continued to be pushed to the margins of the American landscape. Dissociating elements produced a significant shift in the argumentative concepts foundational to a community. As Perelman and Olbrechts-Tyteca have argued, "It is then no more a question of breaking the links that join independent elements, but of modifying the very structure of these elements."⁴¹ The strategy of disassociation allowed figures in Country Life to restate agrarian myth by casting rural citizens as political actors with economic interests because of their virtue, not despite it. Economics, citizenship, equality, and participatory decision-making combined in a mythic connection of agrarian democracy to a modern world.

Country Lifers also sought precedents for their struggle in the history of resistance to authoritarian regimes. Culp compared farmers' struggle with the Boston Tea Party, where farmers "threw tea overboard and even went to the extent of rebelling against a tyrannical form of government."⁴² The implication was clear: when a government ignores farmers, they have no option but to resist. Resistance, even revolution, became acceptable steps when the state had ceased to function in a democratic mode that promoted equality. Rather than see the American state as inherently distinct from authoritarian governments abroad, Country Lifer Lewis Edwin Theiss voiced the belief that people should "judge democracy by what it accomplishes or fails to accomplish for them; for by its fruits even government shall be known."⁴³ Through such dissociation, rural Americans projected a discursive space from which to resist public authority as empowered citizens, despite their displacement to the economic and political margins.

Some within the ACLA were curious to see if the Bolshevik Revolution in Russia would lead to substantive change between peasants and the state. Communism was not immediately rejected by all Country Lifers, but seen as a different governmental form with similar results to those they were experiencing in America. Official discourse from the ACLA recognized the Russian situation a decade after the revolution as different from the promise at the start of the revolution, and another diminishment of farming: "The Communist group now bidding for leadership in world economic life professes friendliness to the peasant, but it becomes increasingly clear that its policy thus far has succumbed to an urban proletariat's desire for cheap food and raw materials. It takes away [the farmer's] hold upon [their] land and offers [them] a collective whose products are disposed of at a price set by an urban proletariat."[44] The accommodation between Communist leadership and an urban proletariat, at the expense of exploited farmers, mirrored conditions in the United States, as described by the leaders of the Country Life movement. Butterfield remarked that the growing political and economic separation between country and city relegated rural communities to the outer reaches of society. With rural communities no longer viewed as essential to the life of the nation, he was left to wonder, "Have farmers nothing to contribute?"[45] For Country Lifers, the city served an important function in American democracy, as did rural areas. The problem became what democracy would look like if rural folk lacked the standing to be full participants. Country Lifer Arthur E. Holt suggested: "There are a good many of us who want to live in a country in which there is no congestion of advantage in the hands of a few. It is absolutely intolerable to us that the American farmer should become a European peasant."[46]

The association between the false promises of government, whether communist or democratic, and the fall of rural people to peasantry continued to be the discursive framework that enabled the dissociation of democracy from the state. "*The working farmers of America as a class have not been represented in any authoritative or adequate way in the groups that have outlined policies nor in the councils that have determined destinies*," Butterfield remarked. "This is not a new situation."[47] The ACLA demanded economic and political equality and a place in decision-making because of the contributions yeoman farmers could make to democratic culture, not merely because they wanted their piece of the proverbial economic pie. Challenging public authority's legitimacy required the ACLA to call for a fundamental change in the way publics interacted with the state.

While the state had reversed the terms of legitimization by attempting to become the legitimization authority, the ACLA resisted this reversal and sought to restore the legitimizing function to citizens. It did so by questioning the nature of a state that marginalized its own citizens and suggested that the cradle of democracy lay in the participation and legitimizing functions that originated with rural citizens.

Cooperation and equality were central to resistance of the legitimation reversal. The rhetorics of Country Life rearticulated democracy with citizens to subvert the present "national policy of economic rehabilitation" that "save[d] the present profit-seeking capitalistic system and the related political democracy."[48] Ernesto Laclau and Chantal Mouffe refer to articulation as "any practice establishing a relation among [discursive] elements such that their identity is modified as a result of the articulatory practice."[49] The example of Country Life demonstrates how dissociation can also have an articulatory function in mythic narratives that, in this case, reestablished democracy with citizens. With democracy located in the attitudes and actions of citizens, the people alone possessed the ability to recognize and animate governmental forms. While the dissociation of democracy and the state challenged the legitimacy of public authority, the articulation of democracy with citizens invented a discursive space from which an agrarian mythic narrative could be reimagined and enacted. The United States was an agrarian democracy, despite its current form as a corporatized state with an interest in profits rather than the well-being of its citizens. The narrative of Country Life expressed allegiance to true democracy even in a state that left citizens exiled in their own homes.

By dissociating democracy from the state, this narrative linked the state with an authoritarian system that, according to ACLA member Gordon H. Ward, had brought about "a Fascist dictatorship over farmers and city workers." Ward noted that American democracy had lost its way as basic rights of free speech, including petitioning the government and advocating for change faded away.[50] By dissociating democracy from the state, Ward linked the state to authoritarian rule, an illegitimate form in a democracy. Calling out this anti-democratic position allowed members of the ACLA to symbolically reclaim their rightful position as the grantors of democratic legitimacy. While public authority had reversed the legitimation process and moved rural folk to the margins, the ACLA continued to act as a public, with members exercising their rights as citizens.

To challenge the legitimacy of public authority, the ACLA public used democratic ideals of equality, independence, self-sufficiency, and liberty to critique the current state of American democracy. At the 1936 ACLA convention, M. L. Wilson focused on the problem of democracy for rural Americans, posing a series of questions that challenged democracy in its current American form. He remarked:

> Our present situation raises before us all sorts of question about the nature of democracy. Is democracy a fixed thing, or is it an evolving, changing idea? Are the concepts of liberty, equality, and fraternity different now from what they were when we lived in a simpler society? Is democracy related to the environment of a people? Did it take one form when we were a nation of frontier farmers, and must it take on different forms now that we have become a complex industrial country with the agricultural frontier gone, and most people engaged in highly specialized activities instead of continuing as members of a self-sufficient family unit such as we had 150 years ago?[51]

These questions were prophetic, identifying how democracy had become a governmental form that exercised power over citizens rather than enable citizen participation. Democracy, for Wilson, was wedded to a capitalist economy that privileged divisions and specialization over self-sufficient communities. His questions rearticulated democratic values for the twentieth century, while identifying rural experience as the link between restored democratic practices and traditions resonant with meaning.

In this formulation, democracy could not be imposed on a people or instituted with a change of political party; instead, it was an attitude embodied in citizen action. Wilson asked if democracy could be faithfulness to a political system and a way of living that must adapt to the conditions of everyday life.[52] The fear of relegation to peasantry could be understood, from this perspective, as motivated by the changing political and economic conditions that made the decline of yeoman farmers possible. However, if democracy should be understood as both an attitude and action that could only occur "from the bottom up" and remained "dynamic and changing, never finished," as ACLA president Nat T. Frame declared, then out of the yeoman-peasant tension would come the possibility of a more vibrant democratic culture.[53]

While dissociation freed democracy from the state, Country Lifers such as Butterfield seized this rhetorical opening to rearticulate democracy as the work of citizens. Democracy's dynamism, he said, relied upon "four great ideas that serve as the underpinning of a true democracy," namely "(1) Individual freedom, (2) Equality of opportunity, (3)

Responsible participation in affairs, and (4) Cooperation for the common good."[54] Such discourse retooled agrarian myth for the twentieth century by reclaiming the status of rural experience as the bedrock of American exceptionalism. Without yeoman farmers, America stood to lose its source of virtue and its democratic foundation.

In the estimation of leaders such as Butterfield, democracy atrophied and could be co-opted by an authoritarian state when it was understood exclusively as a form of government and not a set of practices embodied in the citizenry. The federal government in its authoritarian mode had come to count on a passive citizenry who naively presumed that a governmental form that claimed the mantle of democracy was in fact rooted in democratic practices. Butterfield sought to correct this wrong-headed conception: "Most Americans think of democracy as a form of government in which the people rule through their right to vote. Democracy is something far more than popular election of representatives or even of popular vote on laws and constitutions. . . . We have assumed that the democratic mode of government makes us sure of the democratic mode of working and of living together."[55] Butterfield's challenge to the state's actions dissociated the state from citizens and institutions from democratic practice. His characterization was one of many within Country Life that modified the ruling notion of democracy by emphasizing its importance as a practice only identifiable in the interactions of citizens who share a common space and similar commitments to coexisting equally within it. Agrarians were not merely industrious or virtuous citizens but the ideal democratic participants, capable of inventing communicative practices that align with their position as citizens.

Generating Civic Rhetoric

The ACLA promoted the importance of local communities to the life of the nation. As an ethic of human relations, democracy rooted individuals in their immediate community. "A community gives the individual member a sense of security, a consciousness of having the world at his back," argued Walter A. Terpenning.[56] Terpenning, an advocate for the community commons and "open country," evoked a rich sense of community as the place in which one becomes acquainted with the world. In the agrarian myth as he envisioned it, community is not a place or a collection of disparate individuals but a principle of connection by which we come to know who we are in relation to the land and the others who share in it. For Terpenning, like other advocates affiliated with the ACLA, community stood at the center of an alternative vision of the nation, where rooted people enacted a democracy of equals. In community, the abstractions and inequities of modern life could be corrected as a richer sense of public determination, linked to the collective fate of crops and citizens, emerged.

In this vision, tinged with nostalgia and utopian impulses, democracy formed as citizens developed solutions to social problems through collective practices where all ideas were considered. Country Lifers sought to put these ideals on a practical footing as they circulated texts and engaged in face-to-face discussions regarding what democratic life should be and could become. The ACLA offered a democratic vision that relied upon citizens' embodiment of values such as equality and cooperation in their everyday lives. Only out of citizens' participation in local community could democracy grow and mature. Challenging the prevailing conception of the relationship between individual and land as exploitative, leaders in Country Life saw cooperation as a way of existing in the world oriented by a view of democracy located in and performed by citizens.[57] In building community through cooperative relations, democracy flourished as the land itself retained its life-giving power. Democratic practices bound communities and the environment.

In the rhetorics of Country Life, community and democracy were linked through metaphors of simplicity that oriented democracy in place and time. According to Liberty Hyde Bailey, "When we get away from the soil we begin to get away from simplicity."[58] For Bailey, a movement away from one's local environment would result in rootless disconnection from the world. In an attempt to reconnect communities through care of the land, Frame used his presidential address to the ACLA to recite Bailey's *The Holy Earth*. This poem written two decades prior was relevant, in Frames' estimation, for the present:

> The Holy Earth
> Society must keep
> Its contacts with the earth
> Or it is doomed.
> The farmer is
> Trustee of his soil.
> Let no one use his land
> To the detriment of society.[59]

While farmers were responsible for tending the land for the benefit of the nation, they also recognized a sacred bond to the land as the site of community and democracy. Frame's vision of societal doom triggered by separation suggests that, in its proper form, the relationship between society and the lands is mutually constitutive. When society recognizes the importance of the land, it survives, and the land itself is seen as holy. When this bond breaks, destruction ensues. As in other forms of agrarian myth, the farmer is placed in the lead position, humans and land conjoin, and it is left to farmers to maintain and sanctify the sacred connection. In this position between the land and society, the simplicity ascribed to the farmer is indeed profound.

Simplicity, in such a rhetoric, links community with democracy in an ethic of human relations. Simplicity was a way of living and acting in the world that organized rural life and had the potential to recast the life of the nation. "The simple life," Bailey remarked, "is a state of mind. It is a simplification of desire, a certain directness of effort and of purpose that brings us quickly to a result, and such an attitude that we derive our satisfactions from the humble and the near-at-hand."[60] Bailey's connection between the purposefulness of action and an attitude of satisfaction with one's locality, the "near-at-hand," offered a view of simplicity as an aesthetic of social organization. Simplicity positioned locality as the foundation of relations, for humans and non-humans, as well. Working the land, in this democratic ethos, is not done exclusively for oneself, but is a way of participating in a common good that linked the fates of people. By reorienting democracy through metaphors of simplicity, such discourse generated norms for democratic culture that envisioned how rural people could remake a nation.

Rural people embodied the cooperative attitude necessary for democratic citizenship because of their enduring connection to the land. As G. W. Rutherford remarked, "local democracy is the foundation of democracy. It is in local rural government, then, where the vitality of democratic institutions must be kept alive."[61] Rutherford viewed democracy as practiced in local communities and extending to state institutions, with the vitality of the former implicating the legitimacy of the latter. Focusing on one's locality did not isolate rural Americans from national concerns. Instead, from one's locality a national politics could be formed and sustained in a way that kept citizens in a central position as legitimizing agents. Far from a vestige of isolationism or reactionary politics, simplicity was an ethos that flowed outward. While it may be presumptuous to cast rural citizens at the center of democratic life, the ethic of simplicity offers an accessible motif for others in urban places or various backgrounds or walks of life to connect symbolically. As Edmund S. Brunner put it, "we must recognize neither East nor West, North nor South, white nor Negro, city nor country, owner nor tenant. There must be rather that equality of opportunity for which democracy stands."[62]

When properly grounded in place and citizen actions, American democracy would build across localities, leading to common practices of equality and inclusion in which all parties had a stake. The nourishing origin of democracy, however, remained the rural community. Democracy sprung from, and was sustained at, this intimate level. Without local democratic practices, democracy would fail. While Country Lifers generally viewed rural life as the foundation of democracy, Brunner's vision was broader. He suggested that both rural and urban citizens paying the proper attention to their localities could embody the attitudes and practices necessary for the ongoing production of democratic culture.

In the mythic retelling of Country Life, democracy was "simple" not because it was without complications, but because it allowed direct participation and operated at scale that rendered citizen actions meaningful. Liberty Bailey exclaimed, *"We need the example of simple institutions."*[63] This simplicity was measured by proximity to public authority and the ability of citizens to contribute to the decisions that shaped their lives. Simplicity of institutions ensured citizens' participation in "the machinery of our political life"; they connected people with public authority because "the more direct the institutions the more efficient and enduring they are."[64]

When localities embodied democratic practice a more efficacious collective politics would result. Cracker boxes or cracker barrels evoked local politics as performed by citizens on the open porch of a local country store. While sitting on these boxes, rural folk talked of politics, religion, and society, examining the problems and possibilities, first of their local community then of the region, nation or world. Cracker boxes illustrated "neighborly democracy which in the past has formed the matrix for even the highest individual achievements in political, social and religious thinking ... with the mood of cooperative inquiry."[65] Agrarian simplicity was an affective unification in which "each individual acquired [their] sense of social unity, [their] sympathy for [their] associates with whom [they] worked and played and the primary ideals of kindness, justice and good faith."⁶⁶ Democracy must be a learned practice, situated in one's local community, that ensured the proximity of citizens to authority. M. L. Wilson noted, "I think it is obvious to all of us that this modern development of the town hall and cracker barrel concept has grown out of today's acute need for a revitalization of democracy."⁶⁷

The cracker box was far from the only vehicle for community collective reasoning or democratic deliberation at the local level in the United States. Bailey saw in the American people an inventive genius for constructing "simple institutions" for encouraging direct participation. He finds examples across the landscape:

> The simple native institutions have largely determined the methods and points of view in great geographical regions—the New England town meeting, with its ideal democracy; the southern court-house, with its social stratification; the central-west schoolhouse, repeating the democracy of New England but with a freer individualism; the arid-west ditch meeting, repeating again the democracy, but made applicable by the urgency of a single vital problem. It is doubtful whether a nation of cities could be a democracy.⁶⁸

Each region enacted democracy according to its own character. While Bailey's remark about the inability of cities to serve as society's foundation underscores a Jeffersonian belief in rural democratic exceptionalism, the ACLA took pains to include city dwellers in their vision of a network of localized democracy. For the ACLA, urbanites were the potential allies that Danielle S. Allen has termed "democratic strangers."⁶⁹ She stresses the importance for "democratic citizens generally to develop their capacities for political imagination, particularly with reference to the strangers in their lives."⁷⁰ Allen's provocation is instructive in a study of publics as more than a critique of how institutions exercise power over citizens. Instead, Allen brings us to an understanding of how, according to Jeffrey A. Bennett, "strangers repeatedly actualize the discursive bonds of public life."⁷¹ Country Lifers recognized stranger relationships as a vital aspect of democratic attitudes and actions. As M. L. Wilson suggested, "a great problem in democracy in the United States is to get all classes of people to thinking about their relationships in society and to thinking beyond their own group into the problems of general welfare."⁷² The problems of general welfare did not exist a priori for Country Lifers but were the result of present contingencies encountered by citizens inventing solutions collectively and practicing democracy locally. To disregard democratic strangers would entail abandoning democracy as a way of living, a prospect Country Lifers were not willing to allow. Wilson prophetically cautioned: "We are in great danger that each class in society will develop a feeling that 'if I can have prosperity for myself and my group, even though at the expense of the rest of society, it will all work out some way for the general good.'"[73] For Wilson, like others in the movement, the common good had agrarian roots. It began with caring for the land as a sacred trust and fostered democracy though a complex set of relations involving place, family, community, the stranger as ally, and the nation as a body of rooted citizens. As we developed a feeling for the common good, democracy would be sustained as a fabric of being.

Country Life as Democracy

One way to conclude a study of the Country Life movement is with a list of its tangible accomplishments, including rural mail delivery, crop subsidies, rural healthcare and electrification, and a litany of other federal legislation benefiting rural America, parts of which remain codified in federal and state law and institutions. However, we see the vision and rhetoric of agrarian democracy as among the movement's greatest legacies. In the play of insider and outsider, de-legitimization and re-legitimization, the satellite public of Country Life sought recognition, transformation, and the restoration of distinctive qualities of American democracy. Country Lifers challenged the legitimacy of an inequitable state with a compelling counter-vision of democracy as citizens in action, grounded in locality, but developing an interest in the common good. The ACLA recast American agrarian myth as a way of reclaiming and reorienting American democracy, by citing the actions of citizens living in relation to the land, communities, and each other. In this formulation, the state does not grant political legitimacy to those it favors but instead must be granted legitimacy by citizens who see it as an extension of local communities and an agent of the common good. For the members of Country Life, however, this vision of democracy remained a rhetoric of resistance as they saw the plight of rural Americans and local communities in continuing peril from the diminishments and neglect of an authoritative state. With the decline of rural life, the ACLA saw the decay of the foundations of American democracy. The result of the status quo was likely to be disastrous. As figures in the Country Life movement were prone to point out, the Unites States had begun to emulate the social structures of the old European powers that now lay in ruins after the war. By marginalizing local democracy, direct action, and the patriotic farmer, the United States risked a similar fate, as an elite class consolidated power over a peasantry with scant rights.

By questioning the legitimacy of the state, as well as its role as the legitimator of publics, Country Life sought a reversal of authority to restore the rights of citizens, who they saw as the ultimate legitimator in a democracy arising from community and the land. In this manner, Country Life set new parameters for agrarian myth as rural Americans become not only the model of civic virtue but those most responsible for the sustenance and sustainability of a democratic state. This line of thinking redirected the state propaganda directed toward farmers in World War I, in which the fate of the nation, as well as the prospects for a new world order rising from ruins, hinged on the ability of American agriculture to produce abundance. While American farmers had accepted this responsibility during World War I, after the war farming became just another aspect of the economy to which no special favors were owed. The ACLA sought to persuade others to reflect on the implications of this development. Were farmers simply economic agents or was farming something more? For here the arguments deepened to consider the nature of democracy and practices of equality. What sort of nation was this to be and how could that vision best be realized? While rural people were the progenitors and guardians of democracy, democratic practices were open for others to adapt to their communities. An agrarian vision of democratic practice was a commonplace open to all Americans. While this ideal may remain partly hidden by the continued progression of a state the ACLA resisted as corporatized, authoritarian, and dangerous, it has been summoned by others engaging with agrarian myth as a rhetoric of democracy. The cases in this book testify to the continued persistence of the imaginary of rural democracy, as well as its travails in a nation and world that is predominantly neither democratic nor agrarian. Instead, the myth is an alluring emblem of resistance.

The remaining chapters of part I of this book continue to trace the evolution of agrarian myth begun in the twentieth century. Next, we consider the case of the Southern Agrarians, a group of intellectuals, who, like the ACLA, sought to reverse a decline of rural life, against an adversary now perceived as an industrialized, technocratic and dehumanizing state. Then, we examine the rhetoric of J. I. Rodale, prophet of organic farming, who saw "chemical farming" and the culture around it as degrading the soil and the ancient ways of living on, and through, the land. While both the Southern Agrarians and Rodale add spiritual dimensions and questions of good and evil, sin and redemption, to the discourse of agrarian myth, they also assume the vital links between farming, social life and governance that the Country Life movement explicated with fervor. As we turn attention to contemporary cases in part II of the book, echoes of Country Life abound. The attitude of inclusion in Country Life is born out in the experiences of urban farmers including the largely Latinx community around the SCFarm in Los Angeles. The concept of rural life as the foundation for democracy is at once underscored and negated in the coopted creation myth *So God Made a Farmer*, as the attributes of agrarian citizenship morph in to the qualities of RAM trucks. Finally, Country Life remains a vivid example of the legitimizing counterpublic of united agrarian interests that we explore in the book's conclusion as a prophetic measure for our times.

CHAPTER 3

The Southern Agrarians Take Their Stand

As the Country Life Association reworked agrarian myth to reclaim the yeoman farmer as a democratic figure, another critical social movement was gaining momentum. This time, the political and cultural critique emerged not from farmers but intellectuals who sought to challenge dominant narratives of progress by standing with a history of agrarian regionalism. This movement would not only foreshadow the political balkanization that persists today but would further demonstrate for subsequent generations how agrarianism could be used to generate social resistance. Although many cite Wendell Berry's groundbreaking 1977 book *The Unsettling of America* as pivotal to the genesis of new agrarianism, as well as an ideological basis for modern sustainability movements, Berry builds in part upon the work of a group of writers called the Southern Agrarians.[1] Berry's assertion that repressive political forces in the United States are driven by an obsession with exploitative mechanization extends arguments made by the Southern Agrarians in the 1930s. For Berry, mechanized culture necessitates political resistance in the form of a modified way of life that draws from an agrarian society grounded in local, small-scale agriculture.[2] As we shall see, the Southern Agrarians, too, saw agrarian traditions as potentially counteracting the incursions of industrialism, although they struggled to find an untainted version of the past suitable for scaling-up as the twentieth century progressed.[3]

Consistent with how Ross Singer has defined American agrarianism, both Berry and the Southern Agrarians work within "a philosophical tradition and malleable discursive frame adopting, defending, revising, and reproducing mythic assumptions about the morality of farming."[4] A key transformation in American agrarian rhetoric occurred when the Southern Agrarians embarked on a systematic philosophical defense of the Old South as a bulwark against industrial mechanization. This movement represented an impassioned defense of localized agrarian lifestyles against capitalist hegemony. As this chapter explains, the Southern Agrarian movement envisioned a new form of agrarianism, characterized as a corrective to mass culture, while seeking to infuse its adherents with a diligence, bordering on militancy, that continues to characterize some modern manifestations of agrarianism. This analysis expands our understanding of American agrarian myth in rhetorical space by casting the Southern Agrarians as a fledging social network that, while failing to achieve overt goals, had a prophetic impact on social geographies and cultural identities to come. In this case, this influence has an ironic twist as the conservative Southern Agrarians, rising in reaction against modernity, opened doors to practices of agrarianism now often associated with progressive politics.

Kenneth Burke observed that the intersection between aesthetics and politics provides a unique space for cultures to navigate changing environments by strategically challenging the narratives that form their shared reality.[5] One of the key problems for American agrarianism is that it has historically validated contradictory worldviews; along with the Jeffersonian myth of the independent yeoman farmer working in accord with nature, it has heeded the call to feed the world, as seen in the propaganda of World War I, and conceded to the resultant capitalist expansion. Not surprisingly, the anxiety generated by ever-increasing consumption has spawned a number of utopian countermovements over the past two centuries.[6] In his attempt to resist the seductions of mass society, Henry David Thoreau, drawing inspiration from the agrarian communes of New England, viewed his rural isolation as an act of political purification.[7] The Southern Agrarians differed from their Transcendentalist counterparts, however, by advocating for organized regional resistance to roll back commodity culture. The Southern Agrarians published two primary texts, *I'll Take My Stand* (1930) and *Who Owns America* (1936), to challenge the encroachment of consumption and industrialism before the Southern experience was altered beyond recognition and their proclaimed paradise lost forever.[8] While the Southern Agrarians failed in their project of preservation, by the mid-1930s their efforts to create a Southern separatist movement were transformed into a systemic political critique of the dehumanizing visions of social progress that they perceived to be destroying the agrarian ethic. While they set out to establish Southern sovereignty, these writers foreshadowed the rhetorical parameters for the agrarian resistance movements of later decades, and identified a common enemy.

As one might anticipate, the Southern Agrarians emerged as a reactionary movement. Most pointedly, these writers were responding to H. L. Mencken's satirical characterizations of the South as intellectually impoverished. In response to this cultural attack, movement leaders authored the Southern Agrarian "Statement of Principles" in 1929, drawing extensively from the thought of *Kenyon Review* founder John Crowe Ransom and the poet Donald Davidson. This declaration served as the forward to *I'll Take My Stand*, a series of twelve essays published in 1930.[9] The "Statement of Principles" began by asserting that the South must become a sovereign, self-governing entity—a political philosophy called sectionalism. The declaration asserted that the precepts outlined in *I'll Take My Stand* were not a literary exercise, but an effort that was at once a manifesto designed to foment sectional resistance against the evils of industrialism and a spark for a broader spiritual awakening. They avowed that Southerners "have a filial duty to discharge their own section. But their cause is precarious and they must seek alliances with sympathetic communities everywhere."[10] The statement further declared that "minority communities opposed to industrialism, and wanting a much simpler economy to live by" must create a more humane community and embrace a "genuine humanism rooted in the agrarian life of the older South and of other parts of the country that shared in such a tradition."[11] Opposing this agrarian utopia was the multi-headed serpent of industrialism, mass culture, and consumerism—a pervasive force that the Southern Agrarians sought to arrest. The call for a more natural and authentic space was founded upon the basic principle that the Old South constituted a type of atavistic paradise. This space would serve as a sacred haven for humans to live in accord with nature, providing a model that would inspire others elsewhere.

As the Southern Agrarians reflected upon extant visions of this region, both its literary and political expression, they tasked themselves with generating a new Southern consciousness that linked geography and cultural practice, particularly the reaffirmation of a spiritual relationship with nature.[12] This return to localized farming was imagined as a means to preserve the South as a unique space free from the polluting influences of mass culture, a political ambition that set the Southern Agrarians against economic momentum. For the Southern Agrarians, a heterotopian vision of the South constituted the energizing force behind a new gospel of the soil. Beginning with the work of Charles Griffin, rhetorical scholars have recognized that conversion rhetoric can manifest itself through a variety of forms, but its central tenant is to overthrow a perceived blind or decadent epistemology.[13] Given that the forces of industrial expansion were inescapable, the ultimate legacy of the Southern Agrarian revolution was the creation of a rhetorical heterotopia that animated political resistance to the industrial menace by advocating a return to a spiritual connection with the land and the agrarian ideals.

This chapter reconsiders the Southern Agrarians, particularly the writings of key members such as John Crowe Ransom, Allen Tate, Donald Davidson, Andrew Lytle, and Frank Lawrence Owsley. These authors were unified by a dystopian vision in which the South languished in the clutches of an industrial leviathan that dehumanized its practitioners while seducing them with simulations of nature. The agrarians struggled to disentangle their romanticized Eden from the brutalities of industrialism, a difficult task that forced them to address the internal contradictions of the plantation system. Rather than constituting a rhetorical failure in the agrarian myth, this inability to imagine a true Southern utopia forced the Southern Agrarians to move beyond sectional mythologies, generating a gospel of agrarianism that embraced the localized farm as a militant corrective to modern consumerism. This chapter explores the Southern Agrarians' later writings, in which they argued that the local farmer possessed a spiritual relationship with the land. As apostles of a new-agrarian faith, the practitioners of this pastoral vision advocated for converts dedicated to defending the local, sustainable farm as a spiritual corrective against the forces of materialism. In its radical pastoral mythology, this movement erected the stark boundaries between the symbolic landscapes of corporatized farming and local agrarianism as ideological constructs that continue to shape public debates over food. We turn to the Southern Agrarians not only for insight into the rhetorical inventions and constraints that condition agrarian vision, but also to expand our understanding of heterotopian space as an incubator for political resistance.

Industrial Decadence and the Assault on Nature

In *Ideas Have Consequences*, a work deeply influenced by Southern Agrarian writers, rhetorical theorist Richard Weaver warns of the "Stereopticon." This artificial consumer culture that replaced valued traditions with a dehumanizing relativism was largely a summarization of Southern Agrarian thought.[14] The seeds for this concept can be found in the early stages of this movement's justification of Southern sovereignty as a logical response to the evils of industrial consumerism. These dystopian visions of industrial society were critical to shattering the link between mechanization and progress, providing a counterpoint against which to reimagine their own history.[15] The following section draws from the "Statement of Principles" authored by the Southern Agrarians in 1930, a political manifesto that clearly delineated the decadent nature of industrialism. It also explores Ransom's "Reconstructed and Unregenerate," an essay challenging the assumed, but often unexamined, connection between science and progress. To illustrate the erosion of agrarian space and the degradation in lifestyle, Andrew Lytle's apocalyptic "The Hind Tit" and Donald Davidson's prophetic essay "Mirror for Artists" are used to delineate technological progress as a destructive form of cultural incursion. Together, these essays form a tapestry that Burke would have identified as a mythic rhetoric of purification and return—a dynamic chain of discursive enactments designed to transcend perverse futures for a lost, spiritual origin.[16] As the Southern Agrarian campaign unfolded, their arguments expanded in scope, moving away from economic exploitation to a larger critique of technological society and its capacity to mimic nature—a seduction that led to degenerate lifestyles and necessitated spiritual reawakening.

The critique of the mass capitalist system has led some scholars to view the Southern Agrarians in the light of New Deal politics. Exploring Southern Agrarian rhetoric reveals that this standpoint is incomplete, as these agrarians viewed industrialism and mass society as part of an economic system that not only exploited the public, but embodied evil and the "fallen" condition of humanity. Using classic Burkean God and Devil terms, these writers sought to draw stark boundaries around their idealized vision of the agrarian world and the demonic forces of mechanization degrading culture. This is illustrated when Davidson notes, "Industrialism is quite unconscious that the bargain (which the Middle Ages would have described as a devil's bargain, ending in the delivery of the soul to torment) involves the destruction of the very thing bargained for."[17] Challenging the stipulations of this bargain meant not only battling the oppressive forces of absentee business owners or sharecropping, but the greater forces that threatened the persistence of rural Southern experience. This set the stage for what Lytle described as a "war to the death between technology and the ordinary functions of living."[18] While both the Southern Agrarians and New Deal proponents galvanized around concerns over social and economic upheavals, the Agrarians differed because they viewed mechanization as an extension of an evil that had to be marked, contained, and exiled from the South.

The evils of technology were best expressed in Ransom's writing. He bemoaned the impact of science on humanity, critiquing both the hard sciences' enablement of technological progress and the social sciences' reduction of humans to anonymous machines.[19] He argued that "industrialism is a program under which men, using the latest scientific paraphernalia, sacrifice comfort, leisure, and the enjoyment of life to win Pyrrhic victories from nature at points of no strategic importance."[20] Ransom viewed the sciences as the height of hubris, seducing citizens into thinking that they could become omniscient and thus all powerful. In contrast to the sacred forces that animate authentic agrarian society, forces that emanate from a pastoral relationship with the soil, industrial progress was an attempt to control and thus corrupt nature. This is highlighted in the "Statement of Principles": "Religion can hardly expect to flourish in an industrial society. Religion is our submission to the general intention of nature that is fairly inscrutable; it is the sense of our role as creatures within it. But nature industrialized, transformed into cities and artificial habitations, manufactured into commodities, is no longer nature but a highly simplified picture of nature."[21] In this vision, industrialism is not separate from nature, but worse—a degenerate simulacrum that subsumes and mimics nature.

In refocusing on the simulacra posed by industrialism, the critique of exploitation shifted terms, from economic injustice to a victimization at the hands of a decadent culture. The Southern Agrarians articulated this menace by combining the corporeal and spiritual, writing, "This much is clear: if a community, or a section, or a race, or an age, is groaning under industrialism, and well aware that it is an evil dispensation, it must find a way to throw it off."[22] Here, hyperbolic terms like "evil dispensation" combine with metaphors of a shackled society, terms meant to move beyond economic considerations to the mortal fear of spiritual seduction. Since the Southern Agrarians associated industrialism with consumer society, these forces not only degraded nature through mechanization, but also degraded people by defining them as consumers. In a later work, *Attack on Leviathan* (1938), Donald Davidson argued that the danger of the mass culture came from the promotion of novelty and distraction, leading individuals to relate to their world through commodities.[23] As the spiritual relationship with the soil erodes, the victims of this system are not aware of their own gradual enslavement. The machine and scientific mass production represent a degradation of the spirit by reducing both human labor and products of that labor to mechanical forms without deeper meanings.

For the Southern Agrarians, this evil dispensation was particularly insidious because, rather than allowing Southerners to craft a spiritual relationship with the soil, it reduced the realm of perceptions to the trivial flatness of commodities. Further, this profane mimicry of culture was accelerating. When commodities are produced with alarming speed, consumers are forced to consume at an increasingly higher rate to justify the expansion of production. Thus the "pure idealists of progress must coerce and wheedle the public into being loyal and steady consumers, in order to keep the machines running."[24] The Southern Agrarians used the term "embarrassment" to express the experience of the worker whose labor is supplanted by the machine. By allowing the machine to mediate between the worker and nature, essential components of the worker's identity were lost even as "productivity," the key capitalist measure of labor, increased. Leigh Anne Duck suggests that one of the most poignant aspects of modern agrarian philosophy is its assertion that distance from the land means estrangement from people and objects.[25] For the Southern Agrarians, both knowledge and aesthetics had become bricks in this spiritual prison. They suggested that the educational system was being compromised as young men and women were inculcated in a "false way of life" that trained them to embrace an "inconsequential acquaintance" with traditions, arts, and humanities.[26] The decadent simulacra of industrialism trained its adherents to mediate their reality through commodities, while remaining blind to the depth and resonance of life that they were missing. Despite the endless talk of economic "expansion," the Southern Agrarians saw in industrial capitalism a reduction that only the spirit could quell.

Foreshadowing Weaver's Stereopticon, the Southern Agrarians argued that the damaging effects of these simulated patterns of creation and labor could be exposed in the literature and art of the regions of the country that were fast succumbing to the flattening of experience. Davidson outlined this threat when he explored the artist's dilemma, asserting that the true artist must "share in the general concern as to the conditions of life. [They] must learn to understand and must try to restore and preserve a social economy that is in danger of being replaced altogether by an industrial economy hostile to [their] interests."[27] The work of art produced in an industrial culture was degraded because it was no longer an aesthetic reflection of human experience, and was instead a commodity to be bought and sold in a fad-driven marketplace: "It is now the public which is embarrassed, it feels obligated to purchase a commodity for which it had expressed no desire, but invited to make its budget equal to the strain."[28] For this reason, Davidson warned that the denigration of art was a byproduct of endless expansion of the market system. He suggested that the state of modern art and literature produced in metropolitan centers had lost its humanity, writing that these degenerate artists "run crazily off into briar patches and mud puddles, squealing hideously."[29] When "seeing the world altogether in terms of commodities," the experience of art, however avant-garde, becomes yet another commodity, "a concession to humanity's perfectly unaccountable craving, or as just one more market."[30] Artists caught in the coils of the industrial economy learned quickly only to satisfy the temporary cravings of a consumer incapable of aesthetic judgment. This betrayal of art as means for cultural critique or regeneration, became, for Davidson, the ultimate expression of industrial decadence.

Although addressed in detail in a subsequent section, it is important to note that these writers were not anti-federalist in the modern sense, but sought to challenge a degenerate, simulated reality that transcended regional boundaries. Following anti-federal strands through Southern literature, John Grammar notes that this cultural milieu introduced the vision of the government as a cannibalistic parasite.[31] In recent decades, historians of conservative political thought have suggested that the anti-federalist turn in modern Southern politics owes much to the Southern Agrarian movement.[32] Yet, unlike their modern conservative counterparts, the primary target of the agrarian critique shifted from centralized government authority to the epistemological indeterminacy and decadence of corporatism. From the Southern Agrarian standpoint, the only viable answer to this challenge was a spiritual reawakening that would reanimate idealized forms of history, place, and community, felt by citizens in a bodily form. Yet given the pervasive forces arrayed against them, this goal would prove elusive. Even as they sought to expel the industrial serpent from a mythic Southern Eden, these thinkers would find their greatest rhetorical challenge in reimagining this lost paradise.

Paradise Lost and the Purification of Nature

Once the Southern Agrarians had identified the forces of degenerate industrialism, they faced a more compelling rhetorical challenge in building a cohesive social movement out of their shared history. As farming methodologies changed during the latter part of the nineteenth century and the abolition of slavery outlawed the use of human beings as a biomechanical labor force, the largely rural South would feel the full force of these changes.[33] While the Southern Agrarians initially sought to recover a cohesive narrative of Southern identity, they quickly discovered they were in fact reassembling the fragments of a shattered social system. Earlier readings of the Southern Agrarians suggest that they viewed Southern experience less as a product of history than as an extension of romanticized myth.[34] Certainly, rhetorical theorists have long recognized that myth is intertwined with geographical identity.[35] Southern mythology is poignantly problematic due to internal contradictions between myths of heroism and the reality of systemic violence in the slave system and the manifest evils of Jim Crow.[36] Given this perceived cultural erosion, it is not surprising that the Southern Agrarians struggled to redeem the South as an affirmative space, positioned as a moral corrective to the industrial North. Yet it is precisely this failure to salvage the Old South from its own violent present and past that compelled them to shift their focus from one of sectional resistance to a new alternative in an agrarian imaginary. This section will trace the initial attempt to resurrect the Southern Eden in 1930, with added reference to Vanderbilt historian Frank Lawrence Owsley's "The Irrepressible Conflict," an essay that suggested that Southern history was, by its conflicted nature, disabling. As the Southern Agrarians attempted to reassemble a coherent, regional narrative evaporated, they moved away from a focus on history and instead refigured nature as an oppositional spirituality to the flux of materialist culture—a rhetorical move that would play a key role in the evolution of the agrarian heterotopia found in the progressive agrarianisms that would follow.

First, it is important to understand the rhetorical constraints that inhibited the Southern Agrarians from imagining a coherent Southern utopia. The early years of the movement drew the attention of Seward Collins, publisher of the *American Review*. When Collins later expressed his sympathies with fascism, the Southern Agrarians rejected this dangerous movement.[37] Collins's attraction to the Southern Agrarians likely stemmed from the idea that distinct ethnic populations were spiritually bound to particular geographical locations, a political philosophy sometimes referred to as "blood and soil."[38] The attention of figures like Collins would trouble the Agrarians who were attempting to justify the essentialist links between Southerners and their land without direct reference to racial ideologies—perhaps because they preferred to discuss slavery in an oblique manner. Anxieties about this undertaking are present in Tate's correspondence with Ransom regarding the tension between the plantation system and the yeoman farmer. Tate went so far as to suggest that the group's initial attempt to resurrect their mythic Eden was tantamount to a form of idolatry.[39] He put the question succinctly when he asked, "Why should our tradition compel us to choose anything? Particularly in view of the all but accomplished fact that our tradition is destroyed?"[40] Following a social Darwinian chain of reasoning, why should they resurrect a system that history had seen fit to exterminate? He then asserted that a tradition "can always be defended, but recovery and restoration is a more difficult performance."[41] Similarly, critiquing industrialism did little to regenerate the social order industrialism had destroyed.

Much of the utopian rhetoric of this time period, beyond the work of the Southern Agrarians, responded to socioeconomic crises.[42] These utopian narratives reimagined a lost, idealized past—a golden age that could be resurrected.[43] Ransom initially asserted that a model for this recovery could be found in the provincial antiquity of Europe. He wrote, "unadulterated Europeanism, with its self-sufficient, backward looking, intensely provincial communities," stood as an oppositional force to "the character of our urbanized, anti-provincial, progressive, and mobile American life that is in a condition of eternal flux."[44] This demonstrates the degree to which the Southern Agrarians wished to identify some fixed collection of spiritual values to set in opposition to the placeless nature of mechanistic culture. Yet, unlike their European counterparts, as they excavated their own history, the specter of the plantation emerged. This forced them to project the yeoman myth, and the small, independent rural farm, as the true marker of Southern identity.[45] While this strategy potentially aligned the Southern Agrarians with other proponents of the citizen-farmer, including the Country Life movement, in this instance, the advocacy of the yeoman is fraught with the repression of the plantation system and its prominent role in the South. Here, the South Agrarians attempted to align the plantation with the rise of the capitalist system that demanded commodities such as cotton. Lytle asserted, by contrast, "The farming South, the Yeoman South, that great body of free men, had hardly anything to do with the capitalists and their merchandise."[46] The plantation system, with its murderous exploitation, mimicked the evils that the group rejected, while providing raw materials for the very machines that had consumed the South.

For the Southern Agrarians, this history necessitated a rhetorical division between mercantile culture and the local experience of the independent farmer, but such distinctions often belie the facts. Stephen Holden argues that tracing the roots of slavery and progress back to Jefferson's Monticello reveals a collapse between the human and the machine.[47] While Jefferson is retained in agrarian myth as an embodiment of the citizen as farmer, Monticello remained a plantation, and Jefferson a slave owner. In this manner, Jefferson may be seen as fused with the real and the artificial by using masking dehumanizing labor practice within the myth of the independent American farmer. Given that human mechanization and exploitation resided at the center of their own faltering utopia, the Southern Agrarians eventually had to look beyond historical recovery. Owsley summarized it by stating, "The South either had no history, or its history was tainted with slavery and rebellion and must be abjured."[48] Paradoxically, his answer was to circumvent race by arguing that slavery had no essential connection to agrarian life and was an environmental response to the economic conditions of the nineteenth century. As their rereading of history failed to reconcile the realities of slavery, it became clear that an agrarian South would have to be radically reimagined to provide the solace and redemption these agrarians sought.

Allen Tate, who authored the famous poem, "Ode to the Confederate Dead," suggested, "The Southerner is faced with the paradox: [They] must use an instrument which is political, and so unrealistic and pretentious that [they] cannot believe in it, to reestablish a private, self-contained, and essentially spiritual life."[49] To illustrate this point, it is worth referencing Tate's fictional work, which deeply romanticized the South, but simultaneously attempted to deal with the crippling effects of Southern nostalgia. In his controversial 1938 novel *The Fathers*, Tate followed a literary commonplace of the era by positioning two families on either side of the Civil War.[50] The Poseys align themselves with Northern industrialism and the Buchans with Southern aristocracy. By the novel's end, the Buchan family has been decimated by madness and murder, including the suicide of its patriarch, Colonel Buchan, who cannot face the changing social system around him. The Colonel's inflexibility is juxtaposed with the Posey patriarch, who becomes a slave to trends and whims of mass culture. These concerns are also evident in the Southern Agrarian response to *Gone with the Wind*. Although they admired the writings of Margaret Mitchell for her romantic rendering of Southern experience, the ease with which Mitchell was able to link the plantation model with the modern mercantile culture was troubling.[51] In short, their history did not provide an atavistic paradise, but rather an account fraught with conflict. From this position, even the purification of the Southern yeoman was difficult, as this potentially mythic figure had become too deeply intertwined with the plantation in the modern imagination.

As their cultural inheritance defied easy reassembly, the Southern Agrarians turned increasingly to the concept of nature as a resistant force to the placeless character of industrialism. Ransom suggested that the last bastions against the incursions of industrialism were represented by the small, rural communities that dotted the South, each living in tune with nature. He wrote that the industrial mindset encouraged the "strange idea that the human destiny is not to secure an honorable peace with nature, but to wage an unrelenting war on nature,"[52] adding: "[Humankind] is boastfully called to be a natural scientist essentially, whose strength is capable of crushing and making of [their] desires the brute materiality which is nature."[53] This shift in their writings was subtle, reorienting industrial culture not as the enemy of the South exclusively, but as a threat to nature itself. Tate described the Southerner as a byproduct of nature: "The Southern mind was simple, not top heavy with learning it had no need of, unintellectual, and composed; it was personal and dramatic, rather than abstract and metaphysical; and it was sensuous because it lived close to a natural scene of great variety and interest."[54] Lytle went further, arguing that this sensuous relationship with the land allowed Southerners to live in tune with nature, reading signs in soil and weather. In one passage, he attacked the frivolity of hygiene, suggesting that people from urban areas who sleep on fresh linen or rely on modern plumbing have allowed themselves to become slaves of an industrial leviathan.[55] This association of the yeoman farmer with nature began to coalesce into a regime loosely termed a "routine of living" that these authors associated with the South.

The process of historical reassembly became a rite of purification as the Southern Agrarians attempted to exorcise those elements that defiled their agrarian vision. Even the South's cherished religious tradition was not immune, as Tate summarized: "Its religious impulse was inarticulate simply because it tried to encompass its destiny within the terms of Protestantism, in origin, a non-agrarian and trading religion; hardly a religion at all, but a result of secular ambition."[56] The placeless character of the industrial culture horrified these authors, but at the same time they came to recognize that the serpent had long since infiltrated the South through the guise of economic avarice, both from the plantation system and its relics in racial violence, as well as in the moderate degree of industrialism that had already taken root. Given this failure of history, the Southern Agrarians began to preach a pastoral vision of nature best expressed in localized agrarian space that could defy the spread of industrialism. While this rhetorical vision remained inchoate in 1930, by 1936 the Southern Agrarians had adopted a new strategy: the formulation of an agrarian heterotopia as a corrective space that could, in time, reawaken the spirituality of the nation and overthrow the decadent culture of materialism.

Heterotopian Space and the Genuine Human

As their justifications for Southern sovereignty collapsed under the internal contradictions intrinsic to their shared history, the Southern Agrarians' later writing demonstrated a shift away from sectional resistance, advancing instead an idealized lifestyle that transcended regional boundaries. Scholars have observed that, with the publication of *Who Owns America?* in 1936, these writers began to downplay the plantation mythology to broaden their appeal.[57] Careful analysis of their writings, however, reveals that these thinkers were also confronting the corporate imperatives that rendered regional utopia a virtual impossibility. The following section explores the shifting rhetoric of their 1936 writings where the "routine of living" was transformed into a systematic template for spatial resistance. In contrast to the Transcendental tradition, the Southern Agrarians turned agrarian practice into a form of militant pastoralism that could challenge and contain the expansion of mass culture.[58] From this standpoint, the agrarian life became a means for penetrating the simulated reality of corporatism by connecting the farmer to the authenticated power of nature. In essays such as Ransom's "What Does the South Want?," Lytle's "The Small Farm Secures the State," and Tate's "Notes on Liberty and Property," this militant pastoralism matured into an anti-consumerist gospel in which the independent farm constituted a resistant space where laborers were reconnected with their senses. More importantly, this heterotopian vision transformed the independent, sustainable farm into a counter-hegemonic, heterotopic space that would detach itself from the historical and geographical specificity of the South. Here, the farmer and nature are theorized as being of one essence, in Burkean terms, practitioners become consubstantial with the spiritual projection they create.[59] In this new, purified vision of nature, the Southern Agrarians projected a space wherein the genuine human could thrive.

As noted, the Southern Agrarians asserted that agrarian practice constituted a special form of labor that had an essential link to an authentic lifestyle: "The agrarian regime will be secured readily enough where the superfluous industries are not allowed to rise against it. The theory of agrarianism is that the culture of the soil is the best and most sensitive of vocations, and that therefore it should have the economic preference and enlist the maximum number of workers."[60] They visualized a spiritual awakening rather than a political doctrine. Lytle asserted, "prophets do not come from the cities, promising riches and store clothes. They have always come from the wilderness, stinking of goats and itching with lice and telling of a different sort of treasure, one a corporate head would not understand."[61] The beginnings of a new spiritual alternative in 1936 can be evidenced by their shift from the term "industrialism" to "corporatism" to delineate the forces they sought to resist. More importantly, the shift away from sectional nostalgia allowed this pastoral vision to break beyond the boundaries of the rural South. Since each region of the country could be described as a physiographic zone with its own unique culture, a focus upon localized agrarianism would allow these unique enclaves to flourish.

While some historians have drawn parallels between the Southern Agrarians and their Marxist contemporaries, a key distinction between the two reveals itself in the contrasting themes of the *abstracted universal* and the *authentic local*. Ransom dispensed with any glamorization of machine-based labor, writing that "labor is men laboring. The men who labor are, on the whole, those who are backward in economic initiative and intelligence" and "too helpless and docile to defend themselves."[62] Rather than embrace the burgeoning labor movement, he chided Marxist intellectuals who enjoyed the "thrilling odors from the armpits of men who work with their hands, and they have admired the ox-like strength of laborers, and still more the ox-like herding together in comradeship."[63] In rejecting these intellectuals, Ransom asserted that the peculiarities of the Southern mentality functioned as a corrective to the hyper-corporatized mindset: "Here the Southern temperament discloses a peculiarity which sets the region quite apart from others as a field for industry. Southern labor will not work as fast as other labor. It is even a matter of pride to the laborers."[64] Essentially, Ransom argued that the independent nature of rural Southerners compelled them to work at a pace set by rhythms other than the consistently humming machine. Unlike many Marxists of the time who sought to empower industrial workers, the Southern Agrarians drew correlations between mechanization, materialism, and urban experience that led to the projection of pastoral lifestyle as a corrective to these dehumanizing forces.

In these later writings the concept of nature, now purified of the taint of materialism, can be characterized as a form of salvation. Lytle argued that the rural farmer "does not suffer from spiritual sterilization," which comes from a "disassociation between work and the life of the senses."[65] In the world of the machine, one is called to work by mechanized whistles rather than by nature. In contrast, nature is no "rude lover" who disturbs the farmer's natural patterns of sleep, but instead seduces one into wakefulness by the shifting performance of the morning. He then suggested, "As complicated as the beginning of this day is, it is only one day in a lifetime of years. There is continual variety. There are seasonal changes, the time changes, the imperceptible lengthening and shortening of light hours, the variable weather. The richness of these phenomena defies the hardening of a rigid routine."[66] Thus, the sterilizing impact of the machine culture came from its capacity to divorce humans from their senses—senses that should be ordered by nature, not machines. Standardizing consciousness through the corporate profit model leads to a loss of experiences that are essential to the worker's humanity.

This pastoral, localized vision can be vividly demonstrated in the ways that the Southern Agrarians began to confront the infiltration of consumerism into the domestic sphere. It was Ransom's fear that "agriculture pursued on business principles will always be insolvent, and the class dependent on it will always have an insufficient income."[67] Thus the Southern Agrarian critique of consumerism was heightened as they explored the forces that impelled families to waste resources on purchasing unnecessary products. Not surprisingly, women authors began to appear among their ranks—often viewed as the lynchpin of domestic rural cultures.[68] One of these authors, Mary Shattuck Fisher, argued that true female emancipation can never be found in industrial communities where women are exploited for their labor potential; she described graphic scenes of babies born dead in the basements of tenement buildings. She insisted that only women who come of age in what she called "non-profit oriented" families, whose production and consumption practices do not exceed the bounds of nature, represented true emancipation.[69] This new woman resisted the temptation of defining her status as citizen through consumer excess, thus limiting the impact of Weaver's Stereopticon upon her family. While the advance of the corporate system could not be stopped, this new awareness encouraged matriarchs to respect the shared needs of the common body of the family and to reconsider how their consumption patterns shaped their families, the local community, and the land.

Increasingly, the Southern Agrarians became aware that they were striving to transform the idea of agrarianism itself. One way this is evidenced is in the aforementioned shift in terminology from industrialism to corporatism—the latter a term that began to represent the larger forces of cultural standardization and the inauthentic. Tate warned that "small ownership, typified by agriculture, has been worsted by big, dispersed ownership—the giant corporation."[70] Ransom expanded the agrarian critique by proposing that the North had in fact enslaved itself as a byproduct of destroying the culture of the South—that the South had historically functioned as a limiting device in regulating the erosion of liberty in industrial culture. As the Southern Agrarian political critique evolved from a call for sectional independence into an anti-corporate gospel, the group revised the history of post–Civil War "reforms" as a victory for corporate culture over agrarianism instead of an internecine conflict. Owsley revisited his historical exploration of these injustices but shifted focus to the Fourteenth and Fifteenth Amendments to the Constitution. He warned that "by giving corporations the status of persons, the Federalist Judiciary has prepared the colossal bodies of organized wealth to become undefeated champions of personal liberty."[71] Echoing Owsley, Ransom honed in on corporations as the primary manifestation of the industrial serpent, writing, "in each of these respects corporations have become far more independent of natural law and more ruthless in their competition with private citizens than was ever contemplated in their original statutory authorizations, and they are much more difficult to restrain or control than would be the case if the operations of each were confined to the State which originally chartered it."[72] National and international organizational "persons" that addicted the local populace to their products and exploited their labor required radical correction.

Thus, by 1936, the Southern Agrarian utopia was being transformed into a true heterotopia in which agrarian space, not Southern space, became a model for regulating the scope and quantity of both production and consumption, and a place where the natural human could exist in balance with nature. Ransom, aware that his original production model was not sustainable—as farmers would perform more labor to produce less food—began to call for production capacities set by nature. He defended this position by arguing that to preserve individual freedom, maximum output or corporatism must be set in opposition to the "considerable productivity" of the pastoral.[73] This new resistant space was built around the dictates of sustainability rather than profit—a key component of this new spiritual awakening of the soil. Therefore, the Southern Agrarian assault upon industrial capitalism evolved by degrees into a challenge of the economic systems driven by rampant competition that strained the limited resources of the land. As he attempted to visualize this new heterotopia, Davidson wrote with his characteristic fervor: "The Southern planter or farmer (and not only the Southern one!) gullied and exhausted [their] lands, sold [their] timber, held [their] tenants pinned with the dollar mark, not because [they were] a limb of Satan, but because money had to be forthcoming."[74] The infinitely expansive drive for profit that stood in the way of this spiritual awakening, as this awakening was based on a sensuous relationship with the land that led to cultivation rather than exploitation. By this point, the Southern Agrarians were foreshadowing the concerns of modern opponents of corporatism, particularly the dehumanizing effect of mediating human commerce and interaction through these corporations, which had themselves, ironically, been granted the rights of personhood by laws and the courts.[75] The transformation of the local farm into a site of resistance to corporate greed, environmental destruction, and dehumanizing simulation has evolved in recent decades to become central to the political and economic version of the new agrarianism.

This projected space where individuals lived in harmony with nature was intrinsically linked to the budding concept of sustainability, a core principle that animated the heterotopia projected by the Southern Agrarians. Further, Davidson projected this new pastoral vision beyond the borders of the South as the "diversity of regions rather enriches the national life than impoverishes it, and their mere existence as regions cannot be said to constitute a problem. Rather in their differences they are a national advantage, offering not only a charm of variety but the interplay of points of view that ought to give flexibility and wisdom."[76] Following this chain of reasoning, Ransom also called for "every possible legal assurance to the small independents of their right to compete against the corporations without being exposed to conspiracies."[77] Each of these writers maintained that when local farmers were replaced by disembodied, corporate entities, the soil was perverted and the land exploited by parasitic corporations. As their own vision of a sovereign South eroded, these writers had, perhaps unconsciously, articulated in both loud battle cries and quiet whispers, a new heterotopian vision of localized, sustainable farming—that has reemerged as if carried on the winds of time to later challenge the imperatives of corporatism. This rhetoric hinged on the renewal of the authentic connection between the farmer and the soil, the natural human to an unadulterated landscape, the renewal of a relationship based upon respect, a sense of bodily limits, and the higher purpose of civilization rooted in conscientious cultivation rather than decadence.

Salvaging the Soil from the Blood

In his historical study of genocide, Ben Kiernan ties cases of widespread violence to rhetorical appeals to tradition, embodied in an idealized relation to land and soil. "Racism," he writes, "becomes genocide when perpetrators imagine a world without certain kinds of people in it."[78] This projection of a "purified" land, without cultural outsiders, has spurred the production of such "outsiders" by violent extremists of all stripes: from to Hitler and Pol Pot to the current landscape of terrorists from various ideologies. The Southern Agrarians were acutely aware of their own fraught position within this dynamic, whereby they were easily seen as apologists for the genocide of Africans in the plantation system. Their projected state of yeoman farmers living alongside the plantation system, like an alternative reality in another dimension, remained, at best, another manifestation of an old order that was fading away. As our analysis suggests, the rhetorical strategy that helps their ideas to retain vitality is critique, as they increasingly sought to lay bare another violence, that of profiteering and industrialization. Here, they recast the Southern form of agrarian myth as a nomadic spirit that might find parallels and common company in other places and peoples. While they sought to salvage agrarian myth from the modernists and cultural invaders, it may be argued that the resilience of this myth was their salvation, allowing them to regather a compelling vision of life from the midst of an exposed and ruined order.

In a rhetorical analysis of a localized resistance campaign in North Carolina, Kenneth Zagacki revealed how regional farmers were able to tap into a unique and often submerged form of patriotism to challenge the combined forces of the federal government and large corporations.[79] Local farmers represented a key component of this movement as they sought to protect their land and lifestyle. As political theorists and practitioners alike have discussed, local agrarian space is a persistent and complicated force in American politics because it seems to transcend traditional ideological boundaries.[80] While the Southern Agrarians are often characterized as a doomed, anti-progressive, sectional movement, our analysis reveals that their record is not simple. As advocates such as Berry have grappled with the problem of finding a unique space from which to challenge the dehumanizing forces of corporate culture, they have also sought to restore flagging traditions of localized agricultural life and have drawn on the agrarian heterotopia woven into the American imagination.

While heterotopias have often been categorized as obscure spaces that coexist on the margins of more legitimized spaces, the preceding analysis demonstrates that these counter-spaces can also challenge more dominant geographies. When utopian movements fragment in the face of social or economic realities, their entelechy can be harnessed and stabilized if they are encoded through space. From these new starting points, these places can be reconfigured on a much larger scale—with important consequences for our definitions of humanness. The Southern Agrarians realized that corporatism was inescapable, but some form of cultural practice that adjusted production and consumption to the constraints of nature needed to be cultivated. While Southern sectional movements evaporated by the 1940s, the reemergence and growing energy of natural farming, as well as anti-federalist and environmental campaigns, suggests that the agrarian heterotopia imagined by the Southern Agrarians remains a persistent fixture in American politics. This attests to the importance of recognizing the power of social geographies in charting the rhetoric of resistance movements. Despite the weight they carried, rhetorical traces of the Southern Agrarians have traversed place and time. The Southern Agrarians set the stage not only for the spiritual overtones of modern agrarian movements that would seek to revise food practices to challenge mechanized culture, but also their fervor. As we shall see in case analyses to come, agrarian myth remains a passionate concern, on which the fate of communities, marginalized people, and American democracy turns.

In part I of this book, we examine the historical precedents for today's revival of American agrarian myth, beginning with the figure of the yeoman citizen-farmer at the heart of Jefferson's republic. In World War I, agrarianism was propelled as an American mythos and scaled up in war to enforce a vision of benevolence. The Country Life movement sought to restore the agrarian and community foundations for democracy and political legitimacy. In the Southern Agrarians, the connections between farming, social life, and spirituality become explicit, as these thinkers strove to preserve and then project an agrarian ideal in resistance to modern industrialism and its spiritual bankruptcy.

Today's revival of agrarianism still broadly relies on a rhetoric that blends militant pastoralism and spiritual awakening. As this radicalized pastoral has been resurrected, modern food and environmental advocates have transduced the Southern Agrarian critiques of corporate materialism and now focus upon local, independent farming and the return to a more human scale of living as the salvation of culture.[81] As Zagacki's analysis reveals, anti-federalist citizen-farmers can often find extraordinary resonance with the American imagination when confronting the massive, abstracted forces of corporate government. For many within these social movements, pastoral visions drive both production and purchasing choices, asserting that non-mechanized, agrarian, heterotopic spaces remain a persistent site for resistance. As we shall highlight in chapters to come, this agrarian heterotopia remains deeply ambiguous, even paradoxical. While agrarian myth remains a powerful discourse of resistance that can connect parties around a common visualization of what it means to be an American, this unifying potential makes agrarian myth a tempting mark for co-option. Increasingly, Americans are called upon not only to recognize or rally around agrarian myth but also to interrogate how this myth is being used, to what end, and by whom.

Next, we look toward an additional precursor of today's agrarian revival in the work of J. I. Rodale, the original American advocate of organic farming. As we shall see, Rodale's rhetoric extends the spiritual rhetoric of the Southern Agrarians through jeremiad, the rhetorical genre of lamentation, as he offered a long-running complaint about "chemical farming" and the diminishments of agriculture and social life that it has produced. In the ancient practices of organic farming, Rodale sought a corrective for the falsities of the prevailing agricultural order, fit for a pragmatic, democratic people.

CHAPTER 4

Rodale's Jeremiad Inspires the Organic Movement

In the middle of the twentieth century, a new type of heterotopian science gained traction in the United States. A growing anxiety about an increasingly mechanized society led to a desire to mobilize around agrarianism as a corrective to the artificial consumerism and toxicity of modern society. In the May 1942 inaugural issue of *Organic Farming and Gardening*, the magazine's founder, Jerome I. ("J. I.") Rodale, offered one of his many prophecies: "One of these fine days the public is going to wake up and will pay for eggs, meat, vegetables, etc., according to how they were produced" and how much "they will save on doctor bills."[1] Rodale recognized that, despite narratives that heralded it as a triumph, agricultural industrialization was not simply an efficient means of production that increased yields while reducing labor. Functioning as technological and corporate colonization, industrial agriculture had precipitated the movement of millions of Americans from the country to the city, where many became disconnected from the origins and contents of the food they consumed. Gravely concerned with the decline in the quality of food, soil, and public health, Rodale made it his life's work to educate the public about what he viewed as the sacred agrarian science of organic food and farming, and its role in creating a healthy and sustainable society.

Given the organic food sector's exponential growth since that time, Rodale's advocacy for organic agriculture for over three decades during the mid-twentieth century seems more visionary today than it did to most Americans at the time. Although regarded as the first prominent champion of organic agriculture in the United States and an early, outspoken voice against industrial agriculture, Rodale's critiques of the agricultural use of synthetic chemicals (e.g., DDT, ammonium nitrate fertilizer) in the early 1940s have received far less scholarly attention than similar critiques circulated roughly two decades later in Rachel Carson's *Silent Spring*. This chapter examines Rodale's emergence as the leading figure in the early United States organic agriculture movement as a significant moment of mythic American agrarian resistance that still reverberates in the food movement today. We first offer a historical and biographical sketch of Rodale's emergence as a widely recognized prophet and guru of United States organic agriculture. Using this sketch as context, we then look to contribute a deeper understanding of how Rodale's discourse accounts for his influence on food and farming. We present a close reading of Rodale's most influential work, *Pay Dirt: Farming and Gardening with Composts*, published in 1945.[2] As a prophetic vision for a modern grassroots movement of agrarian farmers and gardeners, *Pay Dirt* sheds light on the changing status of agrarianism in United States culture during the World War II era, and the history of rhetorical discourses found in today's political struggle for ethical food.

This chapter develops the theoretical argument that, as the first book on organic agriculture to be written and published in the United States, *Pay Dirt*'s impact can be attributed in part to what we call Rodale's lay-scientific style of prophecy and his combination of the rhetorical genre of the ecological jeremiad with the mythos of American agrarianism.[3] As we show, Rodale's rhetorical vision consists of two main components. First, demonstrating his own privileged access to cutting-edge science, including his own organic farming experiments, Rodale argues for the scientific and factual superiority of organic agriculture over the dominant industrial model of what he terms "chemical agriculture." Second, and crucially, relying on the jeremiadic form and its narrative dualism of good versus evil, Rodale invites values-based rhetorical identification and grassroots mobilization. Translating scientific discourse into a lay discourse on morality, Rodale adapts the mythic narrative theme of God's creation of the citizen-farmer as the caretaker of a covenant with nature. The broader narrative around this heroic figure is that because industrial-agricultural sins and evils have violated this covenant, organic agricultural methods must be widely adopted before it is too late. To contextualize our close reading, we situate Rodale within the rapidly changing agricultural environment of the early to mid-twentieth century. Additionally, we explore the theoretical concept of the American ecological jeremiad informing our inquiry.

An Organic Prophet Emerges

Accounts of J. I. Rodale's emergence as the nation's prophet of organic agriculture commonly refer to *Pay Dirt*'s reception as Rodale's defining moment of recognition and arrival. The book has been described as a "breakthrough of sorts" that, as Rodale himself noted, led to invitations to speak all over the country, and to his nickname, "Mr. Organic."[4] Numerous newspapers and magazines reviewed the book, which between 1945 and the early 1960s was reprinted fourteen times.[5] Despite *Pay Dirt*'s positive reception among the general public and by some scientists, many scientists and corporate interests reacted with hostility.[6] To rebut Rodale and other critics of chemical agriculture, Monsanto, a leader in the agricultural chemical business, produced a public-relations response titled "Plain Talk, Pesticides and the Environment," asserting chemicals as invaluable to fighting world hunger.[7] Just as industrial agriculture advocates continue to use similar arguments today in defense of the use of chemical pesticides, genetically modified seed, and factory farming, the rhetorical frames that Rodale develops in *Pay Dirt* are commonly articulated among organic advocates today.

Daniel Gross notes that, as *Pay Dirt*'s notoriety led Rodale to accept his newfound role as the voice of "the organic cause," his writings took on an increasingly "crusading" and even "evangelical" tone.[8] *Pay Dirt* may be best understood as part of his broader rhetorical career of fighting on the behalf of food, the environment, and nutrition. Rodale's status as an organic prophet emerged through his deft recognition of how the status of the farm reflects the status of democratic society. In previous chapters, we have discussed the Country Life movement and Southern Agrarians along parallel lines. Similarly, Rodale's rhetorical leadership was influenced and inspired by a group of intellectuals promoting an alternative to industrial agriculture in Europe beginning in the mid-1920s.[9] During that time, Rudolf Steiner developed and promoted an ecologically regenerative approach to agricultural soil science based on the use of natural compost fertilizers instead of chemical fertilizers. Steiner's approach became known as *biodynamic agriculture* (recognized by many today as organic agriculture).[10] Steiner, a recognized philosopher, literary critic, architect, social reformer, and instrumental figure in natural medicine, blended natural science and the study of spirituality.[11] Although most advocates succeeding Steiner did not share the same fervor for mysticism, traces of it can be found in today's holistic discourses of organic food and natural remedies.

Steiner and his colleagues in biodynamic agriculture had been motivated by the mass production and rapidly increasing use of chemical agricultural substances enabled by World War I.[12] Brian K. Obach writes that although the biodynamic movement only began to spread to the United States during World War II, "the United States had its own agrarian movements prior to the development of organic philosophy" that had already begun to contest the emerging industrial agricultural order.[13] The Populist movement of the late nineteenth century articulated anti-corporate grievances against the rising urban and global market monopolies of banks, railroads, and wildly unstable farm-commodity market prices.[14] Additionally, in the early twentieth century, a back-to-the-land movement sought agriculture-based social reforms to counter problems of urbanization and to restore ethics of self-sufficiency and egalitarianism.[15] Also, prior to all these movements, the first half of the nineteenth century spawned food and dietary reform movements including vegetarianism, temperance, and non-processed food advocacy.[16]

In their own ways, each of these movements directly responded to changes in agriculture and everyday American life brought about by modernization's vision of efficiency and convenience. In the 1830s, food and nutrition reformer Sylvester Graham famously questioned efficiency and profit-making shortcuts taken in the production of wheat and in the preparation of bread in commercial bakeries.[17] Graham's critiques of adulterated food emerged as the combined forces of scientific advancement, the expansion of the geographic frontier, and growth in the national population that moved the nation closer to an industrial agriculture system.[18] In the second half of the nineteenth century, American agriculture's first revolution entailed more than transitioning from manpower to animal power. The passing of the Morrill Land Grant Act, the Homestead Act, the Pacific Railroad Act, and a bill to establish a Department of Agriculture (the USDA), as well as innovations such as the steam-powered tractor and thresher, altered food production and consumption in the name of expanding the nation's scientific and geographical frontiers.[19] Despite later protests by agrarian Populists and back-to-the-land reformers, by the end of World War I a second agricultural revolution was underway, involving the application of synthetic fertilizers and later, pesticides. Combined with a widespread transition from animal power to gas-powered mechanization, these changes prompted resistance from organic agriculture advocates such as Rodale.[20]

Rodale and the Agricultural *Zeitgeist*

Interestingly, Rodale's foray into food and health advocacy was not a result of scientific training or an agricultural background of any sort. Rodale was an urban businessman whose dream was to leave his office in the city for a "scientific" farm in the country.[21] When he moved his family and his manufacturing company from New York City to rural Pennsylvania in 1930, his intent was to blunt the harsh impact of the Great Depression on his business. Toward this end, and driven by his own personal passion for knowledge, Rodale founded and gradually expanded the Rodale Press as a side business, and eventually began publishing magazines on food and nutrition, including *You Can't Eat That*, *Health Digest*, and *Prevention*.[22] In 1940, Rodale and his family moved to a farm in the rural community of Emmaus, Pennsylvania.[23] There, Rodale's interest in publishing on food and health topics merged with his experimentation with the organic farming methods he had found when looking for material to publish in his magazines.[24] After World War II, the farm became an experimental research site for organic agriculture initially named the Soil and Health Foundation. Later relocated to Kutztown, Pennsylvania, an expanded and renamed Rodale Institute remains in operation today.[25]

Rodale's publishing and farming experimentation in organic agriculture led him to direct contact with the group of European intellectuals who would shape his rhetorical leadership for the rest of his life. Rodale published magazine articles by Rudolf Steiner's student, soil scientist Ehrenfried Pfieffer, whose 1938 book *Biodynamic Farming and Gardening* was translated into at least five languages.[26] When Pfieffer moved from the Netherlands to rural Pennsylvania after the start of World War II to create a model biodynamic farm and training program, he and Rodale met for the first time.[27] Rodale was most impacted, however, by British agronomist Sir Albert Howard, world-renowned for his research on soil composting.[28] Through personal letters, Rodale corresponded with Howard, who later served as associate editor for Rodale's popular magazine, *Organic Farming and Gardening* (now titled *Organic Life*), and wrote the preface for *Pay Dirt*.[29] These figures and others, including British scholar Lord Northbourne, who coined the term *organic*, as opposed to *chemical*, agriculture, and Lady Eve Balfour, author of the 1943 bestseller *The Living Soil*, provided Rodale with the scientific basis for his emerging efforts to disseminate ideas about organic agriculture in the United States.[30]

Given the strain that the war put on the food supply and the subsequent planting of backyard gardens by an estimated 80 percent of the United States population in 1943, Rodale's increasing devotion to food and health advocacy was especially timely.[31] Andrew F. Smith writes that when Rodale's *Organic Farming and Gardening* went into publication in late 1942, it was "singularly suited to the zeitgeist."[32] Little synthetic chemical fertilizer was available during the war, making the moment opportune to promote organic gardening.[33] However, in the name of maximum production, the war also led to powerful political alliances between big agribusiness interest groups and the federal government, resulting in programs favoring the former.[34] As we will highlight, although organic and industrial agriculture proponents both deployed a discourse of maximum short-term production during wartime, *Pay Dirt* attempted to build on the success of the grassroots victory-garden movement in World War II, to keep citizens actively producing local and healthy food, and to reorient the productionist rhetorical frame toward long-term needs. With the war coming to an end and a new dawn on the horizon, Rodale looked to seize the moment to launch an organic agriculture counter-revolution.[35]

While Rodale was an entrepreneur who dabbled in various ventures before committing his life to agriculture, as Suzanne Peters observes, Rodale's role as organic prophet was that of a "popularizer" rather than "innovator."[36] A prolific author, dedicated researcher, and small-scale farmer in his own right, Rodale embraced the title of guru and, as journalist Eleanor Pereyni described him, the nation's "Apostle of the Compost Heap."[37] Biographer Carlton Jackson's *J.I. Rodale: Apostle of Nonconformity* describes Rodale as an "evangelist" and "one man movement who may be most responsible for today's increasing awareness of food facts and fallacies, from farm to table and all stops in between."[38] As one of the gurus of his era who often placed stock in anecdotes regarding the positive effects of his own experiments on his farm and on his own body (for example, via diet and natural remedies), Rodale was sometimes compared to other food and health gurus, including Sylvester Graham and Bernarr McFadden.[39]

Later in his career, Rodale explored other means of social advocacy. After several years of losing money on his publishing house, he eventually became a person of considerable wealth, which only expanded his available means of food and health advocacy.[40] As an example, Rodale purchased a theater in New York City, where thirty food- and health-themed theater productions he authored or produced in the 1960s were first performed—some of which were later staged across the country.[41] Biographer Carlton Jackson notes that while known for his cool demeanor, Rodale directly and fearlessly responded to opponents ranging from industrial agriculture proponents to New York drama critics to the Federal Trade Commission (FTC) and the American Medical Association (AMA).[42] The FTC and AMA took Rodale to task for what they viewed as unsubstantiated nutritional and health claims. In a nationally publicized hearing against the FTC, Rodale fought the case and eventually won, but the hearing's focus on Rodale's lack of scientific credentials only bolstered a trend in the discourse of his opponents across his career.[43] Over the years, Rodale's "non-expert" advice on personal nutritional and health practices drew more controversy than his advocacy for organic agriculture, but may have affected how some audiences viewed his credibility regarding agriculture.[44]

As this sketch of his emergence as a popular prophet for organic agriculture suggests, Rodale's advocacy emerged during times of rapid modernization defined by industrialization, the rise of big business, urbanization, and scientific advancements. Although the industrial agriculture project was already well underway, many questions remained about how to transition the food system in the postwar era. Rodale's advocacy in *Pay Dirt* consists of an evolved discourse of the jeremiad both to call this transition into question and to suggest a new path. In the next section, to foreground our analysis of this work, we explicate the American ecological jeremiad's discursive form.

Ecological Jeremiad

Rodale's advocacy for organic agriculture can be characterized as asserting uncompromising critiques of industrial farming as well as warnings regarding impending punishment that would be visited upon those who practiced it. Thus, Rodale's words embodied classic features of the American jeremiad, a genre of contemporary discourse modeled on Puritan political sermons inspired by the biblical prophet Jeremiah. The jeremiad of seventeenth-century New England consisted of public and ritualized "castigations of the people for having defaulted" on their bond with the Lord.[45] As a genre within the prophetic rhetorical tradition, the jeremiad warns the people that they are in a dire situation resulting from their violation of God's covenant, and that in order to escape dire consequences they must act urgently to restore it.[46]

The prophet of the jeremiad assumes the persona of what James Darsey labels a "servant" and "divine messenger" who "speaks for another." The prophet speaks truth in a time of crisis, and this truth is not a matter of compromise, but redemption. The prophet's ultimate task is "to restore a sense of duty and virtue amidst decay and venality."[47] Scholars of American public discourse have theorized how modern political figures utilize and adapt the genre of the jeremiad to fit modern situations.[48] David Howard-Pitney notes that such figures have often used the American jeremiad with "familiar myths" (for example, American Exceptionalism or the American Dream) to preserve Americans' self-image and to mobilize a return to foundational values and practices amidst "rapid change" and "unsettling signs of the times."[49] Several studies have identified the jeremiad in modern US environmental advocacy—an ecological jeremiad.[50] Making the case that the jeremiad has been fundamental to the US environmental movement, John Opie and Norbert Elliot argue that the dire warnings of modern environmental visionaries, ranging from John Muir to Rachel Carson to Al Gore, follow in the prophetic rhetorical tradition of the Puritans.[51] Dylan Wolfe describes the American ecological jeremiad as a secularized rhetorical form emerging from a "sacred messenger of a divinely ordained nature, one who has the vision to 'speak for the trees,' and who warns of imminent dangers if nature's covenant is not urgently restored." In jeremiadic discourse, "nature is given divine credence, a stronger sacred tone than is necessary in the secularized modern jeremiad."[52]

As we illuminate in the analysis that follows, as Rodale crafted a modern American ecological jeremiad for the United States' organic food movement, he targeted audiences who would take up the mantle of the ecologically conscious and scientifically minded citizen-farmer. This jeremiad directly invokes the mythos of American agrarianism, which appeals to the virtue of a society as cultivated and preserved by land-owning and self-sufficient citizen-farmers endowed by God as Earth's caretakers.[53] As noted in the introduction to this book, agrarian myth affirms farming as both the most virtuous way of life and an occupation on which all others depend. Further, while historically taking various forms in popular, literary, and political discourse, agrarianism traces back to ideals and virtues of citizenship in ancient Greece, reflects the popular majority status that farmers once held, and has long mediated the cultural separation of the country and the city.[54]

Agrarian myth helped shape Rodale's balance of the implementational and evocative jeremiadic strategies described by John Opie and Norbert Elliot. Implementational strategies follow a methodical and plain rhetorical style and present systematic procedures for actions to restore the broken covenant. Conversely, both more abstract and rousing than implementational strategies, evocative strategies offer moral judgment through reference to biblical, poetic, and non-scientific sources. Jeremiadic evocation places emphasis on the wonder of nature and the need to protect it.[55] Examining these strategies, it becomes clear that *Pay Dirt*'s primary purpose of translating scientific knowledge for practical use by private citizens led Rodale to place heightened emphasis on systematic implementation; however, this technical explication unfolds within a mode of prophecy that evocatively stirs the reader to action.

Sacred Science, Sacred Nature

In *Pay Dirt*, Rodale assumes the persona of a messenger of an ancient and sacred agrarian science of the soil. In the book's preface, Sir Albert Howard provides personal testimony to a kind of conversion experience that Rodale underwent in relation to how he views his own relationship to food and the soil. Howard, who emerges in the book and in other Rodale writings as a high priest of soil science, affirms for readers that Rodale embodies the ethos of a true servant of the organic gospel. Howard writes that he appreciates Rodale's audacity to address the serious problem of chemical agriculture. Despite having no agricultural experience, Rodale had, "courageously acquired a farm, learned how to get it into fertile condition, and then observed the results of compost on his crops, his live stock, and afterwards on himself and the members of his family." Howard adds that Rodale thus "took his own advice" before offering it to others. As told by Howard, Rodale's story of taking up a new life loosely correlates with biblical narratives of conversion, which tell of witnesses such as the apostle Paul communicating with, then being devoted to, a divine revelatory voice that others could not hear directly.[56]

Rodale's persona as a lay prophet of scientific agrarianism develops from here. He draws on references to published scientific findings, interactions with scientists and experienced farmers, and occasional first-person observations from his experimental organic farm. However, rather than further support existing conventions, Rodale engages in subversive rhetorical invention. Specifically, Rodale draws upon the noted repertoire of resources to craft a jeremiadic call to action. This call hails the nation to avoid the grim future that chemical agriculture has in store.

In her study of modern scientific-prophets such as Rachel Carson and Robert Oppenheimer, who model a "hybrid of scientific-prophetic ethos," Lynda Walsh notes that such visionaries often live an "ascetic or marginal lifestyle" by their own volition. These leaders attempt to "demonstrate privileged access to knowledge beyond the public ken," and "use that demonstration to engage the polity in a dialogue about its covenant values."[57] Rodale's asceticism included moving to the country and engaging in an unconventional form of agriculture, as well as practicing and promoting a diet that avoided staples such as milk and white bread in favor of sunflower seeds, Hawthorne berries, and bone meal.[58] Additionally, among other demonstrations of the prophet's privileged access to scientific knowledge found in *Pay Dirt*, Rodale describes his close correspondence with high priests of organic agriculture (especially Howard, Pfieffer, Balfour, and Northbourne) and frequently cites their scientific arguments.[59]

In these obscure texts, Rodale found a vision of the sacred that served as the basis of his personal conversion experience and his enlightened prophecy. *Pay Dirt*'s ecological jeremiad emerges from what these texts describe as the agrarian simplicity of "Nature's Law of Return," which mandates that farmers return to the soil what they took from it.[60] This uncompromising doctrine of nature's divine design is based on the principle that "nature consists of interrelated interlocked life-cycles."[61] Rodale explains that the law is not solely an invention of intellectuals but results from their close observations of compost-based agricultural methods used since ancient times. Rodale writes of the ancient knowledge revealed in F. H. King's *Farmers of Forty Centuries*, an early twentieth-century study of organic farming in China that influenced Howard's work on similar phenomena in other countries. Sharing King's journey to enlightenment across the world, Rodale tells his American audience that "the Chinese get enormous yields by using composts and other organic materials," adding that in contrast to smaller yields in the United States, King witnessed "yields of wheat in China of over 100 bushels an acre as not uncommon."[62] In another example, Rodale states that without the help of chemical fertilizers, "certain varieties of grapes that have been growing for centuries in Indian and Persia still retain their productive potency." This conservation of varieties differs from grapes in countries such as France, where, Rodale says, the use of chemical fertilizers spurred a constant need to introduce new varieties.[63]

Further sacralizing the ancient science of agrarianism, Rodale highlights that no modern chemist can reproduce indigenous seed varieties developed through disciplined subservience to nature's covenant over the centuries: "How many times do we read that . . . our scientists are sent to Russian and Central Asia to introduce healthier varieties from there, ones that do not wilt as soon as the wind blows from the wrong direction?"[64] Rodale's appeal to ancient organic agricultural practices affirms the agrarian idea that the soil alone provides for a person's independence, but only if nature's covenant is protected. Appeals to agrarian independence appear not only in the form of exemplars from the past, but in the form of lessons learned from great mistakes as well. Rodale uses examples such as a historian's account of the decline of Rome as an outcome of the death of agrarian farming through "absentee ownership and one-crop specializing." This model replaced an agrarian model of balancing crops and livestock, as well as times in which the owner of the land completed most of the labor and "grain was sold to the cities for articles not produced on the farm."[65]

Rodale's jeremiad in *Pay Dirt* sanctifies organic agriculture as the "true" science of ancient civilizations, not only through references to scientific findings verifying it but also through evocative and mythic language of good and evil. The jeremiad presents a dualistic narrative struggle between nature's innocence and essential goodness on one side, and chemical agriculture's reckless poisoning of nature's covenant on the other.

Specifically, *Pay Dirt* pits the sciences of soil ecology and plant biology as invested in long-term sustainability against the shortsighted pseudoscience of applied agricultural chemistry. Showing his disfavor for nonorganic, man-made chemistry, in particular, Rodale continuously prophesizes about how it violates the Law of Return: "If soil conditions get out of hand either because of intrusion of foreign elements (certain strong chemicals, for instance) . . . it is then more difficult to grow plants the way nature intended."[66] The result is the death of the farmer's "allies," such as beneficial fungi: "Where any one item in nature's cycle is disturbed, it will be found that others are automatically affected."[67] On the ongoing chemical agriculture project of trying to mimic and subdue nature, Rodale states this much: "American farming, on the whole, is . . . carried out as if the soil were a sort of mine, without thought being paid to the possible harnessing of our miracle-working soil microbes."[68] The chemist's epistemology inevitably disrupts the intricate complexity of nature's sacred "miracle-working" powers.

To expose how chemical agriculture is poisoning nature's covenant, Rodale further invokes evocative jeremiadic strategies. Appealing directly to the sacred nature of this covenant and farmers' burden to protect it, *Pay Dirt* warns that when farmers "violate fundamental agriculture principles" and have "sinned against the land," they are "visited by crop failures."[69] This prophecy continues as follows: "In the long run, this new field of endeavor [chemical and monoculture farming], if not curbed or controlled . . . may extend the single-crop technique of land-mining with . . . attendant evils of soil exhaustion and erosion."[70] Before it is too late, farmers must seek atonement for their sinful chemical agricultural practices. There is already some evidence of this evil beginning to appear: "We are beginning to recognize the disastrous nature of farming malpractice that creates erosion and dustbowls."[71]

Rodale alludes to the direct consequences for human health as well: "We cannot go on forever treating the soil as a chemical laboratory and expect to turn out *natural* food. What we are getting is more and more *chemical* food. Instead of eating live matter which can readily be absorbed by the body, we are consuming food which is becoming more and more artificial."[72] As part of his case, Rodale laments that despite FDA regulations regarding pesticide residues on fruit and vegetables, actual residues are often twice the allowed limit, some residues remain present after washing, and some are inevitably absorbed into the fruit or vegetable.[73] The evangelical fervor of Rodale's argument is elaborated in dire predictions regarding the wrath awaiting the nation if farmers do not heed the call of nature: "These evils are not inevitable of natural things. If we, as a nation, permit the practices to go on, we shall richly deserve the consequences such as those predicted by the prophet Micah: the land shall be desolate because of them that dwell therein, for the fruit of their doings."[74] As we highlight below, Rodale continues to address farmers directly, pinpointing how their violation of nature's covenant is bad for their farms as well as compromising the nation's legacy of agrarian virtues. Further, as we will suggest, *Pay Dirt*'s agrarian commitments continue to shine through as Rodale indicts the farmer's practices without questioning the essential goodness of the character of farmers.

A Totalizing Threat

Rodale's jeremiad explicated several intertwined threats arising from institutions of modernization long held as antithetical to American agrarianism, notably, big business and technical solutions reflective of shortsighted thinking. In conjunction with his evocative jeremiadic critique of chemical agriculture and its supporting institutions, Rodale seeks to inspire and mobilize current and potential citizen-farmers who, like him, "believe in farming as a way of life" despite its joys being "almost completely lost in practices through the . . . havoc created by chemicalized agriculture."[75] Here, Rodale appeals to the fundamental American agrarian principles of agriculture as culture and agricultural living as the root of happiness rather than just another occupation or sector.[76] Rodale envisions a nation of agrarian citizen-farmers and, in this respect, his jeremiad harkens back to Jefferson's ideal of agrarian communities spreading on plentiful land. In contrast to Jefferson, however, Rodale envisions agrarianism as an ethic of redemption by which the farmer would break from the captivity of agriculture in its dominant, toxic, industrial forms.

While Rodale looks to restore agrarian democracy with the farmer as a proxy, corrupt forces have seriously threatened the ancient and sacred link between citizenship and farming. The chemical farmer has broken nature's agrarian covenant and is in urgent need of atonement. According to Rodale, his audience of "the general farmer," which includes the "professional farmer" as well as "amateurs," part-time farmers, weekend farmers, and gardeners, has been led down the wrong path. Sympathizing with his audience, Rodale offers that the evils of chemical agriculture have led many of these people to be "scared off the land by the intricate technology of chemicalized agriculture." Rather than a life of simple agrarian pleasures, "farming has come to seem, in recent years, too complex, too unrewarding."[77] Many of these people "could have enjoyed rural life and good husbandry, and could have found, like Washington, that farming was their 'chiefest amusement.'"[78] For Rodale, however, organic agriculture, while pleasurable, has a distinctly serious mission: it is a potential revolution toward a more democratic, environmentally sustainable, and public-health-promoting food system. Stressing the urgent need for transformation, Rodale prophesizes that in the postwar period, "every agricultural resource will be strained to the utmost to feed the battered world, and all the small gardener and amateur can produce will help the food supply, at least locally."[79]

Before this revolution can advance, farmers must urgently seek atonement through the adoption of ancient organic agriculture practices. Rodale's implied conclusion is that unlike organic methods, chemical agriculture is unnatural to the farmers' intention—it is external to their innate virtue. Borrowing from the language of the agrarian myth, the problem is not farmers' inherent "goodness of heart," but the temporary and complex propaganda that has compromised the "clearness of head" and simplicity of practice that farming and rustic experience of nature enable.[80] On the threat to the covenant raised by well-intentioned but ignorant farming practices motivated by chemical-agriculture propaganda, Rodale contends that "the use of chemical fertilizers on the land is such a complicated procedure that many farmers do not learn to handle them properly." He continues, "The average farmer doesn't pretend to understand . . . and since [they] can't have the county agent at [their] elbow all the time [they use their] own judgment and [are] apt to suffer by it."[81] Again, the threats to the covenant—technical complexity and the farmer's dependence—are quintessentially anti-agrarian.

In Rodale's jeremiad, these evils are pervasive and institutionalized; they are a totalizing threat to all life. The so-called experts of government and higher education further legitimize this decline of agrarian virtue and nature's covenant. Rodale juxtaposes the farmer's goodness with such institutional collusion: "Farmers often use these strong chemical weed-killers, without knowing that they kill [farmers'] best friends, the soil bacteria, fungi, and earthworms. And when professors of agriculture recommend their use on a large scale, it is sad indeed."[82] Elsewhere, Rodale elaborates on the corrupting influence that the chemical agriculture industry has on related educational and research institutions, which in turn affect the farmer: "Practically all agricultural text-books on the subject of the soil have a kind word to say about the earthworm, but rarely if ever do they dwell on the pernicious effect of chemical fertilizers on these helpful creatures."[83] In this discourse, by rendering none other than the compromised health of the earthworm as symbolic of ideological deception, Rodale follows in the agrarian rhetorical tradition of illuminating troubling alliances of undemocratic collusion and unbridled power across governing institutions.[84] Big business is a problem because a short-term means-to-ends profit motive drives its dangerous experiment with nature, the food supply, and the nation's health—all with the support of governmental agencies.

According to Rodale, chemical agriculture propaganda has instilled in farmers a narrow-minded attitude focused on expediting profits: "Some farmers employ strong chemical fertilizers to speed up the maturity of the crop so as to get to market early and enjoy the premium paid before the market becomes glutted and also to clear the land for a second crop.... Do farmers actually save time by using such artificial methods? By going against nature, they suffer later in the form of plant diseases and insect depredations."[85] Contrasting this rush for profit with the need to work within the limits of nature, Rodale appeals to the timeless and sacred ways of the Law of Return: "We learn from the Old Testament that the Hebrew farmer fallowed every field at least once in seven years. It was part of his religion and he was considered an outcast if he did not do it."[86] In this discourse, the imperative to restore nature is not only scientific but mystically divine. Rodale uses evocative jeremiadic appeals to the Bible to move a wide lay audience and to define an organic science of the soil as not only practical and true, but also sacred and moral.

Rodale responds critically to a host of other agriculture practices inspired by chemical agriculture. Further addressing the false prophecies of maximum short-term farm profits, Rodale speaks against the alarming rate at which tenant-operated farms continue to spread across the country. He contends, "Farm tenancy is responsible for many evils in agriculture," affirming that it violates what Chester Eisinger calls the "freehold concept" central to the agrarian myth: the political principle of landed property ownership in wide distribution.[87] Against a movement for an agrarian democracy based on the permanence of organic agriculture and a landed stake in society, "the average tenant is here today and gone tomorrow. [The tenant] doesn't build up [the] land. Many tenants actually sell their manure. They violate all the rules of good farming; they plant too many acres in open cultivated crops such as corn and potatoes which permit destructive erosion.... Why should they do all this? The land is not theirs!"[88] In this excerpt and in others on the topic, Rodale does not directly address the tenant farmer's character, but rather the general premise and practices of tenant farming. In *Pay Dirt*, this rhetorical framing approach holds, as Rodale tends to avoid direct confrontation with the farmer's exceptional, God-given character. As such, Rodale's jeremiad leaves intact the mythic image of the farmer as beyond moral reproach and identifies the cause of farmers' sins as outside influences.[89] It may be argued that leaving the farmer's character intact is rhetorically necessary for maintaining the effectiveness of the fundamental jeremiadic tension between crisis and renewal. If the farmer's character is seen as internally flawed or shrouded in doubt, the potential for heroic agency is undermined: the jeremiadic narrative would crumble and agrarianism could no longer be seen as a pathway back to a happier state of national virtue.

In addition to attempting to secure identification with the farmer in these ways, Rodale translates scientific discourse into appeals to common sense and agrarian practicality. Rodale appeals to the commonsensical idea of prevention and long-term thinking over the reactive use of chemical remedies. *Pay Dirt* identifies antibiotics and other modern medical treatments for sick farm animals as well-intended but merely reactive transgressions on nature's covenant. On the rapid rise of farm animal disease noted by the Animal Veterinarian Medical Association, Rodale writes, "The approach to the problem by animal doctors and scientists is, of course, curative rather than preventive." He elaborates, "But this is a negative approach. These diseases are merely an indication that something is causing trouble in the animal" and medicine is not going to get rid of the cause."[90] *Pay Dirt* presents a thorough indictment of the rise of what are now called concentrated animal feeding operations (CAFOs). Rodale writes that in the name of efficiency and profit, "in the past quarter century or so, chicken raising has become . . . an assembly line sort of production." Amid the rapid spread of disease in these "factories where the hens are penned up in batteries or tiers of individual cages" and fed low quality food, the so-called experts recommend vaccinations and synthetic hormones.[91] Offering a prophetic conclusion and a jeremiadic call for common-sense action, Rodale concludes, "These diseases are warnings by Nature that something is wrong and that it is necessary to start rebuilding from the soil up."[92] Thus, disease among animals is the farmer's punishment for trying to take shortcuts around the honest farming that nature intended.

Rodale's rhetoric draws upon this implied agrarian standard of nature-as-measure of the health of the farm and the community in other ways as well. For instance, he points to the health of wildlife and the broader ecosystems in which farmland sits as indicative of the relative virtue of a given farm. Rodale affirms that it is time to "consider an important, often neglected, asset of a well-run farm. The wildlife population is a good indication of whether or not sound farming practices have been followed."[93] Rodale also contends that despite the "great role that birds play in insect control" in natural ecosystems, chemical agriculture harms them, only to make the need for chemicals seemingly greater: "Soil pollution is probably one reason why bird-life is greatly reduced on farms where heavy use of chemicals and poison sprays is customary."[94]

Similarly, *Pay Dirt* laments that when it comes to insect control, "our valuable allies the toads are disappearing from farms."[95] Another problem is that farmers are disrupting the balance between agriculture and nature by "removing hedgerows and stonewall fences, where small brush grows, to increase the size of their fields and save time cultivating." This practice discourages bird and wildlife havens, increasing destructive insect life.[96] Removing natural borders of trees and hedges may also hinder the health of the soil: "They are very effective as windbreaks in keeping the drying winds down and the temperature of the ground up . . . important too for the land in preventing wind erosion."[97] Corporate farms of thousands of acres, which cut down trees to operate tractors more efficiently, are encouraging this destructive practice.[98] All of these changes in the landscape are evidence that "the balance of nature is upset."[99] The implication of Rodale's critique of the destruction of a natural agrarian harmony of biodiversity is that the farmer must atone for these sins. As expounded upon in the next section, Rodale gives significant attention to the action steps needed to achieve this ultimate objective.

Restoring the Covenant

The final component of Rodale's ecological jeremiad consists of the presentation of action steps and further words of inspiration. Rodale identifies citizen-farmers, not the institutions of chemical agriculture, as uniquely possessing the virtuous agency required for revolutionary change. He states that the "land can come back into 'good heart'—a secret that only the compost farmers know."[100] Drawing explicitly upon agrarian myth, Rodale identifies the farmer as "the true custodian of the nation's land."[101] Affirming that there is still time to act, he concludes, "Badly eroded, worn-out soil will not recover over night, but fertility *can* be restored." As citizens upon whom the nation's virtue is modeled, farmers and gardeners must give back to the soil what they have taken from it.

In his attempt to outline implementational strategies for the farmer, Rodale prophetically warns farmers that they should not give into the temptation of man-made, chemical shortcuts, for dire consequences are imminent: "No acceptable substitute for animal matter has as yet been devised, nor is there anything as simple to handle without fear of disastrous consequences due to burning, overdoses, and all the other complications. I feel quite certain that the fertilizer formula of the future will consist of plant matter, animal matter and finely ground up rocks. This is Nature's way."[102] Offsetting the grim evocation characterizing such predictions, Rodale's attention to restoring the covenant is methodical and detailed. To support his optimistic vision for a better future, Rodale attends to everyday, practical methods to be used by farmers and gardeners. To deliver these implementation strategies, he largely uses a plain, assertive, and methodical rhetorical style. He draws upon Pfieffer's and Howard's studies to list "thirty-six reasons why compost farming is superior to farming and gardening with artificial fertilizers," then provides a host of tips for achieving optimum results.[103] He also lends farmers and gardeners direct, step-by-step advice on using compost: "The instructions which follow represent the *Howard Method*, also known as the *Indore Process*. This consists in mixing vegetable and animal wastes with earth and water." The instructions cover several key categories in the process: ideal location for the compost heap, how to prepare the location, what green matter to use and in what proportion, how and when to turn the heaps, as well as explaining the need to avoid chemical activators recommended by many government agricultural agents. Grounding his argument with emerging scientific evidence, Rodale addresses how to spread compost and how to use manure with specific crops, such as corn: "Some farmers apply their manure or compost only when they plant corn . . . figuring that it leaves a residual value for the three or four remaining years in the rotation. Experimental work done at the Ohio Agricultural Experiment Station seems to show that it is best not to use all of the manure with the corn."[104]

Rodale notes that as farmers begin to adopt these organic techniques, they would do well to exhibit patience, display humility, and diligently heed "Nature's Law of Return."[105] Reminding farmers that, "Mother Nature has a way of caring for her children," Rodale urges humility in the face of wonder: "Nature is an experienced soil chemist, as well as soil biologist. She works deftly and harmoniously."[106] Despite the temptation to exert control to achieve short-term gratification, ancient agrarian virtues of patience and appreciation for nature should not be violated: "We must not expect perfection. There will always be some destructive insects. There will always been some disease on a small scale. They are Nature's way of eliminating the weak and perpetuating the stronger, hardier strains. But we don't have to take measures like those of primitive surgery."[107]

Restoring nature's covenant therefore means exerting the discipline to entrust in nature's slow pace or sometimes frustrating designs.

Additionally, *Pay Dirt* invites farmers to look to the past for timeless and sacred agrarian wisdom. Rodale directs farmers to the example set by the most methodical "old-fashioned farms," which, unlike those with disastrous soil erosion, have achieved great success. He encourages modern farmers to start by thinking about their farms' general design, stating, "The greatest virtue of old-fashioned farms was that they were self-contained and had a balanced complement of both crops and livestock."[108] Like these farms, modern farms would do well to follow the proven practices built around animal manures, green-manure crops, and other animal and vegetable residues.[109] Consistent with his discourse focused on unearthing the farmer's sins, Rodale translates science to a wide audience, urging farmers and gardeners to use agrarian common sense, which is about fitting within nature's balance: today, "farmers often carelessly destroy manure and then spend hard-earned money in purchasing artificial fertilizers, because it seems like less work."[110] To the contrary, farmers must tap the remarkable efficiency of nature's recycling capabilities: "With the aid of this rich organic material, they could produce a far better kind of feeding matter" for their poultry, and hence meat and eggs, than they can purchase.[111] Using his own farming experience as testimony, Rodale elaborates on tapping nature's cycle when raising chickens and other poultry: "The best eggs come from chickens that can run on grass in the spring and summer and have access to outside all year round, supplemented by a diet of properly raised food. On our farm, we feed only food grown on our own place with the exclusive use of organic natural composts."[112] Rodale prophesizes that eliminating the mass-production poultry business will also reduce the increasing number of issues addressed through chemicals. He predicts, "There is a tremendous market waiting to be developed—the production of high-quality eggs by chickens kept under natural, healthy conditions, at a price higher than average."[113]

While Rodale is firm in his call for disavowal of all synthetic chemicals for agricultural use, his jeremiad is also careful to skirt the impracticalities of idealism. For example, he does not expect all his readers to immediately move to the country and explains that he is not suggesting that farmers completely dispense with the benefits of technological advances and modern conveniences. He demonstrates this when addressing the question of whether farmers committed to organic methods should use tractors: "I think we must use common sense. The world is advancing. Wonderful machinery is being developed. . . . Sometimes it is necessary to have a tractor." Rodale adds that "there is no question . . . that the tractor is expediting the mining and exhaustion of our soil, but this can be counteracted by a return to organic farming practices."[114]

Although it would be reductive to characterize *Pay Dirt* as a call to turn back time, it calls for deep reflexivity and identifies as one of chemical agriculture's cardinal sins its abandonment of the wisdom of past generations. As the basis for a new organic farming movement, Rodale appeals to starting "from scratch" at little or no expense, which will recover the virtues of agrarian ingenuity, independence, simplicity, and honesty to the farm. Affirming the importance of the soil and the principle of agrarian freeholding of small parcels of land to reclaim sound agricultural practice, authentic food, and health, Rodale asserts, "Our land, actually, is the basic capital of the nation and should be used in that spirit."[115] Of the path forward, he offers the following: "We do not need to wait for a government program, natural resources surveys, subsidies—or track on another mortgage—to begin."[116] The revolution will begin with citizen-farmers at the grassroots level, eventually challenging federal authority that drives chemical agriculture. But changes are still needed at the institutional level. Among recommendations that Rodale makes for government reform is legislation to foster the renewal of an agrarian democracy of freeholding landowners: "Preventing ownership of more than a certain number of acres varying with the location and productivity of the soil. Vast-acred, assembly-line, single-crop farms should be outlawed, or strictly controlled for fertility maintenance."[117] *Pay Dirt* also calls for further research on organic, compost-based cultivation methods and the formation of an alliance between medicine and agriculture, which includes the USDA, to study how soil microbes—positive bacteria and fungi—can be most effectively used to eliminate the use of chemicals, benefiting both agriculture and public health.[118] Here, as in Rodale's recommendations to the farmer, which in some cases are as simple as allowing hens periods of rest from egg-laying, Rodale appeals to harmony between nature and agriculture as the foundation for actualizing a new agrarianism.

In *Pay Dirt*'s concluding chapter, Rodale brings his jeremiadic call to action full circle, prophesizing a potentially better tomorrow and rallying his readers with words of inspiration. Rodale writes, "We are more acutely aware of the deficiencies of 'modern' agriculture than ever before," and "across the country, there are signs of positive change on the horizon."[119] He adds, "the future of agriculture can be very promising. In almost all quarters of the globe, progressive minds are challenging the commercial farming practiced for the past fifty years."[120] During the expansion of the American frontier, new land seemed infinite, and problems of soil depletion through bad farming practices seemed unimportant. However, those times have passed, and people such as "practical farmers," amateur gardeners, soil biologists, doctors studying nutrition, specialists tracing the origination of plant and animal disease, "conservationists interested in keeping our natural resources available for generations to come," and "leading writers on agricultural problems" are turning to composts.[121] Rodale adds that following the positive lead of the back-to-the-land movement, many people are leaving urban centers with no intention of returning. As we discuss below, before the hopeful vision of Rodale's prophetic words could begin to be realized, those who joined the organic agriculture cause inspired by Rodale would endure years of uphill struggle.

A Legacy of Resistance

J. I. Rodale's rhetorical leadership early in the organic agriculture movement in the United States marks an important moment of modern agrarian resistance against the rise of industrialism and modernization. Rodale's American agrarian ecological jeremiad and persona of the lay-scientific prophet in *Pay Dirt* were instrumental in establishing him as the nation's first prominent advocate of organic agriculture. As we have seen, Rodale's rhetoric links the sacred and mythic tradition of agrarianism to its scientifically proven results. Like the propagandists of World War I, the Country Lifers, and the Southern Agrarians, Rodale links agricultural practices to the virtues of citizenship; however, for Rodale, these values are not distinctively American, but connected to ancient agrarian practices of caring for the land. He calls for a spirit of humility in relation to the "miracle-working" powers that make life possible. In defense of this sacred agrarian tradition, he delivers harsh judgment on the current state of American agriculture and places blame for its sins on the temptations broadcast in chemical agriculture propaganda. Unnatural to agrarianism and nature's covenant, industrialism breeds dependence, short-term and linear thinking, and unnecessary technical complexity, which deplete agrarian virtue and drive people from rural lives into the city. In this jeremiadic narrative, only a return to agrarian organic farming will rescue the food system and the nation from impending doom. Yet, even amid a flight to the city that threatened to depopulate farm communities, Rodale recognizes yearning for natural practices and a resonant life on the land.

With its emphasis on reclaiming the ancient and sacred genius of the natural and the simple, *Pay Dirt*'s ecological jeremiad for organic agriculture speaks to the core of the agrarian tradition. Paul Johnstone observes that one of the fundamental tenets of the American "agrarian creed" is that "agricultural life is the natural life, and, being natural, is therefore good."[122] As Rodale sought purification in a return to the natural, he articulates a second agrarian tenet that marks nature as beneficent. He assumes that a nation endowed with rich soil and plentiful environmental resources will be prosperous if nature is permitted to produce its bounty without synthetic interventions. As Richard Hofstadter explains, "If the people failed to enjoy prosperity, it must be because of a harsh and arbitrary intrusion of human greed and error"—an immoral breach of "natural law."[123] For Rodale, both urban and rural dwellers suffer from the "unnatural" afflictions of the chemical agriculture establishment that extends from corporations to government. This illness, at once moral, social, and physical, results in degraded soils, corrupted morals, social dependence, and poor health, and impacts the nation in its entirety.

As shown in our analysis, Rodale gives detailed attention not only to how farmers have deviated from the natural, but also how to return to the natural to reclaim agrarian virtue and restore the ecological covenant. Unlike other ecological jeremiads such as those of Rachel Carson or Al Gore, which give more attention to illuminating the problem and its causes, *Pay Dirt* gives relatively equal attention to problems and solutions, using what John Opie and Norbert Elliot describe as evocative and implementational jeremiadic strategies in approximate balance.[124] *Pay Dirt*'s deft articulation of these jeremiadic strategies and its acclaimed reception helped to plant some promising roots for the future harvests of the organic movement. While the dramatic expansion of industrial agriculture in the postwar era all but secured a marginal status for organic agriculture, Rodale's contributions to the mobilization of the US organic movement was tireless, formative, and prophetic of rapid growth in organic agriculture in the late twentieth and early twenty-first centuries. Rodale and his European contemporaries in the study and advocacy of compost-based agriculture offered alternative food advocates and environmentalists a vocabulary and worldview to draw from. His influence was crucial in spreading the message of organic agriculture as a living tradition for any to adopt, and millions have as the movement continues to grow.

Rodale's rhetorical leadership over several decades was also instrumental in raising public awareness of food and farming as environmental issues, as well as showing that environmental practices are closely related to health. Further, Rodale showed that agriculture, the environment, and health all have scientific and moral dimensions that affect the relative vitality of the nation and world. The cultural revolution of the late 1960s and early 1970s, in which Rachel Carson's prophecies played a notable part, aided the cause of Rodale and his contemporaries significantly, as the readership of Rodale and his organic agriculture publications increased substantially.[125] Although Rodale's death in 1971 precluded him from witnessing the rapid expansion of the organic movement in the coming decades, he acknowledged with pride that the younger generations had taken to his message and that the movement was growing.[126]

Despite the growth of the US organic movement since Rodale's time, from the perspective of today's organic-food advocates, who often fight for intertwined issues such as food justice, security, and sovereignty, Rodale's anti-chemical ideal for organic agriculture in *Pay Dirt* and his other writings may appear narrow and difficult to realize. For many, financial access to organic food remains limited, even as the consequences of industrial food are recognized. Moreover, many in the organic agriculture movement continue to speak out against what they describe as the corporate co-option of organic food, farming, and product labeling. Industrial-scale organic farms and corporate organic food outlets, key players in the food system today, continue to use chemicals for various purposes.[127] Further, Rodale did not anticipate that his support for use in moderation of some modern farm machinery, such as tractors, might be seen by later readers as "leaving the door open" for larger, more expensive, technical mechanization, which in turn led to the ability to cultivate even larger farms, some of which purport to be organic.

Although Rodale would later go on to confront some problems of the capitalist system tied to food and farming, it is important to recognize that *Pay Dirt* underestimates the role that the political economy of agriculture plays in the future of food. Suzanne Peters suggests that this shortcoming is symptomatic of an apolitical rhetorical style that characterized Rodale's discourse over the years. While Rodale's magazine publications consistently promoted an organic agriculture revolution, they attempted not to politicize, let alone radicalize, it.[128] Although this may have been a strategic business choice to align with science and popularize organic agriculture and food, it has undoubtedly had a lasting impact on the image of organic food since Rodale's time. In Rodale's discourse, the political affiliations of organic farmers and consumers are virtually indecipherable from those of other farmers and consumers.[129] While this may have fostered the image of a reductive, consumerist movement, its effects may not have been entirely negative. Although current popular discourse associates the organic movement with liberals and progressives, organic producers and consumers are politically diverse.[130] Also, contrary to popular belief, consumption of organic food products cuts across class and ethnicity.[131]

Finally, one of *Pay Dirt*'s most significant contributions arises at the nexus of intertwined issues that Rodale raises, including public health, environmental destruction, and nutrition, as well as the political dimensions of agriculture that he does not cover. Rodale's operating assumption is the virtue of democratic participation. He believed that one should not have to be a credentialed expert to ask questions and construct arguments about scientific and moral topics that affect citizens' lives. In this manner, he demonstrates for future organic agriculture advocates the importance of democratic engagement in both the public and technical spheres. Within the context of food and farming that serves as Rodale's focus in *Pay Dirt*, he impresses upon his readers that industrial agriculture does not own exclusive rights to the discourse of scientific improvement. Reviewing the development of indigenous seed varieties to suit regional conditions, Rodale saw that ancient agrarian civilizations sought and achieved scientific progress without chemicals or the destruction of nature. Addressing charges of nostalgia and impracticality with appeals to the time-tested and sacred simplicity of common sense, Rodale's vision redefined science and progress. Today, we live in the thrall of public scientific controversy regarding potentially catastrophes that range from climate change to viral outbreaks to the future impacts of agricultural biotechnology. In Rodale's prophetic critiques of the unchecked power and dogmas of institutional science, we find not only advocacy for better food, but also, and more importantly, strategies for vigorous resistance on behalf of human and nonhuman life.

We conclude the first section of the book with some brief reflections on lessons learned thus far about agrarian myth as a strategy of resistance. As we have seen, this myth resides deep within the American psyche, with its original terms established by Jefferson as a principle for the foundation for a nation embodied by citizen-farmers and their deep connection to the land and their local communities. As the nation expanded, the myth became part of the rhetorical armature of Western migration, which justified the removal of Native peoples from the lands with the claim that it would now be properly occupied and cultivated by citizen-farmers of European descent. In the twentieth century, the agrarian myth was initially mobilized as the symbolic regime placing the United States at the head of a new economic order of abundance. However, the mass production that monetized this myth displaced the citizen-farmer, marginalizing this key figure as another type of industrial worker charged with food production. As the twentieth century progressed, agrarian myth increasingly became a rhetoric of resistance that sought to restore an agrarian vision, seen in myriad forms as political, economic, spiritual and scientific. The ACLA recognized farmers as the nation's foundational citizens and the legitimators of public authority, and saw the apparent decline of the farmer as indicative of the fall of democracy itself. The Southern Agrarians upheld agrarianism as the salvageable core of a lost society, even if they could find no immediate place to apply it that was not tainted with the violence of slavery or industrialism, both forms of systemic oppression. J. I. Rodale, again recognizing the agrarian connection between agricultural practice and societal participation, advocated organic farming methods as a way of reestablishing connection to place, improving soils and foods, and resisting the deleterious impacts on land, nature, and people. Together, these cases show the allure of agrarian myth, as well as the manner in which issues involving farming and rural life have captured the struggles of the nation as whole. Even with an increasingly urban population, Americans have long contested over farming and our way of living on this expansive land.

In the second half of this book, "Threatened Harvests," we examine three contemporary cases that inventively illustrate the ambiguous legacy of agrarianism as an ethic of resistance. In the case of the SCFarmers in urban Los Angeles we find a compelling example of the recasting of American agrarian myth to fit the diverse social forms of a multicultural and urban society. However, these farmers' experiences also reveal the difficulty of instituting an adapted agrarian community within this new milieu, especially when its social form or methods of production are taken as a challenge to vested interests. The case involving Chipotle manifests another type of ambiguity by conflating the discourse of moral purity with the markers of agrarian myth, and envisions a return to better times. Agrarian mythmaking not only helped to propel the brand's spectacular growth but also sowed the seeds of its dramatic fall. As we will suggest, the experience of Chipotle suggests the complexities of deploying agrarian myth to justify enterprises that use large-scale models of production or perpetuate the prevailing culture of fast or convenient food. In the instance of the RAM Super Bowl commercial "So God Made a Farmer," complexity turns to co-option. What appears as a resistance to all things false or phony, in a nostalgic paean to farmers as the caretakers—not only of their farms, families and communities, but of God's own creation—unfolds as a tribute to a brand of pickup trucks and the corporate colonization of the American farm. We recognize American identity not in the throwback Jeffersonian farmer, exactly, but in the rugged and omnipresent RAM that speaks to the consumer longings of "the farmer in us all."

PART II ▪ **Threatened Harvests**

CHAPTER 5

The South Central Farmers Cultivate a Precarious Community

The troubling story of the South Central Farmers (SCFarmers) is now well-known across contemporary US food and environmental movements. Envisioned as an urban "survival garden" for low-income residents of color in South Central Los Angeles, the South Central Farm (SCFarm) became a symbol of hope for a world in which all people, regardless of income, race, or geographic location, have just access to fresh fruits, vegetables, and green spaces. Scholars of food-movement organizing, urban planning and development, race and coloniality, and environmental justice have argued that the SCFarmers' story exemplifies the promise and perils of community gardening as a mode of grassroots empowerment.[1] Building on past studies, this chapter illustrates that the ways and the extent to which this story reproduces, revises, and resists the mythology of American agrarianism—long the dominant frame in public perceptions of agriculture, must be factored into the SCFarmers' rhetorical agency. The mythic reading undertaken in this chapter promises insight on how issues of cultural identity, justice, and urban agriculture figure into the myth today. This reading addresses the possible role of agrarian mythmaking as a tactic of food activism and place-based resistance. Toward these ends, we explore how this case study attests to the conflicted relations between a rising and diverse agrarianism cropping up in places that once seemed unlikely, and the entrenched neoliberal–free market ideology system of political, legal, economic, and cultural power.

Let us begin with the SCFarm's sociopolitical context and significance. Situated in one of LA's most economically impoverished, crime ridden, and racially diverse areas, the SCFarm benefited an estimated 2,000 people.[2] Until 2007, when the fourteen-acre SCFarm was bulldozed, it was the largest of its kind in the country and was one of very few community gardens located in an underprivileged area in the LA region.[3] The SCFarm's considerable material contributions to food justice, food security, and food sovereignty, as well as its symbolically significant location and size, help to explain why its destruction generated such widespread attention. During and after the widely publicized dispute over the land of the SCFarm, media accounts emphasized these exceptional qualities, generating public support by mobilizing the mythic image of the farmer as noble victim whose plight is innately unjust. Carefully considering how this story was (and continues to be) told is crucial to understanding not only the contours and outcomes of the SCFarm controversy, but also how the SCFarmers have become part of the mobilizing folklore of US food and environmental advocacy.

This chapter attends to the SCFarmers' story primarily by examining Scott Hamilton Kennedy's Academy Award–nominated documentary feature film, and subsequent Netflix and Amazon Prime mainstay, *The Garden*. While recognizing some internal differences of opinion among the SCFarmers, the film attempts to encapsulate the approximate perspectives and experiences of most SCFarmers. When the film was completed, some SCFarmers helped promote it by attending and speaking at public screenings.[4] Expanding upon the conceptualization of agrarian myth as a flexible rhetorical form sketched in our opening chapter and illuminated thereafter, our close reading of *The Garden* makes the case that Kennedy's cinematic storytelling contributes to the myth of a new-urban agrarianism as a form of community capacity building. Moreover, our analysis fleshes out some of the enabling and constraining features of the myth as a political form in a twenty-first-century, urban context. *The Garden* articulates what we term a *mythic vernacular narrative*, specifically a Latinx vernacular narrative of urban agrarian resistance. This vernacular discourse testifies to the potential of a diverse, emerging agrarianism that connects modern urban cultures to agrarian traditions around the world.

We derive the concept and the meaning of mythic vernacular narrative in large measure from Kent A. Ono and John M. Sloop's theorization of "vernacular discourse" as "rhetoric of the oppressed" unique to the local communities with whom it resonates.[5] Here we follow the lead of Darrel Wanzer-Serrano, who, using Ono and Sloop's theory, argues that *The Garden* functions to decolonize Latinx vernacular discourse within a "highly racialized context" that subjugates the SCFarmers' ways of knowing and acting.[6] We also draw from Teresa M. Mares and Devon G. Peña's account of the SCFarm as a "vernacular foodscape" imbued with particular ethnic customs, rituals, and heritage.[7] Our case study extends these previous studies' mutual positioning of the SCFarm as a transnationalizing performance of Latinx place-making, which functions as decolonial resistance to dominant discourses of modernity and neoliberalism.

As already foregrounded, by explicating cultural myth's central role in *The Garden*'s visual narrative, our approach differs from these prior studies. The racial and colonialist ideological force constraining this pro-SCFarmers' narrative cannot be adequately understood without accounting for the important role of new-agrarian mythmaking in what Ono and Sloop would call the "culturally syncretic" nature of *The Garden*'s resistive "pastiche." Using Todd Boyd's notion of cultural syncretism, Ono and Sloop show how vernacular forms simultaneously protest some discourses of the dominant culture while affirming others; they are not purely oppositional to the dominant culture and may borrow from it.[8] Ono and Sloop explain that, motivated by a concern for "local conditions and social problems," vernacular discourse reappropriates bits of classic, popular, and dominant American culture to adaptively generate rhetorical appeal and agency.[9] Rather than mere mimicry, culturally syncretic "pastiche fractures culture in the process of appropriating it through imaginative reconstructive surgery."[10] In vernacular pastiche, much like in Janice Hocker Rushing's account of cultural mythmaking as transitioning and evolving public consciousness from old scenes and teleologies to new ones, the social actor selectively integrates old elements into new forms.[11] In other words, like the practice of pastiche, mythic evolution involves rhetorical invention; using or deconstructing existing myths or parts of those myths, and putting new forms in their place.[12]

By inventing a resistive reappropriation of the traditional American agrarian myth to protest today's dominant discourse of modernity, *The Garden* therefore illustrates the practice of culturally syncretic pastiche. Pastiche links the SCFarmers to new food movement discourses critiquing intersectional oppression, while delinking this case from the rural imagery and location of Eurocentric agrarian myth. *The Garden*'s vernacular pastiche demonstrates the productive possibilities of resistive, decolonizing agrarian rhetorics for advancing new food movements. However, it also demonstrates the formidable constraints and risks that new-agrarian counter-myths face, especially when a backlash of colonialist narratives is mounted against them. The case of the SCFarm attests that vernacular narratives may win the battle for public sentiment yet still be trumped by moneyed interests protected under neoliberal political economic and legal structures.

Speaking from a perspective marginalized by these structures, *The Garden*'s vernacular narrative combines a decolonial discourse of ethnic and racial oppression with a class-based moral discourse that has reappeared throughout the history of US agrarian mythmaking.[13] Found in agrarian discourses ranging from Thomas Jefferson's letters to late nineteenth-century Populist movement protests, this moral economic frame of agrarian virtue discursively positions an exceptional people who cultivate the soil against the corrupt ideologies and occupations of those who hold power in the city.[14] In this dichotomous frame, the farmer and other humble allies are usually portrayed as noble and heroic victims of corrupt power-holders and the natural environment.[15] As we have seen, the examples of Country Life and the Southern Agrarians, as well as Rodale's renderings of the innocent farmer duped by chemical agriculture, differently illustrate this theme of the virtuous farmer pitted against wealthy and cynical opponents.

In addition to contextually adapting this class-based frame inherited from past agrarian mythmakers, *The Garden* invites identification through appeals to value principles consistent with those defining traditional agrarianisms, including but not limited to the land as a vital source of connection with family, community, nature, and faith. Reworking the theme of the farmer as noble victim, the film's urban new-agrarian narrative replaces the old protagonist and setting, the White yeoman working a solitary plot, with a Latinx community that builds capacity through collaboration while turning a vacant lot into an urban farm. Tema Milstein, Claudia Anguiano, Jennifer Sandoval, Yea-Wen Chen, and Elizabeth Dickinson suggest that, in contrast to Western discourses of the environment that implicitly represent White affluence, vernacular environmental discourses among marginalized communities foreground a sense of self deeply embedded at the confluence of food, place, and nature. In this eco-vernacular discourse, food is materially and emotionally central to "past and present senses-of-relations in place"; the loss of the land also means the potential death of culture practiced as "ecoculture."[16] *The Garden* presents just such a grounding of community, not only in place but also in time. At the SCFarm, the confluence of land and community was enriched by the enactment of Latinx knowledge and traditions in the new context of urban Los Angeles. Despite the constraints it faced, the SCFarm represents an intriguing testimony of cultural diversity in urban agrarianism that continues to inspire through film. While the SCFarm faced serious obstacles in the dominant culture, the movement of these farmers from the margins to the center of food advocacy and environmental movements allowed their case to inspire an urban agrarianism that continues to rise.

Hereafter, this chapter consists of four sections. The section that immediately follows offers a brief overview of the US urban community-garden movement. The second section reviews the tumultuous history of the lot on which community members created the SCFarm. The third section presents a close reading of *The Garden* guided by American agrarian myth as a theoretical concept and critical heuristic. Finally, in the fourth section we explicate some implications from this close reading of the film.

Diversity and Community Gardens

Although the US community-garden movement has changed significantly since its late nineteenth-century emergence, some of the same societal problems that prompted community gardens then continue to generate enthusiasm about them today. During those times, cities offered the burgeoning population of economically disadvantaged residents the opportunity to grow food in city-owned vacant lots.[17] Although interest in community gardens waned after World War II amidst the advancement of industrial food production methods and an expanded food distribution system, their popularity increased again during the cultural revolution of the late 1960s and early 1970s, a period of significant urban decline.[18] Today, too, urban advocates propose community gardens as part of a broader effort to address deepening urban economic strife, crime, gentrified urban development, and more generally, the need for community-building amidst increasing demographic and cultural diversity.[19]

When the modern community-garden movement emerged in the late 1960s and early 1970s, many community-garden organizers sought to transform vacant lots into green spaces for vegetable plots, sitting areas, playgrounds, and flowers.[20] Vacant lots were often blighted sites for drug dealing and other crimes. The gardens enhanced the attractiveness of neighborhoods and created opportunities for community development.[21] Many of those who cultivated these new community gardens were recent immigrants or African Americans from the southern United States, who introduced their own cultural influences to the gardens.[22] During this period, the federal National Urban Gardening Program provided financial support for gardening efforts in five cities.[23] By the mid-1990s, over 1 million individuals were involved in more than 15,000 organized community-gardening programs across the country.[24]

From the mid-twentieth century onward, low-income communities and communities of color, particularly in urban settings, have been far more likely than other segments of the population to be food insecure or live in "food deserts" where they lack access to high-quality and healthy foods.[25] Over the past half-century, the growth of the industrial system of highly processed foods consisting of dangerous levels of fat, salt, sugar, and other substances has worsened this problem, adding another exigency for urban community gardeners to address. Contributing to a countermovement and working against fast food and corner convenience-store grocery shopping, community gardens promote making fresh, organic, and nutritious fruits and vegetables available for people in need. More so than in the past, then, the community gardens of today are often viewed as part of a grassroots response, or at minimum, a coping mechanism for dealing with not only dire poverty and hunger but other burgeoning consequences of a faltering industrial food system and a global environmental crisis. As Thomas Lyson writes, enacting "civic agriculture," community gardeners engage the food system in a way that not only increases their access to fresh produce, but also has the potential to transform them from passive consumers to active and aware "food citizens."[26]

Public expectations for community gardens have expanded as they are cast as correctives for various social ills. Scholars and practitioners affirm them as tools for fostering, among other positive social ends, food and environmental justice, food sovereignty, food security and public health, environmental and food literacy education, community beautification, and even improved neighborhood real-estate value. There is good reason for most of these expectations. Research suggests that community gardens often deliver many civic benefits: fostering positive social interaction between neighbors of different ethnicities and class backgrounds; functioning as sites of physical activity, therapy, and various educational and cultural events; contributing to air purification; controlling city noise and temperature; offering fauna and flora habitats; helping to improve neighborhood safety and stabilization; driving up rates of home ownership while resulting in fewer vacant lots and buildings; providing a less expensive alternative to city parks; and increasing adjacent property values.[27] Of course, these benefits do not magically appear as soon as vegetables are planted in an empty lot. Many community gardens exist temporarily, and the success and resilience of community-garden initiatives may vary greatly. Furthermore, as the case of the SCFarm shows, even successful and widely popularized community gardens can be precarious if not protected by governance, land-use laws, or property rights.

Throughout the history of the SCFarm, the number of positive effects multiplied but were also precarious and increasingly under threat. The SCFarm had become representative of what Marco Cenzatti, following Henri Lefebvre's notion of space, calls "heterotopias of difference"— "spaces of representation" given shape by the lived experiences of their occupants. These heterotopias "are produced by the presence of a set of specific social relations and their appropriation of physical space," which as demonstrated by the SCFarm is not inert, but rather, participates in the social relations in its own unique way.[28]

Some of the threats to such heterotopian spaces are external. Indeed, the history and fate of the SCFarm can be read as a mythic mediation of what Robert Emmett describes as the "overarching tension" in today's urban community-garden narrative. This tension emerges from the implicit conflict between valuing land as private property and valuing it as a public good that regenerates community.[29] In the case of the SCFarm, this tension erupted into a pitched battle in the public and legal spheres. While the farmers "lost" the key battles and their garden was destroyed, the example of the farm persists in the equally heterotopic space of public imagination. Part of the value of *The Garden* and other discourses around the farm is to encode the SCFarm experience as testimony to the power of food production to build community and change lives. From the standpoint of rhetoric as a form of cultural memory, SCFarm is far from gone.

Before turning to our close reading of the film's vernacular narrative, we provide a contextual backdrop for understanding political dynamics that play out in the film's story of the SCFarm. While featuring some aspects of the background account, the film heavily focuses on the firsthand experiences of those involved in the farm. To more deeply contextualize the SCFarm's past and the film itself, we begin by exploring some of the complexities leading to the dramatic showdown over control of the land that began to unfold in 2004.

Vacant Lot/Contested Land

Prior to becoming the site of the SCFarm, the lot at the corner of Forty-First and Alameda in South Central LA was already a focus of vigorous political contestation. In 1986, the city of Los Angeles obtained the then-vacant lot through eminent domain, with the intention to install a power-generating waste incinerator on the property. As a part of the Los Angeles City Energy Recovery Project (LANCER), the city paid Alameda-Barbara Investment Company, led by its primary investor Ralph Horowitz, approximately $5 million for the land.[30] Thanks to the efforts of a group of residents known as Concerned Citizens of South Central Los Angeles (hereafter, Concerned Citizens), the investment firm never carried out the waste incinerator project. The predominantly African-American nonprofit organization teamed with national and grassroots environmental, slow-growth and public-interest law groups to conduct a health risk assessment and successfully block the construction.[31] This grassroots effort has become known as an environmental justice movement success story.[32]

The land previously designated for the waste incinerator remained vacant until 1992. South Central LA had been the epicenter of widespread rioting following the Rodney King police brutality verdict and the blight in the area worsened. This blight, in addition to poverty and despair, hunger, crime, and other problems in the area prompted the LA Regional Food Bank to ask the mayor to use the land on a temporary basis. Positioned across the street from the property, this nonprofit food-distribution organization reached out to neighborhood residents and facilitated the formation of a community garden on the lot.[33] To begin this project, the food bank distributed what it envisioned as survival garden plots to families with incomes no greater than 150 percent of the region's poverty level.[34]

Two years later, after canceling a plan to sell it to a public housing corporation for the creation of affordable townhomes, the city sold the property to its own Harbor Department for $13.3 million.[35] The department reaffirmed former Mayor Bradley's agreement with the food bank and created a mutually revocable permit allowing the LA Food Bank to continue to operate the site until other plans for the property were established.[36] Shortly thereafter, the food bank notified the dozens of low-income families cultivating the land that it would no longer be able to cover the costs of managing the farm. The gardeners, who were already cultivating hundreds of small plots on the land, responded that they could manage the land on their own.[37]

Primarily composed of Mexican and Central American immigrants, the SCFarmers as a community reflect major demographic changes that had recently occurred in South Central LA. The community had shifted from an African-American majority to a predominantly Latinx population.[38] Under the name of SCFarmers Feeding Families, the group originally consisted of roughly 360 families, including US-born Mexican Americans and people from indigenous diasporas across Mesoamerica.[39]

After the SCFarm formed, external developments, including modernization of the nearby transportation infrastructure, increased the land's potential economic value. This renewed former owner Ralph Horowitz's interest in the site, but negotiations for his development corporation to repurchase the land from the city fell through.[40] In the late 1990s, Mayor Riordan began to discuss the conversion of the site into an industrial park as part of a broader economic development plan—a plan that was endorsed by the still-active Concerned Citizens, who had opposed the incinerator.[41] Still, the garden remained, and in 2001 Horowitz's new development group, Libaw-Horowitz Investment Company (LHIC), officially filed suit against the city of LA for breaching the original eminent-domain contract. Under the contract, if the city had failed to use the site for non-public or non-housing purposes within ten years, Horowitz's company had a right to repurchase the land back from the city.[42]

At that time, the city opted to enter into settlement negotiations with LHIC and, in 2003, agreed to sell Horowitz the property for private business use for approximately $5 million, pending the dismissal of the lawsuit and LHIC's donation of 2.7 acres of the site to the city for a community recreation space in the form of a soccer field.[43] The sales agreement was clearly below the market value at the time, but prominent African-American councilwoman Jan Perry, whose district included the property, approved it. This approval took place in a closed session of the LA City Council.[44] The Concerned Citizens group had been adamant about their desire for a soccer field and supported the settlement. The group argued that not protecting Horowitz's private-property rights would set a dangerous precedent, possibly making it more difficult to establish temporary city land-use contracts in the future. Concerned Citizens also contended that use of some of the SCFarm land for a soccer field would make the property more accessible for the recreation use of all members of the community.[45] The struggle over the SCFarm site therefore became defined in part by division between local groups consisting largely (but not exclusively) of African Americans on one side and Latinx on the other.

In January 2004, Horowitz gave notice to the food bank that their revocable permit would terminate by the end of February of that year. This prompted a series of lawsuits and countersuits between the SCFarmers and Horowitz. Represented pro bono by the progressive law firm of Hadsell and Stormer, the SCFarmers argued in court that their rights had been violated due to the city's "back-room" closed-session negotiations with Horowitz and the below-market value settlement. They were granted an injunction to remain on the land until the case was resolved.[46] In late 2005 the court ruled that the closed negotiations were legitimate and that the land must be turned over to Horowitz. Many SCFarmer families, along with local and national allies, insisted that they would turn to civil disobedience, if necessary, to retain the land and save their garden.[47]

As eviction became imminent in May and June of 2006, protestors from around the region and the country, including Hollywood stars and famous musicians, joined the SCFarmers at the community-garden site.[48] There were reports that Horowitz was willing to accept a sum of $16 million for the property, but attempts by politicians, including Los Angeles mayor Antonio Villaraigosa and California senator Barbara Boxer, to negotiate with Horowitz failed. Reported multimillion-dollar pledges from the Trust for Public Land and the Annenberg Foundation also failed.[49] The Annenberg bid was received after Horowitz's bid deadline, and he did not respond.[50] In June 2006, bulldozers uprooted crops to clear a path for sheriff's deputies, who forcibly evicted hundreds of farmers and their supporters from the land and arrested forty-four for obstruction.[51] Authorities arrested ten more protestors in early July, when a smaller group attempted to stop bulldozers from plowing over what remained of the fourteen-acre garden.[52] A few weeks later, another judicial decision upheld the sale of the site to Horowitz.[53]

Documentary as Vernacular Mythmaking

Understanding Scott Hamilton Kennedy's intention as a director and producer helps to place *The Garden* within the rhetorical frame from which it emerged. Kennedy has stated that his purpose with this documentary was to "tell the story" of this community garden from the perspective of farmers and community members directly involved in the dispute.[54] In an "extra feature" interview with film critic David Poland included with the DVD version of the film, Kennedy clarifies that given that he started filming in 2004, he could not have anticipated that so many dramatic twists and turns would develop over the next few years. According to Kennedy, he became interested in this story early on because it seemed like a "perfect tale of democracy" in action, adding that only later did he develop an affinity for the farmers. This affinity grew to the point that upon finishing editing the film, it became hard for him to watch the tragic ending because there were "so many ways [the community garden] could have been saved." In other words, the intent was not to create a political film or to be an activist filmmaker, but rather to make a film about activism. Yet the affinity that Kennedy developed for the SCFarmers shines through in the film's primary focus on SCFarmers' viewpoints and experiences. As a result, the film functions as social justice advocacy.

Kennedy's preference for grassroots voices over the use of expert commentary or voice-over narration results in a vernacular film about the creation and experience of social truth among a community of people in a distinctive cultural space. The film's scope and cinematic techniques emphasize its situated position in place and time. Rather than taking on an investigative role to determine the truth regarding the downfall of the SCFarm, including what may have occurred behind the scenes, Kennedy has described his intent as embracing ambiguity. Acknowledging that he does not have all of the answers to why the SCFarmers lost their fight, Kennedy describes his technique as follows in his interview with Poland: "I like letting the facts be . . . letting the people use their own words to express what happened." Kennedy adds, "I like the grey areas in a story." *The Garden*'s official website, sponsored by the production company Black Valley Films, maintains a position consistent with Kennedy's. The website states that the tragic outcome aside, the film animates both the promise and pitfalls of American freedom and democracy, as reflected in the grassroots organizing of immigrants who originated from countries in which "they feared for their lives if they were to speak out." The website adds that rather than asking viewers to take a side, the film invites viewers to reflect upon conflicts between American values that linger throughout the film, stating that that *The Garden* "raises crucial and challenging questions about liberty, equality, and justice for the poorest and most vulnerable among us."[55]

Additional production company website descriptions retain an emphasis on the film as an unbiased contribution to rational public deliberation regarding the upholding of core American values. Implicitly gesturing toward the film as a hybrid of what Bill Nichols calls "expository" and "observational modes" of documentary film, Black Valley Films states that *The Garden* combines "the pulse of verité with the narrative pull of fiction."[56] Black Valley Films frames this tension as arising from a naturalistic method of being on the scene while remaining unobtrusive. The value of this naturalistic observational mode is that it captures how the events in focus incrementally unfolded, giving a firsthand account of their impact on the people involved.[57]

In dialectical tension with this naturalistic approach, the film relies on an expository technique that has long been central to the popular public perception of documentary.[58] As noted, Kennedy features direct-to-the-camera verbal commentary and personal testimony from key parties involved with the farm. Despite not using the traditional expository technique of voice-over narration, the juxtaposition of voices maintains an expository emphasis on argumentative logic, perspective, and audience comprehension.[59] By deemphasizing any sense of the filmmaker having extended engagement with individuals featured on camera, this mode complements the observational style.[60] As will be elucidated in the analysis that follows, the naturalistic approach that Kennedy utilizes to construct a vernacular pastiche proves potent for foregrounding the affective experience of place-based resistance. Further, it illuminates how the SCFarmers ruminated through their experiences by using a sense-making approach that was fundamentally oppositional to that of their opponents. The SCFarmers enacted a social vision that placed community and collective rights above individual self-interest.

The Vernacular Garden of Good and Evil

From its first scene, *The Garden* contributes to the evolution of a urban new-agrarian myth grounded in vernacular discourse and imagery. Bringing these new characters and scenes into focus, the opening scene features a farmer leaving his house before dawn to go to work. Unlike the White, rural, male farmer who American agrarian narratives have often mythologized as virtuous protagonist, this farmer is a person of color who lives in the city. Upon his arrival at the farm at daybreak, it becomes clear that this farmer also cultivates a different kind of farm than is typical of agrarian narratives set in rural locations in the United States; he is an urban community gardener, he does not own or rent his land, and he does not sell his harvest. Rather than using a horse, tractor, or pickup truck, this farmer takes a commuter train to work. The film shows this farmer and others diligently breaking the soil with hand tools; planting crops; and, with a look of pride and anticipation on their faces, watering the crops. As children pick the vegetables, the viewer is exposed to a new imaginary of family farming.

As Ono and Sloop explain, vernacular discourse often functions "to upend essentialisms, undermine stereotypes, and eliminate narrow representations of culture."[61] Almost immediately, *The Garden* contests not only dominant representations of immigrants and Mexican Americans as lazy or criminal but also of the American farmer as White and rural. This productive disruption continues to develop as the camera cuts away from the SCFarmers and shifts to an aerial shot of the SCFarm. The SCFarm appears as a green, lush, and visually inviting island in the middle of an otherwise dreary scene. It is in sharp visual contrast with the vast and drab industrial cityscape surrounding it for as far as the eye can see. Beyond the SCFarm, there are few trees or green spaces of any size. The area is an industrial zone of concrete and warehouses, and the only sign of life from above is the traffic on the highways running through it. While the traditional American agrarian myth assumes that country, nature, and farming exist in a place of solitude and separation outside of urban areas, the aerial visual shot inverts this premise; nature and agriculture appear to sprout from the center of a distinctly non-rural industrial zone.

A traditional agrarian moral frame of corrupt urban powerholders exploiting humble rural cultivators of the soil and other "ordinary" people structures the film as a mythic collision between good and evil. The opening scenes described affirm agrarianism as a dignified and virtuous life of cultivating the soil. Consistent with agrarian myth, the farmer rises early to cultivate the land and to reap rewards for the mutual benefit of family, the community, and the land. However, as the film sides with not-for-profit community empowerment in a political struggle against the neoliberal economics of privatization, it complicates the economics-infused morality that has historically defined US agrarian uprisings. Even as it engages the traditional agrarian frame, the film memorably integrates non-European ethnic agricultural practices in the urban setting of Latinx Los Angeles, while retaining an emphasis on agrarian virtues such as simplicity, practicality, honesty, solidarity, and environmental sustainability.

The traditional agrarian moral-economic narrative of corrupt urban elites exploiting humble rural farmers emerges in *The Garden* in both verbal and visual ways. Rufina, one of the leaders of the SCFarm, states that Horowitz, African-American city councilwoman Jan Perry, and the primarily African-American community organization Concerned Citizens have orchestrated an out-of-court "backroom deal" to use the land for a warehouse and a soccer field. Later scenes suggest that Perry may have worked in alliance with Concerned Citizens because they helped her get elected and they remained politically powerful in the community. Visually, interview footage of Perry and other opponents of SCFarmers in the film typically places them against a dim or dark background, in some instances with shadows partially hiding their faces. In one striking image, Perry, standing in the shadows with the night sky behind her, denies that she was personally involved with negotiations with Horowitz and asserts that the matter was litigated in court. However, in the next scene, Tezozomoc ("Tezo"), another SCFarm leader, exposes Perry's dishonesty. Tezo obtains and reads from out-of-court settlement documents describing Perry's direct role in the negotiations.

Adding to this dichotomous moral frame of the honest and humble versus the corrupt and powerful, the film juxtaposes interview statements from new owner and business developer Ralph Horowitz with statements from the oppositional community organization Citizens of South Central. The result of this pairing is the critical portrayal of Horowitz's statements as xenophobic and racist. This framing of Horowitz's remarks suggests that he has stereotyped and demeaned the SCFarmers as poor, unable to speak English, and as feeling entitled to the land.

For their part, the SCFarmers featured in the film charge that racism against Latinx people has led to the eviction. As the conflict between the SCFarmers and the city comes to a dramatic climax, Horowitz explains that his reason for not selling the SCFarm back to the SCFarmers, even for what would be millions of dollars in profit, is that he does not "like their cause" and their "'you owe me' mentality." Consistent in his appeal to the Western colonialist and neoliberal discourse of self-determination and individualist meritocracy, Horowitz also defends his property ownership by appealing to the common sense of a cost-rewards calculus of self-interest: "I have a huge mortgage, real estate taxes, liability insurance, all costs and expenses that any private property owner has."

Aligning with this framing of the SCFarmers' opponents as crudely self-interested, prejudiced, and heartless, two SCFarmers are shown sharing their opinion that the eviction has come "because we're Latino." Rufina states the following about the Concerned Citizens' opposition to the SCFarmers: "They want to do the race thing. They wanted to do the race thing from the get-go, to make this a brown and black thing," in historically African-American South Central. Concerned Citizens' leader Juanita Tate stresses, inaccurately, the high number of undocumented

Latinx people who have moved into the neighborhood. As we expound in the next section, the mythic appeal to exceptional virtue in the modest agrarian life takes on new urgency in the contested, heterotopic site of the SCFarm, where farmers appear as model citizens developing community amid decay. Despite this apparent virtue, however, they are cast as outlaws without title or claim to the garden their labor has made.

The Mythic Connection

As *The Garden* constructs a pastiche of vernacular discourse, it presents the SCFarmers' various, intertwined forms of mythic connectedness as unique, timeless, and vital sources of societal well-being. The film affirms that, as part of the nation's enduring cultural mythology, farming is the agrarian practice of familial, community, spiritual, and environmental connection. Familial connection is one form of mythic agrarian connection to which *The Garden* appeals. Revealing the SCFarm as the basis of familial bonds and familial heritage contributes significantly to the film's promotion of emotional identification with the agrarian myth's narrative of the farmer's plight. Throughout the film, Tezo plays a key role in this agrarian appeal to familial connection. In one scene, Tezo speaks with the SCFarmers' civil-rights attorney Dan Stormer and the legal team representing the SCFarmers. In a voice shaken with emotion, Tezo explains, "For me, it's very personal, because it's something that my dad left to me. My dad is ill, and he can't be with us, so it's personal in that manner." Stormer then states that his team will be writing up a case to obtain a temporary injunction to stop the eviction.

In another poignant scene, with the sun going down in the background, Tezo works in solitude, breaking the soil with his hoe and then carefully positioning his seeds in the soil. Seeming to use this moment of solitude and his engagement with the soil as a source of Thoreau-like agrarian reflection, perspective, and guidance, Tezo, through voice-over narration, reflects on the ongoing land dispute and SCFarm. The camera turns to another farmer, who states, "Tezo's father, he teach me how to grow the plants, like professional. He teach me." The voice-over of SCFarmer Juan Gamboa then emerges with a picture of Tezo's father. Gamboa iterates, "Tezo's father was a very respected person here in the garden. A very helpful person. He was very admired here. We loved him a lot." In this scene, family and community connection commingle and converge.

Another example of how the SCFarmers' vernacular narrative appeals to familial connection is when Rufina and other SCFarmers give city councilman Tony Cardenas a tour of their community garden. Cardenas tells the SCFarmers, "I learned how to farm the land and I got my first blister when I was just five or six." Cardenas also notes the instructive value of the farm to a group of SCFarmers: "These are things that your kids cannot learn in books." As others nod their heads in agreement, one SCFarmer emphatically responds, "No they can't learn this in books. We need to teach them here."

In addition to the agrarian appeals to familial connection, the film animates the SCFarmers as practicing agrarian environmental connection as well. Verbal appeals to this facet of agrarianism manifest, for example, when the SCFarmers receive their eviction notice and respond by questioning why they were not given enough time to harvest crops at the end of the growing season. During the public comment session of the city council meeting, SCFarmer Don Eddie stands, holds up his arms, opens the palm of his hands, and states (in Spanish): "I am a farmer and here is the proof in my hands! To not allow us to stay on that land that has already been cultivated? For what reason?" Not only is Eddie's question logical, his embodied performance, including his reference to his worn hands, emotionally evokes mythic agrarian morality. As Thomas Burkholder notes, in American agrarian myth nature is "beneficent" and its bounty rewards honest labor. Productive citizens who work the land are entitled to its rewards.[62] The breach of this agrarian contract has long provided farmers with a basis for public moral grievance. An injustice to the farmer is also an injustice to the community, whose well-being depends upon the farmer's success more than it does any other citizen. This dependence pertains not only to food, but to the farmer's economic importance as a key patron of urban merchants ranging from bankers to skilled tradespeople.[63] In the agrarian myth, as Richard Hofstadter explains, this means that agriculture, as a "calling uniquely productive and uniquely important to society," has a special right to the concern and protection of government.[64]

The SCFarmers' appeal to this mythic agrarian moral contract also supports the intrinsic value of the land above and beyond its economic profitability. In this discourse, appeals to the farmers' intimate connection with nature and the deep meaning of the land as part of a familial heritage cast the film's appeals to agrarianism as a source of spiritual connection. The film invites viewers to connect with the SCFarmers at an emotional level, but also to reconceive of the issues surrounding the garden through an alternative cultural logic beyond the colonizing epistemology of neoliberalism and modernity. In multiple instances, the film shows farmers in tears describing the SCFarm as a "sacred" place. In a scene following the devastating announcement that the garden would soon be bulldozed, an elderly woman from the SCFarm community asserts that, in addition to a plant from her plot and whatever else she can manage, she will be rescuing a statue of the Virgin Mary from the site. When a group of SCFarm protestors march to the mayor's office, Tezo meets with the deputy mayor and pleads, "It is as if I went to your community and took down your temple, I took down your church. That's what we're talking about. These are sacred things, and you're taking away our way of life." Another SCFarmer encapsulates Tezo's sentiment, succinctly stating, "Because without the land, we are nothing." The SCFarmers' appeals suggest that while the land offers dignity and sustenance to individuals, it is also about more than one person or family, with a value beyond even the SCFarmers as a group. The SCFarmers promote the benefits of their farm for the broader community, future generations and the environment.

These appeals to agrarianism as spiritual connection with the land reflect agrarian myth's capacity to sanctify certain social arrangements as not only timeless but embedded within an embodied and place-based sense of identity and existence. However, the urban new-agrarian message in the spiritual rhetoric of *The Garden* changes the terms of other manifestations of this myth. For the SCFarmers, sustainable, healthy agricultural practices, such as organic farming using biodiversity-affirming permaculture, summon this sacred connection. The spiritual wellspring of agrarianism is open to those who come with an earnest heart and can be tapped by communities in different locations and occasions.

The Garden also suggests that an agrarian environmental ethic can provide a means to recovering the sense of community that is lost to commercial plunder of the earth and its people. Early in the film, Tezo recounts the SCFarmers' positive impact on the local environment and community through the creation of the SCFarm. While the area in and around the farm had been subject to extensive, almost irreparable, environmental decay and industrial blight, the SCFarmers were able to rejuvenate it. Tezo states that the fourteen-acre lot was in such disrepair that "[they] actually had to buy dirt to basically put dirt in here because what was there was actually the foundation of the older warehouses." Similarly, upon learning of the eviction notice, a woman in tears states of the land, "It was dirty, and we cleaned it. We were happy because we had a piece." As the SCFarmers rehabilitate the land into an extensive community garden, they practiced agrarianism as an embodied, material, and place-based connection to environment and home.

Invoking the mythic idea of agrarianism as performing a civilizing function, these appeals to spiritual connection mix with the law of environmental return. In return for the constant and tireless effort that agricultural land requires, the farmer and the community receive tangible and intangible benefits including sustenance, moral fortitude, and community voice and efficacy. For the SCFarmers the community garden provided affordable and healthy food, a strengthened community, and sense of both resilience and resistance as they sought to maintain the farm in the face of neoliberal capitalism and colonial enclosure of property interests. The SCFarm generated a vernacular place-based community that would extend its appreciation for the local cultivation of fresh, nutritious produce and the educational value of urban green spaces into the wider community. SCFarmers featured in the film offer statements such as, "I have learned from this land to be proud," adding that it was a legacy to pass to their children that could not be learned in the wider society. The land instilled, through trial and error, lessons about the virtues of diligence, care, self-sufficiency, and cooperative community. These harvests affirm Wendell Berry's contention that "proper concern for nature and our use of nature must be practiced, not by our proxy-holders, but by ourselves."[65] If we are to reclaim responsibility for nature and for our food, we must overcome dominant Western conceptions of nature as a collection of ideas or set of material resources and see it in the substance of our home and lives.[66]

Prior to creating their urban farm, the SCFarmers' culture predisposed them to a rich conception of the human-nature relationship. *The Garden* shows that the lessons SCFarmers learned from each other and from the land about familial, community, spiritual, and environmental connection reinforce the values of their Latinx heritage. *The Garden* casts the SCFarm as a cultural and community site for the ritual preservation of Latinx ethnic foodways and agricultural practices. These foodways and practices center on the celebration of biodiversity as a key source of a community's vitality and abundance. Latinx urban-agrarian practices, to which the film draws attention as culturally meaningful and civically virtuous, additionally relate the SCFarm to wider social movements. They parallel food advocacy across broad segments of culture for a sustainable and just alternative to the monoculture of the dominant food system.

Biodiversity is a guiding value for alternative food movements, as *The Garden* vividly portrays. Much of the food grown at the SCFarm mirrored and affirmed the gardeners' cultural heritage; many of the plots featured heirloom plants from the gardeners' homelands,[67] some of which are medicinal and used in the absence of health insurance.[68] In a memorable scene, Rufina gives District Fourteen council member Antonio Villaraigosa, who would later become the city's mayor, a tour of the community garden. She appeals to Villaraigosa by presenting the rich biodiversity exhibited there, which included a great variety of medicinal herbs as well as edible produce such as avocado, guava, apple, banana, and papaya. The SCFarmers show deep appreciation for the agrarian practice of place-based resistance grounded in the virtue of biodiversity-as-sustainability.

Ethnicity-aligned agrarian practice provides the generative basis for the SCFarmers' challenge to dominant Western discourses of commodification and nationalism steeped in the infinite manufacture of homogeneous and predictable space.[69] The SCFarm vernacular narrative identifies the community garden as a heterotopic space with intrinsic ecological, cultural, spiritual, and educational value to be protected beyond profit-making potential. It is constructed as distinct from the dominant discourse of modernity that claims space and land as resources for privatization, exploitation, and consumerist-economic growth.

For communities such as the SCFarmers, whose families immigrated to the United States within the past couple of generations, industrially produced, genetically modified, and highly processed food violates ethnic foodways based on organic, fresh, locally sourced, and homemade ingredients. *The Garden*'s vernacular narrative celebrates this foodways heritage grounded in place-based cuisine and biodiversity. Emphasizing ethnic and class-based functions of the SCFarm such as these, Teresa M. Mares and Devon G. Peña describe the SCFarm as a "vernacular foodscape" and "autotopography" for performing identity and reterritorializing agricultural ethnic customs.[70] Some of the heirloom seeds used by the SCFarmers could be traced back 5,000 years in their origins, and the selection and arrangement of edible crops and aromatic and medicinal herbs mirrored that of the hometown kitchen garden, or *huerto familiar*, common in Mexico, Central America, Puerto Rico, Cuba, and the Dominican Republic.[71]

Agrarian knowledge that the SCFarmers brought to the garden also included its internal governance structure. In response to a public statement by Juanita Tate of the Citizens of South Central that SCFarm families were holding a disproportionate number of lots and selling their harvest for profit, SCFarm leaders acted to ensure that this was not the case. Tezo says to the camera, "You're allowed to help your family. You're not allowed to make it a for-profit business. There's a group of people who have a lot of lots. We don't want that to be the focus of the garden." The camera then turns to the SCFarm rulebook and shows a SCFarm leader announcing a few new rules for the garden. He states that failure to comply will result in plots being taken away within twenty-four hours. Rufina adds, "We don't give out plots to two or three people within the same family. That is one of the rules. It's always been that way." Tezo explains that if Perry and others successfully argue that the SCFarmers are using the land for profit, they will certainly succeed in taking the land away. Illustrating the challenges of creating and maintaining an agrarian democracy, internal conflict over plot ownership escalates to the point that a disgruntled SCFarmer attacks one of the SCF leaders with a machete and is arrested. Despite the emotional toll of these events on the SCFarmers, the arrest restores the SCFarm as an agrarian community that values the family and a common good.

Read from the perspective of dominant American agrarian narratives, the democratic land-use procedures and practices that the film ambiguously depicts affirm the Jeffersonian-yeoman ideal of an agrarian democracy in which every citizen, or family, lives in relative self-sufficiency through the cultivation of a small plot of land. Under this ideal, each citizen-farmer has a relatively equal political and economic stake in the success of the nation. However, while the democratic land use at the SCFarm may affirm the myth of American agrarianism, the specific system for governance and distribution of land is modeled on the Mexican *ejido*. The ejido system consists of a governing *junta* council and general assembly.[72] Unlike the Jeffersonian ideal, the ejido does not require smallholding private ownership of agricultural land as the basis of democratic culture.[73] Instead, the ejido aligns with what Thomas Lyson describes as civic agriculture forms such as community gardens in which personal stakes become communal as the food production and distribution process is mutually realized and shared through material and social benefits of collaboration.[74]

The ejido is one of several political devices shown in the film that contribute to the SCFarmers' democratic resistance to modern colonialism and its neoliberal forms. Much of the externally focused democratic action appeals to American ideals of freedom of speech and assembly, due process, and justice for all. For example, when given their eviction notice, the SCFarmers begin to organize a legal defense and public campaign to save their right to the land. The film shows them testifying at LA City Hall, giving tours of the farm to local, state, and national political leaders, as well as celebrity-activists, and holding public rallies. Indeed, in their discourses promoting the film, Kennedy and his film production company directly appeal to the opportunities for free speech and democratic representation that immigration afforded these urban farming families.

As *The Garden* invites viewers to grapple with challenges that new-agrarian conceptions of democracy and freedom pose to society, it links agrarian myth to the current political and cultural moment, and attests to new agrarianism as a promising form of socially resistive "delinking." Walter D. Mignolo explains that "delinking means to change the terms and not just the content of the conversation."[75] The SCFarmers sought to change the content of the conversation about rights, at its foundations, as Rufina states, "I think everyone talks about the rights of the owner. Nobody is debating his rights. But I'm also talking about the rights of the community." However, the SCFarmers also changed our collective conversation about agrarianism through their re-assemblage of agrarian myth into a pastiche of diverse connections, fit to purpose for a place and occasion. Changing terms, Mignolo suggests, requires more than a brief interruption of the status quo; it entails deep resistance to epistemic dominance. At stake is not only the language and logic of the conversation but its link to the dominant ideology of modernity.[76] Delinking requires shifting and revising cultural values to reconfigure the nexus of nature and culture, including human relations to the environment, other creatures, and our place in time. Through this reconfiguration, we expand the possible meanings of economy, community, politics, and life, and enact a place-based "body politics of knowledge."[77] The legacy of the SCFarm, as encoded in *The Garden*, may be aptly viewed as such a delinkage and generative expansion.

Lost Farms, Found Forms

The mythic reading of *The Garden* that has been constructed in this chapter theorizes Scott Hamilton Kennedy's documentary film as a mythic vernacular narrative of new-urban agrarianism. Engaging past studies that situate the SCFarm and *The Garden* as vernacular forms,[78] we highlight that the force of the racial and colonialist ideologies constraining the SCFarmers' narrative cannot be fully understood without accounting for the important role of agrarian myth in the culturally syncretic nature of *The Garden*'s resistive pastiche. The film significantly revises and reappropriates the moral frame of American agrarian myth to pit honest, humble cultivators of the soil against corrupt and greedy power-holders. The story develops within a dichotomous frame in which the sympathetic character of the farmer as noble victim nests within a pastiche of urban agrarianism defined by its mythic connections to family, community, lands, and spirit. This pastiche combines fragments from American agrarian myth, commonplace values from the dominant national culture, and Latinx cultural discourses. The result is a vernacular narrative of social resistance that marks the SCFarm as a heterotopic, temporal realization of social, environmental, and food justice for the culturally marginalized.

In our reading, Kennedy's film functions within an ideological register as a cinematic deconstruction of the dominant discourse of neoliberalism in land-use and property rights contexts—especially urban community capacity-building contexts. Within this critique, the mythic image of the American citizen-farmer undergoes what Ono and Sloop term "reconstructive surgery."[79] The resulting hybrid and culturally syncretic form links the SCFarmers to existing critiques of identity-based oppression made by the food and environmental justice movements. Importantly, this form also delinks the SCFarm experience from the White, rural imagery of Jeffersonian agrarian myth. Despite a lack of epistemic privilege and cultural capital, the film's protagonists articulate a place-based, embodied, and decolonizing discourse, pitted against the city government, property owner, and a community organization that opposed using the land for a Latinx community garden. The film shows these parties conspiring in explicit and indirect ways to end the garden and evict the farmers in a campaign that seemed to combine xenophobia and racism with capitalist greed and political corruption. The empowered farm community was greeted as a threat, branded as illegal, and expelled from the land.

In the absence of compromise, the complicated political-economic and cultural dynamics of city land-use that *The Garden* highlights came down to a matter of what the community and city officials valued most— or at least what its influential dealmakers valued most. Garrett Broad describes this case as a showdown between an upstart discourse that affirmed the land's value as a scene of cultural formation and a dominant discourse that only recognized value in terms of economic exchange.[80] To extend Broad's point, the dominant discourse was upheld, but the discursive field of the SCFarm remained open, even if the farm itself is displaced. *The Garden*'s mythic narrative identified an enduring, and unresolved, tension between food movements that address community needs, such as community gardens, and legal, political, and economic structures that reify property rights and frequently pit the claims of individuals against one another.

This identification of agrarian values with community protection and restoration has animated the agrarian discourse and myth examined in this study, from Country Life, to the Southern Agrarians, to Rodale. Although American agrarian myth has affirmed the virtue of private property as a direct stake in community and the success or failure of government, mythmakers have questioned the value of economic growth, while protesting its effects, including inequitable land distribution and negligent environmental stewardship.[81] Among the SCFarmers, however, this discourse of critique turns generative, to include agricultural practices such as biodiversity, as well as links to social forms and narratives from outside the experience of American agrarianism. While this broadened sense of agrarianism yields new possibilities for action, it may also lead to the entrenchment of interests who perceive this recast agrarianism as a threat. In the case of the SCFarm, the presence of a resistant and generative agrarianism in an unexpected place led to a pitched battle between a fledging urban agrarian community and a network of adversaries with authority and money, with the status quo maintained. As the remaining cases in this book illustrate, however, such an either/or opposition is not the only possible formation when a new agrarianism meets neoliberal consumerism.

Speculation about why, exactly, these farmers lost their farm, despite their adept social organizing and the range of allies they attracted, is beyond the scope of this study. It is clear, however, that entrenched racism barely masked by the veneer of neoliberalism was a key, if not determinate, factor. In this system of racial neoliberalism, law and governance are constructed in ways that protect the implicitly White and wealthy's property interests rather than empower marginalized communities. In this view, the discourse on neoliberal "growth" perpetuated oppressive colonialist legal and planning practices in a state in which colonialism traces back to the Mexican-American War.[82] As others have noted, the legal and political decisions determining the fate of the SCFarm are highly consistent with colonialist and neoliberal practices.[83]

Communities who creatively invent vernacular agrarian counter-myths to advance urban community gardening may also face challenges to their status as agriculturalists. Despite the continuing cultural resonance of mythic assumptions about the morality of farming, the dominant myths of the soil were created in a Eurocentric, White image and typically set in a bucolic, rural place in communities of property owners. As Christine Oravec writes, cultural myths are not universally available bearers of rhetorical agency for anyone who wishes to use them to shift hearts, minds, and bodies toward particular ends. History, dominant generic conventions, rhetorical context, and other factors bestow an ideological quality upon a text.[84] This means that subordinated societal groups face extra constraints when trying to break through an ideology's mythic shield.

As a result, the rhetorical force of the SCFarmers' challenge to neoliberal urban-planning discourse was likely limited as traditional, Western agrarian myths were stretched, turned inside out, and creatively reimagined in a Latinx-decolonial image. Despite the SCFarm's failure to gain a permanent footprint and the reification of neoliberalism, the case of the SCFarmers and their vernacular narrative of community empowerment linking food, culture, and environment continues to influence activists and academics. The SCFarmers demonstrate that rich discourses for advocacy can open ruptures within the Western-neoliberal-colonial formation, even if these are temporary, or patched by entrenched interests. As Oravec has observed, "we might only be able to gauge the power of a dominant mythic ideology at the points where this power is broken."[85] In this chapter, we have contended that the SCFarmers' expert legal counsel and relentless determination aside, the mythic quality of the story that Kennedy and others told generated the bulk of the SCFarmers' political agency and public support. The story is, therefore, instructive to activists seeking to advance new- agrarian resistance among marginalized people or in surprising places. The SCFarm demonstrates the power of mythic, new-agrarian appeals in embodied performances of vernacular place-making.

Further, this case study prompts our attention to capacity-building and how new-agrarian narratives can evolve to connect different communities who see changing practices regarding food and environmental stewardship as vital to reversing years of neglect, degradation, and contamination that have resulted in impacts including urban blight, food ghettos, food insecurity, chronic health issues, and environmental contaminations. The spread of urban agrarianism, especially when practiced as community-building and cultural inclusion, may be one of the strategies to turn this tide. As Clara Irazábal and Anita Punja conclude, "No matter what the change produced by the garden in South LA had ultimately been, there was a substantive rationale, on the basis of environmental justice and planning ethics, that should have provided sufficient grounds for the city to prevent the dismantling of the farm."[86] Perhaps the challenge for food advocates going forward is to build resilient networks that will have both the argumentative effectiveness and critical mass needed for civic leaders to place ethics, social justice, and planning on an equal footing with other claims.

While the case of the SCFarmers raised questions about how agrarian visions can be secured in the realpolitik world of property interests and governmental policy, our next case moves into a different sphere of contemporary experience, the virtual domain where branding and consumerism are recast by social media. We focus on Chipotle's heralded *Food with Integrity* campaign as both an audacious appropriation of agrarian myth by a rising fast-food chain and a morality tale for our strange times. While agrarian myth fueled Chipotle's fantastic ascent, the brand's failures laid bare contradictions that could no longer be obscured by consumer desires or by the brand's characteristic enchantments and charm. As we will suggest, this case leads into a consideration of the abilities, and limits, of agrarian myth when used to grant an air of authenticity to virtual experiences.

CHAPTER 6

Chipotle Brands Agrarian Innocence

During the CNN South Carolina GOP town hall on February 18, 2016, presidential candidate Donald Trump was asked about his well-known fondness for fast food, given the scrutiny on its health impacts in recent years. Trump defended his preference by arguing that fast-food chains such as McDonald's, Burger King, and Kentucky Fried Chicken are safer than local restaurants because of their "cleanliness." He observed that under the threat of great financial loss, national chains have a mandate to uniformly ensure every food item is uncontaminated. He reinforced his point by referencing the ongoing Chipotle Mexican Grill crisis, in which a chain that publicly trumpeted its environmental friendliness had been beset by a series of foodborne illness outbreaks involving noroviruses and harmful bacteria, namely E. coli and salmonella.[1] By following a different path, Trump suggested, Chipotle had violated the conventions that has allowed the fast-food industry to become the basis of food consumption in the United States. Chipotle's strategy had put the public at risk, and now it was paying a heavy price. Not only had Chipotle's brand become an object of disdain and ridicule, the survival of the fast-rising chain was suddenly in doubt.

In a tempest of three months, Chipotle was blamed for illnesses arising from food poisonings across the country. The initial outbreak of E. coli came in Washington and Oregon in late October 2015 as people who had eaten at Chipotle locations grew ill due to a rare strain of the bacteria.[2] By early November, Chipotle had closed 43 locations in those states.[3] Eventually, the initial outbreak impacted 55 people in 11 states, from the Pacific Northwest to Pennsylvania and Ohio. In December, this was followed by a second outbreak of another rare E. coli strain that impacted individuals in 3 states in the Great Plains.[4] E. coli was not the

163

only culprit; norovirus cases included an outbreak among 120 Boston College students, as well as 100 cases in California. A salmonella outbreak consisted of 64 cases in Minnesota.[5] By late December, this cluster of poisonings garnered national media attention. The *Huffington Post* published maps of the outbreaks, concluding that "people keep eating at Chipotle, people keep getting sick."[6] The *Washington Post* described how the "crisis" had arisen from, and shaken, the company's business model, which emphasized the use of organic and local ingredients on a scale not before attempted in the fast-food industry. The *Washington Post* reports cite the president of an industry research firm, who stated that given the complexity of deploying a *Food with Integrity* marketing campaign in a rapidly growing fast-food chain, "the outbreak was almost bound to happen."[7]

What started as a business analysis became a moral lashing, with Chipotle's food-safety crisis transforming into what a Bloomberg exposé, published in late December, described as a "karmic boomerang."[8] This piece featured a memorable graphic of a worker with a shocked expression, clad in a hazmat suit, gas mask, and face shield, using metal tongs to hold an apparently hazardous burrito wrapped in foil. Below this image of industrial exposure, the headline reads: "Inside Chipotle's Contamination Crisis: Smugness and happy talk about sustainability aren't working anymore." The article features the case of a consumer who was contaminated with E. coli and sees this event as not only a serious medical illness, but an ethical betrayal. Noting that he, like others, had trusted that Chipotle was providing "food with integrity" as promised, the subject notes, "We fell for their branding." He also faults the company's response to the outbreaks, citing a sign placed at a closed location in Portland, Oregon: "'Don't panic . . . order should be restored to the universe in the very near future.' "'That felt so snarky,' the subject notes, 'People could die from this, and they were so smug.'"[9]

The crisis was compounded by the failure to track the contaminations to their sources. The apparent inability to provide a full account of the cause of these contaminations has raised serious questions about the "eat local" imperative the chain harnessed in its brand narrative.[10] As the public image it had built through innovation collapsed, the company's finances also deflated. Chipotle had grown rapidly to 1,900 locations, while its stock soared to a market valuation of nearly $24 billion, making it the most highly valued fast-food offering on the New York Stock Exchange. By January 2016, Chipotle had experienced a 30 percent drop in sales, and its stock value plummeted by 45 percent.[11] Legal trouble was also looming as the U.S. Attorney's Office in the Central District of California opened an "unprecedented" criminal investigation of Chipotle. Three days after the announcement of this investigation, Chipotle investors filed a class-action lawsuit claiming that the corporation misled investors.[12] Meanwhile, customers sickened by the outbreaks were suing the company.[13]

While the seriousness of foodborne illness outbreaks like this one is clear, the intensity of these responses exceeds that to most outbreaks in a nation in which food contamination remains prevalent. The Centers for Disease Control and Prevention estimate the number of illnesses per year in the United States from foodborne agents at 47.8 million, causing more than 127,000 hospitalizations and over 3,000 deaths.[14] The cases attributed to Chipotle constituted only a tiny portion of these incidents and resulted in no permanent disabilities or deaths. Nonetheless, except for the Jack in the Box E. coli outbreak of 1993, which, by contrast, left 4 children dead and 178 with permanent injuries, no foodborne illness alerts have done more damage to a single organization in such a short time.[15] Why, then, were these outbreaks so damaging to Chipotle? Why were the responses from the media, food-safety experts, and the public so severe?

Addressing these questions requires attending to the construction of Chipotle as a brand and the place of this construction in the public consciousness. Here we find the reformulation of agrarian myth in the vibrant imaginary of purification and innocence restored. That this bold reclamation of agrarian life was made in the service of a convenience-food brand, and a distinctly media-savvy and sophisticated one, was an irony that seemed charming until people began falling ill. Then, it was a simple step for media outlets, and Donald Trump, to call out hypocrisy. While Chipotle told us it was better than other fast food, it turned out to be worse. The story of Chipotle's branding, however, is more complex and speaks not only to our strange talent for reveling in falls from grace, but also to the deep cultural longing that Chipotle identified, expressed, and sought to monetize as well as fill. While Chipotle's branding may have contained the seeds of its demise, it animated a latent agrarianism as a scene of redemption. Chipotle offered rebirth and restoration, the opportunity to reclaim what we thought had been lost with the turn to industrialism and consumerism, as well as their inevitable progeny: fast food.

This chapter engages Chipotle's famed *Food with Integrity* campaign, a series of media constructions that surfaced deep-seated anxieties about food and its meanings. The campaign readily connected food production with social organization and patterns of interaction. The loss of traditional means of food production and life on the farm had also diminished social institutions and robbed people of "authentic" connections, to other humans as well as nature. Chipotle's branding brought this loss into public consciousness while offering an imaginative retelling of agrarianism as a sheltered place of innocence and nostalgia as the solution. Chipotle identified for its public what was wrong and then offered a story associated with the brand itself as the way to make it right. Buying into Chipotle was a choice of better fast food, but so much more. It was an admission that the culture had gone astray and an endorsement of the "imagination" that held the alluring promise of setting it right, for all of us. When the heroic promise of *Food with Integrity* collapsed, the contradictions of a new agrarianism performed through mass consumption were laid bare.[16] Chipotle's attempts to capitalize on consumer anxieties with the promise of innocence and purity ended in a public storm. The sage was revealed as a snake-oil salesman offering not a proprietary elixir, but fast-food burritos.

This chapter examines the paradoxes implicit in Chipotle's striking articulation of new-agrarian fast food. Before returning to the backlash against Chipotle and the company's attempts to regain its footing, we will first trace how Chipotle's brand was built upon a novel rearticulation of agrarian myth that foregrounds the appeal of authenticity. We begin with an exploration of the ability of discourses of branding to stir consumer desires and generate affective states. Here, we turn to the desire for authenticity as central to Chipotle's branding, but also as suggestive of the anxieties over authenticity, and a fear of artificiality, that no fast-food brand can finally quell. Chipotle's leading-edge branding involved a projected return to an agrarian dreamscape outside the consumer culture of brands, convenience, and mass production. The funhouse mirror of ironies built into this strategy was part of its unquestionable charm. With the mirror cracked, we can better recognize the artifice and explicate its construction.

Branding "Authenticity"

In its *Food with Integrity* campaign, Chipotle claimed the nostalgic mythos of the "natural" in organic and local food and deployed romantic imaginary of an agriculture of simpler times to hail an urban new-agrarian consumer-citizen. Agrarian myth lent Chipotle the comforting narrative and familiar symbols of the humble and noble farmer working bountiful and uncontaminated land to produce "real" food. Through eating at Chipotle, the new-agrarian consumer-citizen would regain a material and timeless connection with wellsprings of clean food and moral virtue. Chipotle asked customers not only to choose certain food products, but also to "buy in" to a mythos of food production and consumption where a better and more authentic world was possible. As Greg Dickinson writes, rhetorical appeals to authenticity aid the consumer with one of the central challenges of contemporary culture: locating oneself in an unmediated sense of time and space.[17] Embodying an imagined authenticity through foodways is a means of achieving moments of coherence, connection, and identification amidst spatialized abstraction and mediated fragmentation. Drawing from Raymond Williams's concept of "structures of feeling," Sarah Banet-Weiser notes that today's brands demarcate affective states, inducing consumers with promises of regenerated selves freed from the weight of history: "Far more than an economic strategy of capitalism, brands are the cultural spaces in which individuals feel safe, secure, relevant, and authentic."[18] *Food with Integrity* similarly provided an alluring landscape where consumers could recognize themselves in a brand laden with meanings.

Examining the culture of "foodies," Josée Johnston and Shyon Baumann theorize that these food sophisticates identify "good" food worthy of their consumption—specifically, with at least one of two qualities: authenticity and exoticism.[19] However, each of these qualities is fraught with tension and ironies. While exotic eating is caught between democratic inclusion and neocolonial exploitation, the desire for authenticity arises from an awareness of the prevalence of artificiality. Americans of earlier generations who ate from the diverse food raised on family farms did not search for "authentic" foods, nor would they have sensed the profound lack that drives such a search. Authenticity, Johnston and Baumann note, is socially constructed and relational: "People understand food as being authentic if it can be characterized in certain ways *in relation to* other foods, particularly inauthentic foods."[20] Among the qualities these authors note as characteristic of foods constructed as authentic is simplicity. In simplicity, food is seen as emerging fresh, directly from a simple way of life. Authenticity thereby becomes an ethic for living as much as a quality of the food itself. This lifestyle, however, is also a construction, determined relationally through contrast with the inauthentic complexities of modernity.

Chipotle's brand can be understood, then, as comprised by artifices of desirable qualities, as exotic, animated worlds generated through constructions of agrarian authenticity, set in contrast to the logic of contamination and seen as permeating not only food but all parts of life. As Maurya Wickstrom notes, the brandscape is a strategic assemblage of symbolic codes to form a manufactured lifeworld of consumer desire.[21] Chipotle's brandscape aligns consumer desire for authenticity with a provocative retelling of agrarian myth, steeped in romanticized visions of a purity of being, reclaimed and untainted from the maw of history. Chipotle invited consumers into a moral frame pitting pure agrarianism in stark contrast to the industrial food system's violence and corruption. While the appeal to authenticity created a popular and distinctive brand in a short time, the claim to a moral high ground of food purity placed Chipotle in a vulnerable position. If the promise of authenticity elevated the brand, it could also undo it, as the brand's inability to live up its mythos caused its well-crafted constructions to crumble as the public awoke from the dream.

The assembled symbol systems known as brandscapes are more permeating than ever, as companies, including Chipotle, eschew broadcast media for digital hyperrealities that stream through a seemingly endless array of formats and devices. To describe the processes driving the broad spectrum of cultural action and marketing among these networks, Eric Jenkins uses the term "animistic mimesis." He suggests that brands are "desiring machines" able to manufacture alternate realities, which in turn constitute "bodies that can achieve affective experiences through an image-object. These virtual bodies turn into consumer habits, as viewers learn that their own affective investments can turn the commodity into a pleasurable experience."[22] Jenkins notes that new media habituate patterns of action that provide the illusion of authenticity, "making visible consumers and articulable commodities as well as templates for the American Dream."[23] These new media landscapes cast experience as the varied manifestations of consumerism. As we increasingly live within brandscapes, it should not be surprising when brands present to us constructions of authenticity, albeit in simulated form. When we buy in to these brands, we are not only choosing a product but tacitly consenting to a vision of the authentic, made tantalizing when it hints of the exoticism that suggests a reality elevated beyond the lived mundane.

Guided by these theoretical insights, this analysis to come situates the new-agrarian mythmaking of Chipotle as a commodified authenticity projected through an exotic imaginary in digital environments. In following the progress of the *Food with Integrity* campaign, we explore Chipotle's reliance on nostalgic narratives and rituals of purification for freeing consumers from the guilt of complicity in the industrial food system. We consider how seductive animistic mimesis intensified this self-purification by creating an illusory space of food revolution while presenting consumers with a model for brand-based activism. Finally, we weigh in on the downfall of Chipotle, while also tracing the brand's attempts to retell its story. Chipotle was a victim of its own discourses, found wanting when judged by the constructions of authenticity that the brand itself produced. Just as intriguing, however, is the brand's dramatic success, as with artifice apparent, it effectively sold the dream of authenticity. Chipotle revealed a fissure in the consumerist psyche that allowed agrarian myth to well up as the fond illusion of a garden lost, nearly forgotten, but now restored, amid the malevolent industry of fast food.

Escaping Corruption in "Back to the Start"

By linking its ingredients to a natural-food revolution that seems to challenge the phantasmal nature of industrial food, Chipotle's famed *Food with Integrity* marketing campaign attempted to generate an ironic media brandscape focused on reconnecting people with nature, each other, and the captivating prospect of "authentic," but still convenient, food. This reinvention process parallels animistic mimesis, in which suffering is overcome by nostalgia. In Chipotle's branding, alternative foodways overcame industrial agriculture's chemical "contamination" to ease consumer anxieties about placelessness, depersonalization, and the dissolution of traditional forms of community.[24] While Chipotle the chain sold burritos with some organic or locally produced components, Chipotle the brand offered a mended world.

The Chipotle revolution first took the national stage during the 2012 Grammy Awards with an animated two-minute and twenty-second video called "Back to the Start," featuring a Willie Nelson rendition of the Coldplay song "The Scientist."[25] "Back to the Start" summoned farmers and consumers to turn away from an industrialized and polluted modern food system and return to a simpler pastoral relationship with the land and the fulfillments of "real" food. Like the film *The Garden*, discussed in the previous chapter, "Back to the Start" introduces a distinctive aesthetic.

While *The Garden* uses a vernacular style to construct a pastiche of images and voices reflecting vernacular discourse, "Back to the Start" introduces an animation style that may be termed "homespun allegory," as deceptively simple characters and settings tell a compact story charged with symbolic meanings. While the Southern Agrarians looked to a gospel of redemption tied to constructions of place and Rodale recast the jeremiad in the service of agrarianism, the animated videos of Chipotle packed statements of moral gravity into the accessible form of animated shorts. As its title suggests, "Back to the Start" is an origin story in reverse, a backward progression to realize a rebirth, redemption, or second chance. Viewers become the consumers of the cathartic purification whereby rightful order is restored.

The video opens with "the start" in its first-run manifestation—a small-scale farming operation of the old school, encoded in memory. Quickly, the operation begins to modernize, adopting industrial farming practices marked as sinister and violent. Animals that once roamed freely are confined as livestock with their entire life cycles mechanized. In a striking sequence, fattened pigs, caricatured as round in the style of the animation, are "processed" into pink cubes that fit neatly into the trailers of a caravan of semi-trucks that roll past the smokestacks of a food factory. Images of corruption bombard the farmer, including a large medicinal capsule; solid goo of feed, waste, or both, coming out of a pipe; and imprisoned pigs. Amid this calamity, he has his redemptive epiphany: he can go back to the start. First demoralized, then disgusted, but now inspired, the farmer liberates his animals from confinement to open pasture; eliminates the use of hormones and antibiotics; and creates a happier, healthy, family farm. He loads eggs gifted happily by free chickens into the back of a retro milk-delivery truck refitted for Chipotle. A sign of wooden plaques at the end of the farm property, and the video timeline, of the type used in bygone times to list different products for sale at a roadside store, instead reads:

> Cultivate
> A Better
> World
> chipotle.com

The journey to this better, agrarian world begins by stoking public anxiety about corporate control over food production and its dire consequences for human, animal, and ecological well-being. The farmer's redemptive performance counters industrialization, appealing to collective longing for an honest food culture grounded in authentic relations. Anxiety is released into this timeless state where the innocent past is projected into the present. Inviting consumers to show their support for this message, including by downloading the Willie Nelson theme song, the "Back to the Start" campaign refines the critical stance of a new online media characterized by an ethos of "sincerity, not irony."[26] At Chipotle, consumers can eat without guilt or guile, knowing that they are contributing to a retro-revolution of consciousness in the convenience-food market. The charming, disarming style of the animation belied the boldness of its mission. Through "Back to the Start," Chipotle announced that it would lead the return of the eating public to the agrarian ethics of simplicity and purity. After the apparent epiphany of "Back to the Start," Chipotle the restaurant chain mirrored its branding, announcing that it would recast its procurement procedures to draw on organic and local resources, when possible.

Beginning with "Back to the Start," Chipotle's *Food with Integrity* campaign capitalized on the longing for authentic experience that rises amid technologic abstractions from processed or genetically modified food to artificial intelligence. As Banet-Weiser attests, brand cultures facilitate relationships with consumers hinging on promises of authenticity and sincerity.[27] In identifying with Chipotle, consumers chose to set themselves apart from the demoralizing mainstream, represented in the animation by the massive prevalence and many diminishments of industrial food production. Chipotle promised an immediacy of meaning, easy realized through a simple choice of brands. While the mainstream might muddle on, Chipotle brought an ambience of specialness to select consumers, while hinting at a moral wisdom and political determination that drove the animation's deceptively simple story. The farmer's inspired "return" to a condition of personal fulfillment and cultural richness is made as a surrogate for a society gone astray. As in earlier episodes of the American agrarian experience as a discourse of resistance, the farmer becomes a mythic figure imbued with a collective tale of loss and reclamation.

As we have seen, the mythic motif of the farmer as a model of morality and defender of democracy is part of the fabric of US national identity. "Back to the Start" applies this commonplace in a novel manner for a twenty-first-century consumer republic. In Chipotle, the restorations imagined by the War Garden Commission, Country Lifers, Southern Agrarians, or Rodale ironically come to pass through a politics of techno-capitalism in which a convenience-food brand assumes the mantle of democratic virtue. Ernesto Laclau and Chantal Mouffe contend that since late capitalism has channeled patterns of consumption into an increasingly politicized economic landscape, resistance entails theorizing the possibilities for pluralized publics.[28] Chipotle attempts to establish a brand identity as "a place apart," through an animistic mimesis projecting a better world as not only possible, but resonantly present for a discerning public.

In a manner reminiscent of Marshall McLuhan's slogan "the medium is the message," Chipotle expressed this specialness by embarking on a marketing strategy based exclusively on digital and social media. This daring move reinforced the bold assertion that Chipotle offered not only a new type of fast food, but a refined vision of democracy as a consumer paradise with a distinctly agrarian flavor. With its charming "retro" aesthetics, this approach targeted younger and affluent consumers by relating emerging trends to a romanticized past. Simultaneously, the reliance on digital and social media squarely situated the brand in a communicative regime associated with participatory democratic practice and resistance to authority.[29] Deploying a vision of the past on the leading edge of the present, Chipotle targeted the coveted young-adult market while summoning brand loyalty as a journey taken together. Beginning with "Back to the Start," Chipotle's branding unfolded as an adventure in time, as the past becomes the future.

With the *Food with Integrity* campaign underway, Chipotle sought to reconstruct its corporate culture to meet the demands created by the alternative hyperreality it was now projecting. Fast-food chains have typically followed the McDonald's model of marketing low cost, often highly caloric food to stoke consumer desire. This way of doing business extends beyond fast food and has become a hallmark of other aspects of the consumer economy. George Ritzer describes this effect as reflecting the "McDonaldization" model of efficiency, calculability, predictability, and technological control, adopted across a wide array of institutions today.[30] Chipotle's campaign addressed this model indirectly, suggesting that, without discarding these principles, another way of doing business was possible and desirable. Addressing consumer anxieties prompted by McDonaldization, but cognizant of the place of convenience eating in the culture, Chipotle projected agrarianism as the sanitization of fast food. While Chipotle offered a road not taken, it would remain in the profitable lane of fast food on a large scale.

The complex relationship of Chipotle to McDonaldization becomes even more intriguing when we consider that McDonald's was an early investor in Chipotle at the start of its rapid growth. The film *Behind the Counter: Inside Chipotle Mexican Grill*, produced by Bloomberg Television, details the history of Chipotle's rise to prominence after substantial investment by McDonald's and the uncertainties about Chipotle's future when the relationship between the companies ended in 2006.[31] Although the work projected an air of independent and objective documentary filmmaking found in some food and farming films of the past, it functions much like a promotional infomercial showcasing Chipotle's history and business model.[32] The film's narrative suggests that a difference in ethical vision and rise of conscience on the part of key personnel at Chipotle led to the split, as if decision makers at Chipotle were synonymous with the farmer in "Back to the Start." Like the farmer seeking redemption, Chipotle personnel interviewed in *Behind the Counter* link animal suffering to the existing fast-food model. In an era, quite literally, after McDonald's, they sought to provide a purified space to cleanse themselves, while still providing the convenient and relatively inexpensive food consumers demanded. The company, then, followed the progression in "Back to the Start," heeding conscience to acknowledge past culpability and setting out bravely on a path of its own.

The rest of the story is more prosaic. While Steve Ells launched Chipotle with money borrowed from family and friends, McDonald's made a major investment in Chipotle, soon becoming its largest investor. This capital fueled Chipotle's dramatic expansion. By 2005 McDonald's decided to make Chipotle a publicly traded company and divested its ownership stake for considerable profit.[33] Although Chipotle's tale of moral longing is compelling, its history was also driven by strategic business decisions, including those that led to its branding of "food with integrity" to distinguish itself in the crowded fast-food market. These twin narratives, of Chipotle as a place apart and Chipotle as a business like many others, have continued to run as parallel ledgers throughout the history of the brand. By projecting itself as not only unique but culturally important, Chipotle called attention to its brand as a symbolic landscape only partly related to the burritos it serves. With the contaminations, however, this projection snapped back, revealing a fast-food chain caught in the tangles of its own pretentions.

While the Chipotle brandscape connected corporate history and consumers to the simple story of a farmer, it also sought to fulfill agrarian myth for new times by projecting the brand as beacon of democracy. The aesthetic of simplicity that characterized its campaign was paralleled by its attempt to build community and support for its ideology through direct response via digital technologies. Consumers were encouraged to share and comment on Chipotle's videos as a means of both spreading the message and showing they had "bought in" to the brand's vision. With media and form mirroring content, Chipotle promised to ease the tensions of history through a quintessentially American utopia of consumption as a moral good. Agrarian myth, ever at the ready, was drawn on to legitimize this strange and unworkable ideal. The mismatching of ethical food and agrarian utopias with the fast-food industry is poignantly reconciled by the spirit of the farmer in "Back to the Start," who knows that the infernal cacophony of industrial food production, as Willie Nelson croons, "does not speak louder than my heart." In the first rendition of *Food with Integrity*, Chipotle became the crossroads where agrarianism, fast food, consumer choice, and democratic virtues appeared to coincide. As the campaign continued, Chipotle sought to fuse consumer desire with agrarian simplicity through the ritualistic tale of a scarecrow.

Purification in "The Scarecrow"

Two years after "Back to the Start," Chipotle extended its vision with the viral animated video titled "The Scarecrow."[34] Viewed over fourteen million times on YouTube, "The Scarecrow," like "Back to the Start," combined nostalgic music and slick animation to narrate an escape from contamination to a purified agrarian society. The video tells the story of the scarecrow, a worker at a fictional meat-processing plant called Crow Foods, Inc. It opens with a pastoral scene of a barn, hilly farm fields, and a scarecrow on a pole, but as the camera pulls back, we find the scene is a faded mural on the side of the Crow Foods plant, with billowing smokestacks around it. As he reports to work through a dark passage and keyhole door, the scarecrow is positioned at a conveyer belt that parallels the food product line, with other scarecrows spaced at regular intervals, each with a crow on the shoulder. Apparently, the workers, too, are made by industrial production. The assembly line machine stamps packages with the label "100% Beef-ish," while behind the plant, chutes deliver processed food products to customers in front of a facade adorned with images of baby chickens and the words "All Natural." The scarecrow's work yields an insider's view of the grisly operations, as chickens are injected with growth hormones and cows are stored in stacks of metal crates too small to accommodate their corporeal forms. The scarecrow is cajoled by the crow, who functions as a surveillance device to keep the workers on task and in line. While the workers are meant to be zombies doing the bidding of the crows, our hero the scarecrow reacts differently; he bears the burden of violence and degradation in his sad eyes and drooping straw. His affect reveals an inner light that is not yet extinguished.

It is not accidental that the figure of a scarecrow is used in this context, as it references L. Frank Baum's persona of the scarecrow as a farmer in search of a brain in the *Wizard of Oz*. The soundtrack song, "Pure Imagination," sung here by Fiona Apple, also holds a key place in the collective memory, as it was initially recorded by Gene Wilder in his role as Willy Wonka in the original film version of *Willy Wonka and the Chocolate Factory*. In this Chipotle video, the scarecrow summons his brains and its powers of "pure imagination" to reinvent the food system that he knows is based on death and has tried to steal his own life. Dejected by the scenes he has witnessed at Crow Foods, the scarecrow journeys away from the factory on a commuter train. As he leaves the city, we witness an arid landscape, reminiscent of the Southern Plains in the Dust Bowl era. No food is grown here. Instead, a billboard for Crow Foods dominates the landscape. The scarecrow is again saddened to learn that he, or another like him, is represented in the giant image as the avatar of the diabolical Crow Foods. The scarecrow's identity seems defined by industrial food production, which, as the video shows, not only causes animals to suffer, but renders the natural world lifeless to make products that do not resemble food.

As in "Back to the Start," "The Scarecrow" turns on a moment of epiphany. The scarecrow returns to a tiny farm that remains green amid the arid landscape. This time, it is the plucking of a red pepper from a vine that leads to personal and social transformation. With the pepper in hand, the now-smiling scarecrow turns to the city looming in the distance. He harvests ripe vegetables, packs them in wooden crates, loads an antique pickup truck, and heads for a marketplace that he will re-create. We see him chopping produce in a kitchen and preparing a beautiful burrito, while Fiona Apple sings, "If you want to view paradise, simply look around and view it. Anything you want to, do it. Want to change the world? There's nothing to it."

The choice of soundtrack is enhanced by Apple's delivery, which emphasizes the word "pure" each time it is sung. The creators of "The

Scarecrow" have noted that Apple's mournful voice was more successful in conveying the sadness of the modern food culture than the glib and upbeat Wilder who popularized the song.[35] The video performs a quest for purity to supplant the wasteland of Crow Foods. Irony functions in the service of sincerity as the fantastic foods presented in Willy Wonka's garden of desires are recast as burritos. Still, like Wonka's products of imagination, through which dreams are fulfilled, the shift in the consciousness of the scarecrow to devote his labor to purified agrarianism reflects a utopian sensibility. At Chipotle, healthy, pure food is reinscribed as the product of longing, parallel to Wonka's magical candy.

In a cityscape that is now brightly colored, the scarecrow establishes a vendor stand between the mechanized chutes where Crow Foods dispenses its unearthly products. His spirit restored, the scarecrow is finally able to perform his namesake function and shoo away the pesky spy-crow. As we have seen in agrarian episodes throughout this book, the farmer in American agrarian myth has long been a trusted model of morality and democratic citizenship.[36] In this instance, the scarecrow stands in symbolically not only for farmers, but other people working in the food system as well. He grows the food, harvests, transports, and cooks it, and finally becomes the outlet through which this new/traditional food reaches the public. The changes projected by the scarecrow's imagination are revolutionary and systemic. Over his food stand, a banner echoes the message from "Back to the Start": "Cultivate a Better World."

Through this story of "The Scarecrow," Chipotle further develops its mythic brandscape as the illusory landscape of nostalgic memories and longings, where—drawing on the critical lexicon of Eric Jenkins—the digitalized materialization of bodies can achieve pleasurable experiences through an "image-object" of conscientious consumption.[37] Here, animistic mimesis adopts agrarian myth as a template for reclaiming national values of authenticity, honesty, and democracy, while connecting with nature through food. As Jenkins notes, animistic mimesis has a powerful impact on the consumer consciousness because "in the sparks or moments of intense affect, there is often a de-subjectifying effect, a sense one has been lost or displaced."[38] "The Scarecrow" attempts to snap consumer consciousness out of the permeating dystopia of mass-produced food by showing that this system defines not only how we eat, but also how we work, how we live, and who we are. The scarecrow experiences hard truths about food as a crisis of identity that deprives him of even his most basic self-defining act of scaring crows. Similarly, his remaking of food constitutes a rebirth of his sense of self and his social role. The city and earth are also recast and made to thrive upon the realization of the value of food. While "Back to the Start" boldly casts Chipotle at the vanguard of a new-agrarian revolution restoring an authentic basis in farming, "The Scarecrow" goes further: our collective fate depends on following the path of the scarecrow and reclaiming identity as the co-creators and recipients of agrarian burritos. The viewers are no longer a consumer demographic, but citizens of the Chipotle Republic.

The humble style of the animation and air of innocence found in "Back to the Start" and "The Scarecrow" masks the audacity of these implied claims. Viewing "The Scarecrow," we are to infer that the birth of Chipotle represents the first salvo in a food revolution that will generate an alternative heterotopia and eventually supplant the industrial machinations of mass food culture. The virtues and charm of the characters, however, allow these claims to be advanced with an air of humility on a humane and intimate level. While "Back to the Start" evokes liberation through reformed farming practices, "The Scarecrow" tells its story of salvation by revealing contaminations and then unveiling the space of pure imagination where innocence can be reclaimed and a systemic new beginning is possible. The video allows consumers to alleviate guilt and the burdens of their previous choices to experience an uncanny rebirth. Chipotle not only aligns with organic or local food movements but transfigures them in a morality play that projects a reality cleansed of history in the form of purified food. To update the famous line from Wendell Berry, eating at Chipotle (or in the white SUV after driving through), is a sanctifying act.

While the shift from consumer ethics and liberation to a distilled sense of purity may seem subtle, it is crucial in recognizing the nature of the brand's later travails. Beyond attempting to foster brand identification or allegiance, Chipotle ensconced its image in a larger narrative of food orthodoxy, including the discursive poles of harmony, connection, and idealism that Tema Milstein finds in Western environmental dialectics.[39] Chipotle positioned itself as a social cause championing small-scale farming, humane treatment of animals, environmental sustainability, and the public's right to clean foods. For Chipotle, the contradictory message rang clear: this is not a fast-food brand, but a cultural revolution in which old models are ritualistically purged and convenient consumption is authentic and freed from guilt. The decision to forsake corrupted forms of communication for participatory social media magnifies this strategy of division. It recalls Guy Debord's analysis of the society of the spectacle as operating though simulations that not only supplant realism but displace practices of critical resistance.[40] At the end of "The Scarecrow" there is nothing left to resist as the world has been made new by pure imagination. Instead, we are given the spectacular burrito as a consumerist sacrament.

In his analysis of the cultural debris surrounding Disneyland, Jean Baudrillard suggests that the hyperreal is "a space of the regeneration of the imaginary as waste treatment plants are elsewhere, and even here. Everywhere today one must recycle waste, and the dreams, the phantasms, the historical, fairylike, legendary imaginary of children and adults is a waste product, the first great toxic excrement of a hyperreal civilization."[41] For Baudrillard, the hyperreal represents a form of cultural contamination inherent in modernity. Mary Douglas, however, suggests that discourse and ritual actions can be used to draw distinct boundaries between insiders and outsiders, moral orders and the taboo, purity and filth. Dirt is "matter out of place"; it facilitates condemnation of any idea that may cloud or contradict cherished classifications.[42] For Chipotle, these contradictory claims about purification and waste become inextricably linked. "Back to the Start" and "The Scarecrow" sought to separate the brand from the contaminations of the industrial food system and reify a purified agrarian alternative that will absolve consumer guilt and tie personal and cultural identities to the fate of a revolution. Yet Chipotle's status as a fast-food outlet remained unchanged. Even as its brandscape projected a new way of eating, Chipotle the chain was deeply embedded in an accelerating culture of mass scale where convenience cannot be compromised and shareholder value reigns supreme. This contradiction was not easily resolved in substance, so it was left to the imagery and discourse around the brand to mask, assuage, and mediate it. Chipotle's revolution would be simulated.

Given this precarious position, it is little wonder that Chipotle sought validity and protection in the shelter of related discourses. Commenting on Chipotle's media savviness, Matthew W. Ragas and Marilyn S. Roberts contend that the brand moved from agenda-setting to an approach that may be termed "agenda-melding."[43] Here, rather than generating or projecting a resolution to issues, the brand targets a demographic with an existing set of concerns. Chipotle harnessed the food and environmental movements to prop up its status among change agents. It also sought to capitalize on the "soft" attraction of many members of the public to these movements by offering a convenient, easy way to demonstrate their allegiance. Eating at Chipotle constituted a statement without commitments or work. In a paradox that is normative and profitable in a consumer culture gone hyper, Chipotle positioned itself at a juncture where anxieties over consumption could be quelled by further consumption, of the right kind.

Love and Contamination in *Farmed and Dangerous*

After the success of "The Scarecrow," Chipotle attempted to capitalize upon its niche in marketing food consciousness and consolidate its position at the head of a food revolution. Its media campaign produced a mini-series in a documentary-style satire entitled *Farmed and Dangerous* that aired on Netflix in 2013.[44] The series dramatized the efforts of organic farmer Chip Randell to challenge the food industry, codified as a company named Animoil, that purposefully introduced toxic elements into the food system for profit. Randell is an "off the grid" Romeo who must woo a well-heeled, capitalism-oriented Juliet away from the system of corrupt food production, while leading a digital campaign against Animoil and industrial food. This series attempted to link Chipotle to food activism and colonize the anger and frustration that natural-food advocates held toward the industry. Stewart Lockie notes that while neoliberal political orthodoxy affirms individuals regulating their own behaviors as consumers and entrepreneurs, proponents of alternative food networks see civic virtue regarding food as better expressed in social solidarity and coordinated actions.[45] In *Farmed and Dangerous*, the character of Randell becomes a nexus where food advocates recognize themselves in a fellow sojourner committed to the cause. However, Randall's efficacy as a social organizer spreading the word about toxic food masks his status as a fictional compilation of traits, set in a narrative told to benefit a leading provider of fast food. The satiric nature of the series and the exaggerated virtue of Randall and evil of Animoil highlight this sense of artifice. As the ultimate advocate for food purity, Randall is an engineered construct in a mediascape where the humble virtues of earnestness and proper conduct stand out.

Even as Chipotle summons the constellation of discourses related to food and the environment, the universe of *Farmed and Dangerous* is self-referential, as if Chipotle now has the cultural debate under containment. *Farmed and Dangerous* details the attempts by Randell to expose a plot on the part of the Animoil corporation to develop a synthetic food supplement for cows called PetroPellets. As the name suggests, PetroPellets are made from petroleum products. According to Animoil, they promise to set a new floor for the expense of feeding cows for market. However, they generate an unfortunate side-effect—the cheaply fed cows occasionally explode. When Randell obtains pirated footage from Animoil's security cameras, videos of exploding cows flood the internet as he attempts to expose the corporation's plan to poison the food supply. Buck Marshall, a public-relations expert, or corporate shill, who sets out to defend the industrial food complex, counters this effort. While Randell's character is naively honest and wholesome, Marshall is deceitful and cynical to the point of sending his own daughter, Sophia, to seduce his nemesis. Marshall embodies the perceived evils of modern agribusiness with rationalizations such as "now people only die because of what they eat, not because they do not have enough to eat; so that is an improvement." Much of what Marshall says and does is intended as a form of dark humor, but his character embodies what many food advocates feel is an accurate and revealing portrayal of an industrial food spokesperson. The whole cultural battle that pits advocates for food alternatives against the supposed evils of industrialized food is embodied in these two representative figures.

The line between reality and simulation within this bubble was further blurred as Chipotle ran marketing spots along with the series, featuring the Marshall character as an industry spokesperson responding to Chipotle's "Scarecrow" campaign as naïve propaganda. Marshall mispronounces the restaurant's name and suggests that labels for GMOs are unnecessary because they "often point out things that people do not need to know." He further argues that labels unfairly attack big agriculture, as labeling will do to them what it did to big tobacco. In promoting the pure and natural, Chipotle wrapped itself in layer upon layer of artificiality. When Marshall refers to people who drink organic milk as "homogenaphobic" the efforts to use humor as means to stimulate deeper tropological impulses come clear. Industrial food producers are both toxic and clueless, and the Chipotle consumer is in on the serious nature of the joke.

Like a moral anchor in this unstable irony, Randell resists Marshall's efforts to seduce him into renouncing the campaign to purify the food industry. As the series progresses, the audience identification shifts to the authentic love interest that develops between Randell and Sophia, with Sophia coming to understand the error of her father's ways as she falls in love with the farmer and his simple honesty. She becomes the surrogate for the once-thoughtless consumer who now finds redemption and happiness in a lifestyle underpinned by pure foods. Again, Chipotle promises a revolution of the spirit, this time enacted through the rituals of love.

The real and unreal mingle complexly in the world around *Farmed and Dangerous*. While presented hyperbolically in the series, the concept of PetroPellets hits close to home as reports of traces of hydrocarbons in human food spread among food advocates as well as in scientific studies.[46] While Randall's steadiness and honesty are "over the top," they buoy food advocates to face an industry whose intent may not be too distant from poisoning for profit. Although "Back to the Start" and "The Scarecrow" masked bold claims in digital animation with a homespun feel, *Farmed and Dangerous* finds its moral center in the rightness of advocacy for healthy food within a world gone mad. With Marshall's moronic denials, the enemy is shown to be dysfunctional and incoherent. The commingled powers of truth, love, and agrarian virtue have the enemy on the run. A new paradigm of food, health, and solidarity seems at hand, displaced only by the thin veneer of satire. In *Farmed and Dangerous,* Chipotle is but a beat removed from proclaiming itself the progenitor of an earnest new world founded in the morality of purified food.

The Backlash against Chipotle

At the time of this authorship, several years removed from the outbreaks that sickened several hundred people and cost Chipotle revenues and standing, the cause of these incidents remains unknown. With different pathogens leading to the illnesses, the problems may have arisen from different sources in the supply chain, including local suppliers. However, like other fast-food chains, Chipotle acquires much of its produce and meat from a few vendors on a large scale. The unusual difficulty in pinpointing the source of the outbreaks despite internal and regulatory investigations, as well as the rarity of the strains of bacteria involved, have led some commentators to suspect that the incidents may have been the work of industrial sabotage or foul play. But some experts have maintained that a lack of conclusive findings is not unusual in investigations, especially in mass-produced food.[47]

The number and variety of outbreaks at Chipotle has been commonly viewed as evidence of an institutional or systemic failure within Chipotle. Dr. William Schaffner, infectious-diseases specialist at Vanderbilt University Medical Center, noted that incidents of two strains of E. coli and norovirus at Chipotle led to the question: "'Is this a coincidence or is this a systematic problem of food-handling distribution at Chipotle?'"[48] "There's a problem within the company," Michael Doyle, Director of the Center for Food Safety at the University of Georgia, told a Bloomberg reporter.[49] When reminded of the emphasis Chipotle placed on the use of organic, local or non-GMO foods that did not use antibiotics, Doug Jones, a retired professor and editor of the leading food-safety blog *BarfBlog* replied, "Blah, blah, blah. . . . They were paying attention to all that stuff, but they weren't paying attention to microbial safety."[50] The Bloomberg piece ends with a dig at Chipotle's demographic as well as the brand's attempts to evade reality: "Millennials may discriminate when they eat, but bacteria are agnostic."[51]

While critical discourse was prompted by the public-health scare involving Chipotle, the target of the criticism became the brandscape Chipotle had so carefully crafted. As Banet-Weiser suggests, projecting a brand as a culture entails a degree of ambivalence characterized by flexibility of movement, unpredictability of articulation, and the potential for destabilization.[52] In the case of Chipotle, this precarity solidified as a singular critique: while Chipotle had crafted a public image of agrarian authenticity born of pure imagination, it had forgotten the reality-driven requirements of food safety. While celebrating its claims to the future of food, Chipotle was brought down by neglecting the workaday practices that more pedestrian vendors were obliged to follow. Like a tragic hero, the brand was laid low by its fatal flaw of fantastical pretensions.

While this indictment may seem too neat as it turns Chipotle against itself, it extended a critique that began before the outbreaks. Journalists, food advocates, satirists, and corporate agricultural interests had all begun to question the *Food with Integrity* campaign as misleading, misinformed, or fatally idealistic.[53] As Pete Weber stated, "Chipotle is playing sort of a dangerous game here. Along with targeting branding-averse millennials, "The Scarecrow" is pretty obviously a play for the slow-food and farm-to-table crowd that still occasionally likes fast-ish food."[54] As Chipotle marketed a vision of authentic food that did not compromise convenience, the food industry's ingrained supply chain practices led to a reality with compromises. Try as it might to position itself as revolutionary, Chipotle remained mass-scale fast food.

The comedy group Funny or Die used the Chipotle medium of choice, video shared through social media, to present a satiric animation pointing out the simulated nature of Chipotle's imagined agrarian dreamscape.

Their satirical video offers a poignant critique of Chipotle's marketing by reframing Chipotle's scarecrow to address one of the brand's chief values: honesty.[55] Funny or Die's video "Honest Scarecrow" represents culture-jamming, a form of anti-corporate advocacy that recontextualizes marketing messages.[56] In this case, a retelling that purports to relate the truth about Chipotle jams the message of authenticity and purity that "The Scarecrow" presents. "Honest Scarecrow" replays the original video but adds a few well-chosen textual annotations and changes the lyrics of "Pure Imagination." The remake opens with a clear delineation of the side-taking implied in "The Scarecrow": "Not Us" an annotation says with an arrow pointing to a Crow Foods billboard; as the scarecrow appears he, too, is annotated with an arrow, as "Us." This revealing choice of pronouns suggests the bond of unity that Chipotle calls for from the public, a consubstantiation reminiscent of Kenneth Burke. This assumption is underscored by the altered lyrics to the song: "Come with us and you'll be in a world of pure imagination." This world, however, is not that imagined by the agrarian entrepreneur scarecrow, but that of "an ad made for you by a giant corporation." While the original sought to use its artifice in the service of an ideal of imagined authenticity, the satire lets the images play while erasing the pretense, "We'll begin, drop you in, to a great high budget animation. What you'll see will be pure manipulation."

Having surfaced, the artifice implicit in the original version becomes the focus of attention as we see how it plays on the emotions. The sadness of the scarecrow becomes our own. As the scarecrow ascends to the top of the building where the boxed cows are housed, the plaintive voice sings, "If you want to cry, here's where you do it." As those longing for authenticity in an unsatisfying culture, we are easy marks, "It's what you want to believe." In a specific reference to the questionable claims made by Chipotle, the scarecrow's authentically produced and lovingly made burrito appears with an annotation and asterisk: "Burrito GMO Free!*" The disclaimer follows: "kind of . . . in some places." As the story of the scarecrow ends, the unity of Chipotle and consumers becomes a directive, "Let's engage with our brand and be sure to follow us on Twitter." Chipotle is "us" and a "brand" at once, a contradiction advanced in the strange chanting with which the song, re-voiced as "Pure Manipulation," concludes: "Inside your brain, download our game, engage our brand, this is our name." It is a song of love and possession for connoisseurs of custom burritos: I am Chipotle and Chipotle is mine.

Lessons in Agrarian Branding

In November 2017, approximately two years after the initial outbreak, Steve Ells, the founder of Chipotle, resigned as the company's chief executive. While the company had put a comprehensive food-safety campaign in place, its stock remained trading at less than half its share price before the incidents. Financial analysts cited the role of hedge-fund manager William Ackman, who had a 10 percent stake in Chipotle, as pushing for Ells's ouster. The company announced that it would seek "a turnaround specialist" to replace Ells.[57] In February 2018 the search concluded with the hiring of Brian Niccol, the outgoing CEO of Taco Bell, as the new CEO of Chipotle. A chain famous for cutting its own path in the fast-food world would be led by the former head of a company known for "Mexican" food with no claim to authenticity. The *Denver Post* noted the irony: "Looking to boost confidence in its 'food with integrity,' approach to quick-service dining, Chipotle Mexican Grill is turning to a man who most recently sold foods with names such as Doritos Locos Tacos."[58] In contrast to its previous brand image, Chipotle was now making conventional business decisions for the usual reasons, such as hiring a successful CEO from a rival brand as a signal of stability to squeamish investors.

In its marketing, Chipotle responded to the outbreaks and criticism by attempting to graft safety onto its previous image of mythic-agrarian purity, suggesting that it was a place for food that was not only "fresh" but "safe."[59] As they sought to regain their footing, however, Chipotle also launched a new animated video, "A Love Story," which tells a classic tale of love eclipsed, for a time, by avarice and pride, with a return to true, lasting passion, as well as authentic burritos.[60] In this video, Chipotle not only recasts its fall from grace as a mistake in judgment but rehearses the grounds of its forgiveness and redemption, as the power of love shines through in the production and sharing of food. The story begins with longing as the large sad eyes of the male lead sits at his homespun orange juice stand, dreamily regarding his neighbor as she squeezes lemons for her lemonade. The soundtrack is the Backstreet Boys' "I Want It That Way," which begins, "You are my fire, my one desire." While the male fantasizes about snuggling with his fellow entrepreneur, he notices that his stand is not attracting customers, while she is serving a line of thirsty people. He imagines the solution: branding. With this orange juice sign on fence posts across the neighborhood, business booms, and a rivalry is born. The two characters build a succession of stands, restaurants, and chains while the Backstreet Boys croon, "Ain't nothin' but a heartache (Tell me why); Ain't nothin' by a mistake (Tell me why)."

The plot turns on an incident when the woman lead investigates the artificiality of the sparkly drink her chain is now selling. She enters a darkened production facility where vats marked "lemon flavor" and "lemon color" are linked to hoses. Robots prepare fake food with equipment that includes rows of microwaves and an automated arm, which adds a substance called "Sparkle Flavor Booster" to artificially frosted tacos and burgers. When she attempts to turn the line off, the machinery arises to expel her as an intruder. She is coiled in a hose and thrown down a chute marked "Trash" that drops her onto a pile of trash bags in an alley.

From here, the story unfolds as a tale of heterosexual monogamy, channeled through authentic food. Apparently at the end of his own failed struggle to set his business right, the male lead bounds down a trash chute just down the alley. They find each other anew: "You are my fire, my one desire." However, Chipotle provides what we now recognize as its characteristic gesture: redemption through the reclamation of purity and authenticity. Having confessed their love, the characters end the rivalry and competition for profits that led inevitably to the industrial production of fake food. Instead, they join in a heartfelt collaboration to regain their ethical standing and no doubt jumpstart their mutual futures. They create burritos of love sold from a food truck, no doubt to an audience of urban hipsters who are receiving the signal that it is acceptable (and safe) to include Chipotle among their favored food choices.

While "A Love Story" rehashes the brand's vanguard status, it also can also be read as an apology. Chipotle's stunning growth is revealed as an errant path taken for the wrong reasons, which led to its food poisoning incidents. While the video does not show customers becoming physically ill, the products generated at the horrific facilities that supplant the homespun lemonade and juice stands are poisons composed of materials that are not food. While the "progress" of the rival brands shown here is clearly the wrong course, the judgmental public, too, may have made a mistake in dismissing Chipotle too hastily. A second chance would allow the company to express its true intentions, which have always been present in its branding. After the fall, when all has been trashed, the brand and its public can recognize each other and embrace their mutual love as "one desire." Aptly, the video ends with direct appeal to a new consumer loyalty program: none of "us," whether Chipotle or its public, will stray or betray again.

As corporate branding of authenticity using digitally mediated and mythic-agrarian imagery, this love story seems an attempt to secure Chipotle's utopian brandscape after the radical shocks of the food-safety incidents. This strategy's effectiveness remains unclear. At the time of this authorship, Chipotle has remained in the ambiguous position of a major chain that, although maintaining an enviable market share, has taken a "hit" and is searching for a way to reconnect with consumers. In this difficult moment, it is also evident that Chipotle's new-agrarian brand-making has had positive implications in the United States food system. The influence of the brand is evidenced across the food retailing landscape. While the rapid growth and stock valuation of Chipotle may have been a bubble that could not be sustained, the overall market for alternative fast food is growing in the wake of Chipotle. Other convenience-food chains branded as healthy, authentic, or ethical are on the rise, including brands selling juice, smoothies, salads, and quick Asian and Middle Eastern fare. Chains such as Panera Bread have faced similar struggles with sourcing organic and local food but have also grown steadily and made profitable use of social media.[61] Such brands have also drawn lessons from Chipotle and have typically been more circumspect with their claims and less bold in their marketing. Perhaps mass-produced fast food will never be entirely "authentic" or 100 percent "safe." If the public wants food that is healthy, organic, local, and safe, is it willing to scale back in convenience? Are there ways to "scale-up" local production without the overpromises of Chipotle? Convenience brands continue to explore the parameters of what they can deliver safely and profitably.

For the purposes of the present study, the case of Chipotle leaves questions about the status of agrarian myth. How can this myth be made operational in a consumer society without obscuring, or twisting its links between people and the land, and agrarian work and democratic practice? While the myth may seem excluded from consumerist experience, Chipotle capitalized on a deep longing for agrarian authenticity that the dominant food culture makes more acute. The success of Chipotle in unleashing agrarian longing reveals a permeating sense of loss.

Who has legitimacy to speak for the agrarian in consumerist America? In the brandscape of Chipotle, the reform-minded brand appropriates and retrofits American agrarian myth in a way that nonetheless operates on a mass, industrial scale. Perhaps by necessity, the agrarianism it presented was freed from the workaday world of farming in a perilous economy and set in a timeless realm where nostalgia was not tied to the past but projected as an aspirational future. In the next chapter, we examine a different use of agrarian nostalgia in the case of a RAM truck Super Bowl ad. Here, the past does not animate a revolution but supports a consumerist status quo that ironically foretells the future as freed from the limitations imposed by mere humans. From charming animations, we move to powerful and intelligent machines that tease an agrarian impulse ever-ready to surface in the American consciousness.

CHAPTER 7

RAM Mechanizes God's Farmer

It begins with a stark, rural landscape with a light covering of snow. Low hills are visible in the background. In front, there is a fencerow of tall, brown grass and a single cow. Vertically, the horizon fills four-fifths of the frame, as if the land is a compacted layer where the heavens come to rest. This winter sky is a single, seamless cloud hanging on the landscape. It shrouds the view, but also colors it with luster like a pearl. The audio matches the scene with white noise. Is it the sound of machinery unseen? An approaching storm? As if no further titling or credit is required, a single name, "Paul Harvey," appears in black, typewriter font. We know then that the famous voice, long emblematic of a popularized form of nostalgic American conservatism, will soon emerge.

Within the first few seconds, Harvey bellows, "And on the *eighth* day." The scene shifts to the black and white image of a rural church below a swirl of clouds. The cross stands out, weathered but resilient, above the windows and door. The camera cuts to an aerial view of a farmhouse and outbuildings in plowed fields of straight furrows, intersected by provisional roads, unimproved but sufficient for a rugged truck. Harvey then adds, "God looked down on his planned paradise and said, 'I need a caretaker.' So God made a farmer." A farmer now appears in another still image, walking through a fence gate, between tire tracks, into a snowy open landscape reminiscent of the opening scene, with his dog bounding at his heels. The scene turns to another farmer—a sepia image of an aged man leaning on a rail, in a denim shirt, battered vest, and cowboy hat, with white hair and mustache, his face and neck weathered and wrinkled. His squinting gaze is locked in study of something beyond the frame. Here is the face of the farmer, a visage to love and respect, of a man still standing, and willing, after the scars of many years.

This is the opening of RAM truck's (formerly Dodge RAM) "So God Made a Farmer" commercial, which originally aired during the 2013 Super

Bowl and became an immediate cultural sensation. The 2013 Super Bowl was one of the most watched television events of all time, with an estimated live viewership of over 108 million people. Since its initial airing, "So God Made a Farmer" has generated roughly 23 million additional YouTube views.[1] The Future Farmers of America also hosted the video on their site, with RAM pledging to donate $100,000 for every 1 million views. Within a few days, this goal had been exceeded.[2] Just over two minutes long, "So God Made a Farmer" offers a sentimental tribute to the American farmer and rural farming communities. The ad combines images and words steeped in the values of family, work, and resilience. It also, of course, associates the RAM brand with this distinctly American vision.

Harvey, the late talk-radio legend and conservative political columnist, originally delivered "So God Made a Farmer" as an address to a Future Farmers of America convention in 1978.[3] However, its earlier history unfolds almost as a meme in the agrarian consciousness. Harvey produced an homage to farmers as a "Point of View" column for the Gadsden, Alabama, *Times* in 1975; however, the sentiment, and some of the phrasing, is taken from a letter to the editor in the Ellensburg *Daily Record* in 1949 that provides a "definition of a dirt farmer." Tex Smith, the contributor of the Ellensburg letter, was apparently not its original author. Smith's letter is a verbatim rendition of a piece attributed to Boston B. Blackwood in the *Farmer-Stockman* of Hartshorne, Oklahoma, in 1940.[4] While Harvey added his distinctive oral delivery, the common themes of the farmer's character, with its resilience, good cheer, ingenuity, and faith, as well as the inherent difficulties of farming, have journeyed forward from the earlier renditions. "So God Made a Farmer," commonly attributed to Harvey, is more accurately a collective text, similar to a folk tale, emerging from a deep repository of ideas and images regarding farming and farmers. This renders its appropriation by a commercial brand, and its incorporation into a subtle visual motif of dehumanization and the rise of machines, poignant and revealing.

As we have seen in other cases examined in this book, American agrarian myth is resonant, ambiguous, and thus useful to an array of parties for various ends. Moments of agrarian mythmaking as divergent as the anti–big business rhetoric of the Southern Agrarians to the rise of Chipotle illustrate the myth's ideological-discursive plasticity. As agrarian myth functions in the service of widely varying political-economic and cultural projects, it does so through an equally broad array of sites and discourses. Yet food studies tend to overlook how social actors not directly or intentionally participating in food movements, and who are not typically defined as contributors to the political economy and culture of food and agriculture, contribute to images and myths about food, agriculture, farmers, and rural life. These "indirect" texts play a role, however, in the production and maintenance of popular-cultural notions of family farming heritage.[5] "So God Made a Farmer" is a compelling and impactful example of this type of text. As the agrarian tropes of its images and solemn words grant authenticity to RAM and its audience, its function is normalizing. The potentially resistant power of agrarian myth, seen by parties as disparate as Country Life, Rodale, the SCFarmers, and Chipotle, is absorbed in the figure of the pickup truck, as the essential American object that any in the audience may aspire to acquire.

In "So God Made a Farmer," we find an instructive moment of mythic rearticulation within a contemporary commercial trend of branding rural and rustic authenticity. The noble farmer is cast as the heroic embodiment of moral virtue, who defines the best of what Americans can be. In this chapter, we weigh what happens when anti-agrarian forces radically reinvent the meaning of agrarianism for political or economic gain, stripping it almost entirely of its anti-industrialism principles. Tarla Rai Peterson describes this problem as ideological co-option made possible by the "mythic permutation" of American agrarianism.[6] Peterson traces the creation of a self-contradicting narrative hybrid in which industrial agriculture's myth of progress coopts features of the traditional agrarian myth to advance the industrial agriculture agenda.[7] As this chapter will show, "So God Made a Farmer" radically permutates agrarian myth by emphasizing and reframing the "authentic" virtues of three topoi (that is, commonplaces and conventions of appeal) of American agrarianism originally theorized by Jeff Motter and Ross Singer: place, practice, and solidarity.[8] Generating romantic and nostalgic appeal toward rural family farming in the American Heartland, RAM sells "authentic" rusticity and the values long associated with it: community, hard work, honesty, simplicity, independence, family, and faith. As RAM untangles agrarian myth from the philosophy of American agrarianism as a form of agricultural and environmental ethics, it defends the cultural status quo from progressive political forces. While paying tribute to the spirit of the farmer as a binding force in national identity, RAM displaces pressing questions about the future of food, farming, and rural communities. "So God Made a Farmer" not only aligns farmers with pickups, it does so by tapping national cultural identifications that, as the ad states near its conclusion, bring out "the farmer in all of us."

We turn now to unpacking American agrarian topoi as critical heuristics for mapping mythic permutations in "So God Made a Farmer." These agrarian topoi should aid scholars attempting to account for the role of rhetoric and cultural mythmaking in the legitimation of food and agriculture interests, identities, and discourses. In the present study, they allow us to identify and critique how and where the RAM commercial appropriates commonplace images and motifs of agrarian experience. We begin with a survey of the historical roots of agrarian ideals and the disparate visions that comingle in the topoi of American agrarianism. We then conceptualize each topos in turn and examine "So God Made a Farmer" in its liminality and luminescence.

Agrarian Topoi

From Jefferson onward, agrarianism has often enjoyed a privileged status as an ultimately authentic, and authentically American, experience. Through techniques of rhetorical association, agrarian mythmaking has long allowed for granting an air of authenticity to a range of phenomena, including commercial products. Sarah Benet-Weiser writes that this capacity to transfer agency is particularly powerful when marketing experiences that "have been historically understood as 'authentic' and positioned as outside the crass realm of the market," such as creativity, politics, and religion.[9] Motter and Singer posit that mythic American agrarian discourses authenticate and morally legitimize specific food and agriculture identities and interests. This occurs as social actors variously configure the meanings of and relations between three topoi: place, practice, and solidarity.[10] As commonplace themes and conventions of appeal, argument, and function, these topoi of American agrarianism trace back to the writings of Jefferson and his contemporaries. Agrarian topoi play an organizing role in the invention of verbal and visual appeals, narratives, and arguments regarding the relationship between nature, culture, food, and agriculture in the United States. Thomas Jefferson and the generations of agrarians following him have often situated smallholder farming practices as the source of citizens' rootedness to a rural place, from which local and national political, economic, and cultural solidarity could sprout. The agrarian ideal served a new nation by connecting farmers to their land, with a deep interest that related them to their community.[11]

As Paul B. Thompson stresses, however, the American democracy that Jefferson imagined was prompted more by his desire for a robust political system than for environmental sustainability. A strong, democratic citizenry would counter the potential for anarchy or disarray that his contemporaries feared for democratic societies. Landowners with a vested interest in maintaining societal structures, Jefferson thought, would resist the excesses of democracy, while the rich natural recourses and abundant land of the North American continent, as well as a low density of population, offered the possibility of a nation of landowners who farmed the land and could make society self-sustaining.[12] Jefferson was further convinced that a nation consisting of a small-scale family-farming majority would guard against the ills of urban manufacturing and the culture around it, including the economic inequality, unemployment, and abandoned farmland that he had observed in Europe.[13] For Jefferson, the land was the basis for an economy that concentrated resources and labor on farming. Even before the Louisiana Purchase doubled the nation's holdings, he wrote, "We have now lands enough to employ an infinite number of people in their cultivation."[14]

The agrarian topoi that we conceptualize below manifest Jefferson's specifically American vision, but originate with ancient civilizations. For example, ancient Roman agrarian law, or *Lex Agraria*, also emphasized the just distribution of small plots of land across classes.[15] It should also be noted that Jefferson's vision was only one of several competing visions of American agrarianism and political economy circulating contemporaneously, and that agrarian visions have changed significantly since his time.[16] Still, in key ways Jefferson's vision has endured, not only in its continuing influence on issues of farming, community, politics and statecraft, but also through its contribution to the cultivation of "Americanism" as a distinct type of relation binding places, practices, and communities. We illustrate this influence in the analysis below. The corporate advertisement that we examine portrays the farmer as the perennial American, striving daily to overcome obstacles and stacked odds for simple but profound reasons: family, community, nature, God, independence, and dignity. In the imaginary of this rural society, undivided, propelled by trucks where once it had horses, Jefferson's vision is sustained, as if the citizen-farmers that he envisioned are still the first among us. The stability Jefferson sought in agrarian democracy is fulfilled in this commodification of farming in the marketing, and interests, of powerful industrial machines born of automation and manufacturing.

Embodied Place

Place is the topos most intimately connected to the Latin root *agrarius*, or, "pertaining to the land."[17] This ancient topos invokes farmers' rootedness in a place of their own and their sense of interconnectedness with local ecosystems as key bases of personal and societal virtue. As we have noted, for Jefferson, connection to land was fundamental to agrarian society, citizenship, and democratic virtue, as well as the best hope for a stable republic that would endure. Others have extended this ideology to develop its spiritual and religious overtones. In the United States, land fertile for agriculture has held deep meaning for those who come to occupy it and has often been viewed as sacred. Yet the selective and problematic frontier imagery of Frederick Jackson Turner and others described in the introductory chapter should function as a reminder that westward European settlement in the name of the sacred must not be romanticized. Indeed, American mythmaking has too often distorted or hidden the conquest, violence, theft, and genocide involved. For this reason and others, the trope of land-as-sacred must be critically examined from multiple angles. This trope has been the topic of several well-known studies of myth in American experience, culture, and art. In *Agrarianism in American Literature*, M. Thomas Inge writes that the earliest European settlers promoted the idea of the sacredness of this new Eden or new Arcadia of lush and unspoiled abundance. The New World offered a "paradise regained" for renewing innocence through the purity of unspoiled nature.[18] Roderick Frazier Nash's *Wilderness and the American Mind* traces conceptions of nature among the earliest settlers, including the idea of land as a sacred place that invokes fear, nostalgia, and serenity.[19] Henry Nash Smith's *Virgin Land* outlines how the land-as-sacred trope compelled nineteenth-century immigrants to settle the West and believe in the superiority of this "garden" over other land.[20] Leo Marx's *Machine in the Garden* describes the popular perception of a special relationship between God and human beings who occupied America during the twentieth century.[21] Together, these studies suggest that the land-as-sacred trope has long been central to the mythologizing of agrarianism as an environmental and political project of place-making. This conception of land as charged with spiritual meanings still animates modern visions of resistant agrarianism that relate to environmental movements. Thus far, our book has explored similar connections between agrarian place-making and the sacred in the gospel of redemptive tradition sought by the Southern Agrarians as well as J. I. Rodale's organic agriculture jeremiad against industrialism's assault on rural, place-based culture.

Inflecting agrarianism as a basis for modern environmentalism, the topos of place marks the perceived sacredness of the land's gifts and the human responsibility to act as stewards for future generations. As Wendell Berry eloquently states in *The Gift of Good Land*, "The life of the place comes in as food, returns as fertility, comes in as energy, returns as care."[22] Agrarian mythmakers emphasize care as a way of life, particularly as an environmental ethic. The land and earthly biodiversity are viewed as ancient and irreplaceable sources of vitality, wisdom, and community in a modern world of fragmentation and alienation. In his study of the connections between American agrarian philosophy and environmental sustainability, Thompson notes that agrarian experience rooted in concepts of place has been proposed as a corrective measure to the isolation and fragmentation of modern experience.[23] Further, environmental discourse often deepens this sense of place as a corrective measure by envisioning the earth as an intertwined collection of living systems or, as in the Gaia hypothesis, a single living organism. Such visions and critiques denounce crude discourses of post-agrarian society that reduce nonhuman earthly life to inert, voiceless, controllable, and expendable objects.

Looking ahead, it remains unclear how a new agrarianism of place-making will fare amid what some fear are post-agrarian environmental effects that may be permanently out of control. Anthony Giddens traces the environmentally devastating and dehumanizing effects of "globalization from above" and the institutions that guide it (for example, the industrial food system). These effects, which stand in stark contrast to the logic of a new agrarianism, have motored a "runaway world" of communication technologies, multinational corporate business forms, and international political governance.[24] The role of socially resistive agrarian experience in this complex picture remains open-ended. Is it the victim of massive changes such as globalization? Is it the hero that offers a route to more sustainable forms of living? Alternatively, is new-agrarian resistance a meaningful contributor to a mixed future, as imagined by the term *glocal*, as complex hybrids of places on different scales?[25]

As we suggest in this chapter, "So God Made a Farmer" resides at the hub of such ambiguity. We approach the ad by considering it as a romantic reimagining of agrarian place-making for a vast and relatively "placeless audience." In this manner, the ad "re-places" Americans under conditions controlled by RAM and corporate interests more generally. To see how this happens and gauge its meaning, we need to first extend this inquiry into agrarian place-making. This involves recognizing how the now-dominant, urban-consumerist sensibility toward place supplants the notion of place with space. As past scholarship illuminates, heterogeneous places become built environments of corporate-capitalist spaces—spaces minimally distinguishable from one another from one locale to the next.[26] Corporate-capitalist interests rhetorically encode this generic space-making as progress in the form of new economic "development." This rhetoric of space-making can be traced back to the Enlightenment and the origins of European colonialism, when perspectival conceptions of space were linked to linear time. Paul Thompson explains that developments such as social contracts and individual rights, important as they may be, added to this picture to form the basis of an economics that valued property and scale, giving birth with the aid of technological advances and cheap fuel to a large-scale industrial farming, as well as the other manifestations of an urbanized, globalized, consumerist, hydrocarbon society.

Agrarianism, however, follows a different dynamic that derives not just sustenance, or a sense of the sacred, but politics, economics, and patterns of reality, from an enduring and complex relation to place. In Jefferson's discourse, the topos of agrarian place as a politic emerged in the connection between the land as the basis for virtuous action and the farmer as the embodiment of place. With an ownership stake in land that provided for family and community, Jefferson's farmer exercised an ethic of caring that emanated from the land to bind a nation of like-minded citizens. For much of the nation's history, despite many changes, agrarian interests adhered to defend non-extractive and relatively self-sufficient food economies. To the twentieth century, Jefferson's republic of farmers was recognizable in the realities of American life.

Today, the scene has shifted, with most US residents living in suburban and urban areas, and earning a living by working for others outside of the home. While it is commonplace to view these demographic changes as alienation from the ecological system from which food derives, the full picture of American experiences and identification with agriculture is more complex. For instance, while federal farm statistics from World War II onward tell a powerful story of "consolidation" that left most farm production to a few large farms, a majority of American farms remain small, with about 53 percent with annual sales of less than $10,000.[27] While some percent of "micro" farmers are new to the scene, others are the descendants of farming families who continue to hold onto their land and identify to federal agencies as farmers. This persistence of rural life is an underreported story, but one that RAM knows well: Americans of many stripes come from farms or identify with farming, and relate to rural culture as memory, or at the least nostalgia, if not daily lived experience.

Despite this broad popular identification, the politics of rurality have an uneasy history. Thomas P. Govan shows that eighteenth- and nineteenth-century political discourses faulted agrarianism's politics of place for its "communist" land redistribution sympathies, even as farmers have at times been fierce defenders of private property ownership.[28] Most agrarian discourses have occupied a position between these positions, with agrarians defending a limited or lean notion of private property rights, with various caveats. Eric T. Freyfogle writes that agrarianism defends not "investment or capitalist property" or expansive properties, but rather, "family property" on which one lives and labors, and which links to a moral reputation.[29] In such an agrarian politic, land fosters well-being rather than wealth, not only for oneself but for family and community. From a Jeffersonian perspective, this rich sense of meaning is essential to maintaining the nation's economic and environmental assets, as well as its political stability. As a stalwart defender of democracy, freedom, and environmental sustainability, agrarianism cannot be reduced to dollars and cents; the farmer's balance sheet cannot account for the life of the place and the value of all it produces, intimately or spiritually, as well as economically.[30]

Place therefore marks the political importance of agrarians as permanent, landed citizens, rather than sharecroppers, farmworkers, or peasants.[31] In the United States, the idea that having one's own agrarian place is the fundamental basis of personal and political virtue pervaded colonial meditations on farming and society.[32] Noting that the colonial "freehold" concept is synonymous with agrarianism and agrarian myth, Chester Eisinger argues that a commitment to the "freehold tenure of land" and "wide distribution of landed property" has been the basis of all US agrarian movements.[33] The basic principle of agrarianism as freeholding, explicated in the works of eighteenth-century authors, is a person's "natural right to land." Eisinger writes that, "through ownership of the land, the individual achieves status and self-fulfillment. Ownership gives him [them] dignity and a place in society." Moreover, "happiness and security result from [the] ability to support [oneself] and [one's] family."[34] Under the freeholding ideal, it follows that "the good political society must provide for the uninhibited development of the farmer."[35] Beyond simply having a landed stake in the society, the freehold concept asks that the law give special favor to the farmer. To operate on behalf of society, farmers must control their political destiny.[36]

Abstractly, "So God Made a Farmer" appeals to this conception of agrarian place. However, the compelling, nearly forgotten world that the commercial conveys is produced through the use of myth as a mask for partial truths. RAM sells a mythic image of place-based, farming identities, but its cause and the ad's ideological function are in most ways anti-agrarian. Below, we develop this argument by critically reading the RAM ad's rendering of the three American agrarian topoi described above.

RefRAMe: "Naturalizing" Place

"So God Made a Farmer" extends what is for many a recognizable, mythic tale in United States and Christian culture: the creation story from the book of Genesis. Harvey's voice, also familiar, unfurls like that of a small-town minister or minor prophet who perceives and channels the mind of God. Brant Short contends that as a major agrarian mythmaker during the second half of the twentieth century, Harvey stood as one of rural America's most trusted and authoritative thinkers on a variety of issues.[37] During a radio career spanning four decades, Harvey told folksy, heartwarming stories and lent his political opinions to as many as 22 million listeners. As Short elaborates, Harvey was widely known for celebrating the "importance of the common person," as well as espousing agrarianism as an "'ideal' rural culture."[38] From the pulpit of "So God Made a Farmer," Harvey's stance is cannily commonplace. Like a trusted second-in-command or farm foreman, his mission is to figure what to do about what the boss hath wrought. On the eighth day, the farmer enters as caretaker, in a position that requires literally "taking care," including treating with honor, and maintaining the sacred stature of, God's own creation. Like an unfallen Eden, the family farm, natural surroundings and rural community extends outward in concentric rings of concern to which the farmer journeys in his trusty RAM.

Applying the philosophical terminology of Albert Borgmann, Thompson explicates farming as a "focusing" experience that overcomes the isolation and fragmentation of modern experience by binding the self and identity to an occupation connected to community and place. In "So God Made a Farmer," we see a spectral range of caring in the mythic figure of the farmer as the embodiment of a place that extends from the farm to civic duties. As Harvey tells it, "God said, 'I need somebody willing to get up before dawn, milk cows, work all day in the field, milk cows again, eat supper, then go to town and stay past midnight at a meeting of the school board.' So God made a farmer." We are shown here a succession of images of rural places: a farmhouse in the first light of dawn, with its first floor already alight; a barn with a modified RAM parked in front, the truck's bed filled with hay; a bare winter field with deep curving furrows leading to a grove or metal silos; a wooden building with an American flag mounted inside, seen through the window. This brief tour demarcates the farmer's world as deeply rooted in natural rhythms, connected to place, animals, and family, with a distinctive, democratic view of civic virtues. The meeting of the school board becomes not an "extra," but another of the farmer's habitual activities, as "natural" as milking cows, plowing a field, or looking after some other aspect of home.

The farm demands an incredible physical work ethic that takes a toll on the body; during planting time and harvest season the farmer "will finish his forty-hour week by Tuesday noon and then, paining from tractor back, put in another seventy-two hours." But this is the simplest part, for the farmer must also be able to overcome hardship and loss with a spirit of resilience and reverence: "God said, 'I need somebody willing to sit up all night with a newborn colt and watch it die, then dry his eyes and say, 'Maybe next year.'" The images from this sequence shift from a watchful farmer sitting near an open barn door; to a saddened figure with his head supported by a hand squatting outside a barn; to a farmer alone in prayer inside a small rural church, leaning over the back of an empty pew, bare head bowed, eyes closed, hat held gently in his hands.

The ad appeals to diversity of race, gender, and age among farmers, but in a manner that does not displace the prominent, central figure of a middle-aged (or older) White, male farmer. The progression of places described above ends with the repeating chorus, "So God made a farmer," but shown now is the first image of a woman: in this case an older White woman with a wry smile, wearing a woolen wrap over a ball cap, paisley scarf and gloves, holding the handle of a farm implement. She is at once weathered and fashionable, sophisticated but unfazed by hard work or the privations of farm or ranch life. As the ad unfolds, the diversity fans out. We see an African-American man in wading boots sitting in the back of a pickup with the hatch open, deep in concentration, checking a cable or line. A young girl, White and blonde, stands proudly with her arms crossed in a sunburst in a field of young crops, again with the refrain "So God made a farmer." Latinx people also belong in this vision of farm and family. The image of worn hands is one of the ad's recurring motifs. We see farmers' gnarled and folded hands indicating strength, open hands holding a baby chick evoking care and love, then suddenly a pair of brown hands holding a collection of newly picked colorful peppers. The next image shows a mother and son behind a vegetable stand selling strawberries and chorizo; behind them, a mural shows a farmyard scene with a mustached Latino man and a horse, and another man with a well bucket. The American farmer, it seems, relates to his or her peers elsewhere. Agrarianism binds the segments of society and extends beyond borders to wherever farmers reside.

This apparent diversity, however, is belied by the ad's predominant images of brawny, rugged, White men and the repetition in Harvey's narration of the masculine personal pronoun "he." It is as if the others have earned the right to appear in this space by their ability to act like the prototypical farmer, the experienced, worthy, bellwether White male. This figure is known not only for the range of his activities, but his resourcefulness and thrift; God made the farmer because he needs someone "who can shape an ax handle from a persimmon sprout, shoe a horse with hunk of car tire, who can make a harness out of hay wire, feed sacks, and shoe scraps." This farmer recognizes that in this creation, nothing goes to waste, and all items have multiple uses. While the farmer is notably strong and rugged, exhibiting in abundance the traits we associate with masculinity, he also is capable of delicacy born of sentiment and love for family, the farm, and God's creation. God needs "somebody strong enough to clear trees and heave bales, yet gentle enough to yean lambs and wean pigs and tend the pink-comb pullets." This mythical, caretaking farmer is willing to "stop his mower for an hour to splint the leg of a meadowlark." No task is too demanding or delicate for God's farmer.

Allow us to elaborate on this point. RAM appeals to the American farmer's exceptional work ethic and caretaking ethos not only by showing them engaging in physical labor, but also through close attention to their bodies. This reveals the effects of a lifetime of hard work, but also suggests the moral centeredness and resilience that allows the farmer's labor to endure. Farmers appear in the ad dressed in worn and torn work coats, blue jeans, flannel shirts, dirty baseball caps, and cowboy hats. Several of the farmers' faces are wrinkled with character. Some farmers' fingernails are chipped and have dirt under them. Other farmers wear scrapes as apparent badges of honor, reserved only for those who work outdoors in daunting conditions. Rather than looking directly into the camera, most farmers shown in the ad look into the distance or focus on the work that they are doing, while their weatherworn expressions exhibit quiet dignity and humility. The ad affirms that it is work, after all, not money or public attention, that connects farmers to place as caretakers who bear the weight of responsibility on sturdy shoulders.

"So God Made a Farmer" enacts a moral order with the farmer as a larger-than-life figure at its center. While the ad appeals to a mass audience, it operates through an exclusionist and subordinating logic. RAM only validates its ad's viewers to the degree that they are able to see at least some part of themselves in the visual narrative presented. Interestingly, as portrayed in this ad, place and the environment are somewhat subjugated, muted, and inert. While the church may call the farmer into a spirit of reverence and provides a safe harbor to mourn a loss, much of the sense of place in the ad is established though backgrounds and landscapes, empty of presence or animation except for the farmer in various guises. These backgrounds are often grainy or blurred, as place recedes into unreality, useful in directing our gaze to the other aspects of the image. The ad's portrayal of agrarian place aligns with Jean Retzinger's description of how it is depicted in Hollywood film throughout the twentieth century. Retzinger writes that the fictional film set in rural US locations often foregrounds "character and a generalized notion of the land" as a "peaceful pastoral setting." Such settings are used to support the "virtues of private property ownership and economic perseverance."[39] In "So God Made a Farmer," RAM's repeating, static images of rural-agricultural place give way to the farmer and the practices and rituals of work. In this manner, it also becomes the inanimate home of machines.

Consumers are provided a bridge between their lives and that of the aspirational figure of the virtuous farmer through the presence of a silent partner, the RAM truck. The truck is present in the ad as a means of transport for people and materials, a platform for working, and a place for rest. Trucks are shown holding farm materials, supporting farmers performing work at heights, and giving farmers a convenient place to sit to perform certain tasks. Even when trucks are not shown, their traces are evident in provisional roads and tire tracks. Other farm equipment is also shown, with an emphasis on its size and power. In one of the video's most arresting images, a farmer rests on the blade in the front of a harvester, in a peaceful posture, with his arms resting on his legs and hands folded. The eye is drawn to the elements of the image around him that enact a frightful symmetry. The massive tires of the equipment, two on either side of the farmer, are also temporarily still and the same size as the farmer. The cab of the combine rises steeply behind, its chair empty as if no driver is needed, its headlights like two pairs of alien eyes. The sky roils with a threatening storm, as if it alone is alive. It is a gothic image of the industrial sublime, tense with a latent violence that seems about to cut loose, once the driver returns to the cab or if the heavy equipment restarts on its own.

While the ad reaffirms the imagery of agrarianism as embodiment of place, the static images, the flat portrayals of landscape, and the presence of trucks and industrial equipment evoke instead a disembodiment of place, where machines supplant humans as the active presence. The audience's connection to both place and the farmer is framed and mediated by the presence of industrial farming. As Robert Goldman and David R. Dickens state, advertisers utilize rural images to sell products through an association with core American virtues. Such ads offer "a celebration of past virtues in such a way that it appears that these virtues might *rematerialize* via the consumption of the advertised product."[40] Adding to a long history of appeals to the American virtue of agrarian living, this strategy appears today not only in commercial discourse, but, as we alluded to in the introduction to this book, also in political rhetoric.[41]

We read the RAM ad as an instance of nostalgic association and co-option of agrarian myth. Despite its acclaim as a rendering of modern agrarianism, RAM's appropriation of American agrarian myth in "So God Made a Farmer" functions primarily as a way of reproducing rather than resisting anti-agrarian values and ways of life. Through provocative images and narration of a way of life distinguished from urban experience, and often thought to be outmoded or passing, the ad satisfies consumer longing not with the promise of transformation, but through a reification of the status quo. We may aspire to the work ethic and moral fiber of the farmer, but the means for reaching the farmer's focused presence, even in our scattered lives, is readily available and already here, in the tools of the hydrocarbon economy upon which our lives depend. Like the Super Bowl, the ad celebrates American identity and exceptionalism while aligning history with the present and projected future. The resolute farmer, like the unfazed quarterback, is the backbone of our cultural perseverance, with the game marking another year as surely as the rituals and work around a farmer's annual crop. The ad is thus an act of "naturalization" in which contradictions are resolved by reference to places, processes, or identities that remain unchanged and largely unquestioned. This rhetoric of naturalization transcends potential discord by claiming permanence by design. Despite its appeal to the distinct qualities of agrarianism, RAM presents a flattened sameness upon which Americans may depend.

RAM's naturalizing rhetoric also elevates rural-agricultural experience to the stature of myth. To borrow from Roland Barthes, myth's powers of naturalization stem from its appearance as "depoliticized" and seemingly "innocent speech." Myth postulates itself, its heroes, and its values as "unquestionable," "eternal," and guided by "natural justification" rather than contingent motive and "historical intention."[42] In "So God Made a Farmer," RAM's mythic narrative naturalizes its conception of place in way that thwarts the risks of appearing political while affirming a conservative political project that supports the status quo, while equating the hydrocarbon economy with the gift of God's creation. We may almost hear the next set of verses from the Book of Harvey beginning with a nod to the sponsors both heavenly and worldly: "On the ninth day, God saw this farmer needed a friend, with massive power and reliability beyond what a horse or mule could muster, so God made RAM." Such a framing relegates criticism to the sidelines while granting corporations an open field, regardless of environmental and human harms.[43] This framing is shored up by the ad's co-option of agrarian virtues such as social solidarity and the practice of resilience. By drawing on the resources of agrarian myth, RAM consolidates its position as the preferred brand of a political economy that masks its problematic qualities in naturalizing rhetorics of connection and endurance.

Rooted Solidarity

As other scholars have begun to demonstrate, romanticized conceptions of agrarianism as reflective of a lost innocence are commonly used to appeal to today's predominantly metropolitan population.[44] While this nostalgia lends a sense of shared cultural rootedness, pride in humble origins, and faith in progress, it can also distort the ethic of political solidarity on which American agrarianism was based, as well as viewpoints on systemic issues in agriculture today.[45] Historically, however, the rural agricultural community's deep bond with a life close to nature, memories tied to its geographical place, and concern for its own well-being have provided bases for forms of social solidarity. Actualizing on familial, local, national, or global levels, agrarian solidarity can be conceptualized through tropes of harmony and defense. That is, agrarian people connect to place and each other in a manner that emphasizes harmonic relations to land, nature, and community, while also defending their way of life, whether as stewards of the local or caretakers on a larger national or global scale, engaged in feeding the world.

As both affirmation and defense of a way of life, the ecologically embedded quality of agrarian solidarity supports what conservationist and ecologist Aldo Leopold calls a "land community" mediated and maintained through a "land ethic."[46] Leopold, whose ideas have influenced new agrarians such as Wes Jackson and Eric T. Freyfogle, expands conventional Western definitions of community to include all earthly beings. In the land community, the human is a "plain member and citizen" living interdependently with other species, rather than a "conqueror" superior to and separate from more-than-human life and elements, including soil, water, plants, and animals.[47] For Leopold, outdoor experience and contact with nature are generative of the ability to act upon other life with an ethic beyond self-interest. Working from an expanded, human-nonhuman conceptualization of solidarity, Leopold advocates for "biotic," organic farming,[48] urging farmers to unite as leading conservationists.[49]

Agrarian voices like that of Leopold invoke visions of solidarity that contest, implicitly or explicitly, the culture of industrialism that fosters alienation, fragmentation, exploitation, hyper-individualism, and hurried living. Carrying on Leopold's vision of farmers as conservationists, Wendell Berry laments that today's dominant agricultural system not only carries the consequences of agrarian life but impinges upon traditional forms of farming and rural life, destroying agrarian solidarity. Industrial agriculture and its heroic individualist myth of technological and business progress assaults the notion of agriculture as culture; it is "a disaster both agricultural and cultural" predicated upon "the generalization of the relationship between people and land."[50] What Berry calls the impacts of agriculture as "exploiter" of people and nonhuman life is fundamentally at odds with an agrarian tradition of agriculture as "nurturer" of "health."[51] He sees agrarianism today as not simply a futile or romantic reaction to global capitalism, but a locally sensitive tradition of "care" and a storehouse for reclaiming a sustainable cycle of life.[52] Berry reminds us that economics at its root is *home* economics, based on the management of the household as connected to the ability to know and live on the land where one resides.[53] One of Berry's contemporaries, Gene Logsdon, similarly shows how the economy of industrial agriculture driven by global markets breaks agrarian connection to place and social solidarity by demanding a rugged individualism that drives farmers to compete and push others out of the market. Further, the forces that drive markets, from supply and demand to interest rates and subsidies, may not be compatible with those upon which responsible farming depends—such as the pace of nature or the regeneration of land.[54]

Against the forces that would supplant or compromise it, new-agrarian mythmakers such as Berry and Logsdon defend the topos of solidarity grounded in the fusion of local economies, rural values, and family culture. Agrarian solidarity is also rooted in a sense of place and an ability to live in a deeply conscious relationship with the natural world. Family is often a starting point for such solidarity, as it involves the laudable but often daunting task of preserving and passing down the land to the next generation.[55] Beyond family, social ties and deep companionship created through neighborly interactions, an ethic of hospitality, and lending a helping hand in times of need are also integral to agrarian solidarity.[56] Additionally, solidarity consists of interdependent economic and political action, such as buying farm and household goods from local rather than corporate chain businesses, supporting a local or regional agricultural cooperative, participating in community supported agriculture, or petitioning lawmakers about an upcoming farm bill.[57, 58] These actions attempt to preserve agrarian solidarity by deepening roots, maintaining harmony, and adaptively defending agrarian traditions and interests in the present to safeguard the future.

In contrast to the more mundane solidarity of everyday life, in which, for example, a farming family lends a hand to a neighboring family to harvest wheat before the rain, overtly political agrarian solidarity has on occasions been impassioned, dramatic, and actualized on regional or national levels. Articulations of this political type of solidarity pervade US rural history. They range from the Whiskey Rebellion of 1794, in which farmers on the frontier of Western Pennsylvania revolted against taxation; to the Grange movement from the late eighteenth century to the years after the Civil War, which formed against monopoly capitalism and railroads; to the National Farmers Organization, formed as a farm price-focused collective bargaining group in the mid-twentieth century.[59] As demonstrated by the late nineteenth-century Populists, the World War I home front, and the Country Life movement, agrarian solidarity sometimes entails close collaboration with non-farm groups and their allies, such as factory workers, women's rights advocates, educators, clergy, and merchants.[60]

For agrarians, appeals to solidarity may be used to protect the exceptional nature of farming culture amid social or economic changes or during disputes with urban merchants or politicians. However, they are also the means for uniting with parties to gain sympathy or form coalitions when the circumstances or issues call for it. In the case of the latter, solidarity appeals stress mutual dependence between farmers and merchants, who in a rocky economy experience "boom and bust" together.[61] As we have observed, appeals to solidarity may also operate through projection, such as in the case of the Southern Agrarians' search for an anti-industrial community unsullied by history. The same is true of the SCFarmers' ability to connect with outsiders—those people inspired by these humble farmers' capacity to create community through the sharing of land and produce. These supporters, like the farmers themselves, believed in the mythic possibility that the farmers could overcome the odds stacked against them. Like the agrarian topos of place, such mythic visions of solidarity have been adapted across discursive formations and rhetorical frames, including those that work for as well as against the humble farmer's interests.

As writers including Leopold, Berry, and Logsdon have shown, living in connection to place requires a community-based way of knowing. This way of knowing fends against the danger of being isolated and lost in the philosophical, economic, and social abstractions of modern life. This same agrarian solidarity toward place-based flourishing manifests in advocacy for organic agriculture, food sovereignty and justice, farmers' markets, urban farming, local grocery cooperatives, fair trade, farmworker rights, and everyday appeals to support family farms and the farmer. Like other aspects of agrarianism, solidarity, too, is subject to romanticism, idealism, and commercial exploitation. To examine the precarious nature of the topos of solidarity, we turn again to "So God Made a Farmer," where agrarian solidarity is reframed through the mass appeal of Americana.

RefRAMe: "Surfacing" Solidarity

Agrarian relations within the local community, as described above, extend from the economic, political, and the ecological, to the everyday, the social, and the familial. The RAM ad portrays these dimensions under the auspices of corporate culture. RAM's rhetorical configuration of the farmer-place relation defines the farmer as a physical, ethical, and spiritual figure, and more importantly, a living hero. The farmer-hero's wide appeal is tied to what RAM correctly identified as this exceptional citizen's capacity to invoke the nostalgic realization of Americanness among audiences. Images of a local church, a simple farmers' market, a family praying before dinner, and a father and son walking side by side on the farm rehearse a familiar narrative about rural life that appeals to an idyllic and close-knit sense of belonging, as well as spiritual solidarity. The farmer's attention to, and hard work within, the community is captured by the late-night school board meeting and visualized by the image of "Old Glory" hanging gracefully inside a care-worn meeting house. The farmer's sphere of care emanates from the local farm to the global society, as we conceive, for instance, that his labor may help to feed those he has never met. The striking image of the farmer leading his family in prayer from the head of the dinner table shows a binding element in this fabric of unity. Family and spiritual solidarity merge in the figure of the farmer as the servant of God, who, as Harvey preaches, made the farmer to ensure care for creation.

While the ad exhibits solidarity in place along the farmer's radius of care, it also evokes solidarity in time as the farmer strives to sustain the farm as a precious gift to be handed down. This farmer, Harvey says, "had to be somebody who'd plow deep and straight and not cut corners"; but even this commitment to performing the job correctly was not sufficient, for the farmer has to also be "somebody who'd bale a family together with the soft strong bonds of sharing." Sharing means not only providing support or guidance for family members or others during difficult times but also preserving the family's farming heritage, as a collective life to be shared through time. This solidarity of generations is shown in the ages of the people in the images, who range from elders to the young. The senior, White, male farmers, however, exercise the ad's sense of gravity. While the generations may spiral outward in space as well as time, they are drawn back to the home place of the farm, where the older people still reside and where the younger people sense, and are often told or shown, that they will always belong.

RAM also structures succession in a conservative-ideological fashion. The farmer must be "strong enough to rustle a calf, yet gentle enough to deliver his own grandchild." Despite this reframing of the farmer as midwife, the logic of inheritance is ultimately masculinist. As the ad moves to its end, the rapid switching of images slows, and we focus on the face of a middle-aged, White farmer, with an old hat, beard stubble, and crow's feet beginning to show around his deep brown eyes. He looks directly out at the audience, the slightest of smiles on his closed lips. The image preserves a poignant moment. This is a farmer, Harvey says, "who would laugh, and then sigh and reply with smiling eyes when his son says that he wants to spend his life doing what Dad does." At the phrase "smiling eyes" the image of the farmer fades to black for a dramatic pause, as the pace of the ad edges closer to stillness. The last image of farm life in the ad now appears; it is also held for the longest duration of any on the screen: eight seconds. The farmer's son appears. He is a clean-cut, young man, set starkly at the front of the image, with the background blurred behind. The son wears a tan jacket over a collared shirt, his eyes look fixedly to the distance, already in the half squint that his father, and probably his grandfather, perfected. He holds his hat against his chest in a gesture of reverence, for the flag, God, and the future that he embodies.

"So God made a farmer," Harvey repeats a final time, with his voice slowed for maximum effect, and the image shifts again, this time to a RAM truck parked on a dirt drive, in front of low, long farm buildings, perhaps for chickens or cattle at sunrise. Is RAM the true target of succession, has it inherited the farm and its earth? Do we now know "our place," only through the mediated presence of the truck? For the first time since the words *Paul Harvey* at the opening, text appears on the screen, "To the farmer in all of us." This dedication hangs over the RAM for a long moment. In a culture in which meanings have become increasingly fragmented and fleeting, farming, as portrayed here, appears to offer a storehouse of stability that links us to our ancestors. This linkage is made, however, not through knowing the past, but acting as consumers; we are asked not to remember, but to buy in.

RAM selectively molds the meaning of solidarity in other ways as well. As solidarity to family, community, and God are prominently featured, the political dimensions of agrarian solidarity remain absent. Notably, we are told that the ad appeals to the "farmer in all of us," but the ad does not mention that working farmers and people who live in small, rural towns are a shrinking segment of the population. The farm is portrayed as the life work and legacy of farmers and farm families, but missing are the vast consequences of the "go big or get out" industrial farming system. Moreover, as the ad memorializes the farmer's prodigious work ethic, it does not list among the many daily activities the "off farm" work at full-time jobs that today's American farmers often perform to survive and produce enough income to afford to farm. The emphasis on family solidarity and the absence of solidarity tied to policy and broader systemic issues parallels Jean Retzinger's description of how the portrayals of rural life in popular fictional films diminish the struggles of rural experience through nostalgia.[62] The RAM ad, however, takes this diminishment a step further by relocating farming as a state of being that any of us may imagine for ourselves. While actual farmers are endangered or operating in manners that simplistic portrayals fail to recognize, farming as a simulated morality is readily available for any to claim.

"So God Made a Farmer" is not a re-statement of deep agrarian values for a new generation, but a surfacing of these values as a skein of images and words that free-float for easy access in the consumer market. By using a series of images in succession, the ad takes on the quality of a collection of memes, packaged for a contemporary media landscape; for a quick communique about rural life, we turn to RAM and its easily replicated moral moments. In this manner, the ad is not a statement about farmers or Americans more generally, but instead a nostalgic, visual, and narrative assemblage of American artifacts. As such, like the Super Bowl itself, the ad condenses and brands cultural forms known simply as Americana. RAM has reframed rural experience as a collection of poignant images, fused in a quasi-spiritual narrative by a voice from another era. Rather than sanctified by a God who hovers just off-screen, the ad is sanctified by the consumer good present throughout.

RAM reframes place and solidarity in a manner that simultaneously confirms and overturns their value in American agrarian myth. A similar dynamic occurs with the topos of practice. As the analysis below suggests, defining practice as an act of resilience affirms its value in agrarian life, while establishing agrarian resilience as a standing reserve of consumerist imagery. While appearing to revive the flagging traditions of agrarianism as a missing piece in American experience and an object of consumer longing, RAM absorbs the energy of agrarian resistance and positions agrarian practice as preservation of social order. RAM discursively reaffirms agrarian myth's continuing, sentimental resonance in a culture with little place or time for agrarianism.

Universalized Practice

While agrarianism is, in part, about connecting people to place and to each other in solidarity, it also a mode of cultural and agricultural practice. Agrarians perform daily work, typically in alignment with seasons and the vicissitudes of weather, crops, and the other variabilities of place, to provide food for the present as well as maintain the viability of land into the future. As we have seen in cases such as Country Life and the SCFarmers, this practice may also extend to the arenas of politics and cultural discourse, as agrarians strive to build a network of solidarity to advance, or defend, agrarian causes or places. Even within these spheres, however, the topos of agrarian practice, tightly woven with agrarian place and solidarity, is grounded in a normative vision of everyday embodied habits honoring an ethic of interconnectedness. These practices include socially and ecologically responsible agricultural cultivation and husbandry, as well as cultural, political, and economic practices that preserve agrarian solidarity and place.[63] Those living off the farm, as well as on it, may also partake in practices that recognize or promote agrarian values, include producing food in home gardens or cooking or eating in ways that draw from local, sustainable, or fair-trade food and agriculture. Similarly, urbanites and others may support local environmental conservation and preservation projects or engage in artistic activities that promote the public's appreciation for local and regional food cultures.[64]

Although agrarian cultural practices and agricultural practices are mutually constitutive, and not always distinguishable, agricultural practices are primary in this relation; everything else flows from them. Forms of agricultural practice are the basis for agrarian living in all its emanations. Beginning with Jefferson and his contemporaries, American agrarian mythmaking centered on the idea that working one's own land to grow food and fiber provides an experiential basis for moral development and the creation of community, as well as the advancement of the young nation. Although this premise still stands encoded in myth, the decline of the rural population beginning in the latter half of the nineteenth century evolved the concept and revealed its nomadic possibilities. Agrarian mythmakers adapted their narratives of virtue to urban and suburban culture, and slowly, to more diverse cultural communities. Today's new-agrarian voices recognize that agrarian practices need not occur in rural places only but may happen anywhere that an individual or community uses the available resources to conduct agrarian actions. These emerging practices that promise to reshape the agricultural landscape include urban gardening and foraging, community-based agriculture, and rituals of conscious food consumption that support nearby (non-industrial) environmentally and locally sustainable farms.[65]

Potentially, however, agrarian practices do more than help people to grow or eat well-grown foods; they also support cultural meanings around the principles of connection to place and familiar or community solidarity that we have described above. Agrarian practices of connection are, in effect, ways of manifesting embodied place and rooted solidarity. Whether described through rhetorics of labor or of those denoting "mindfulness" or "holism," the agrarian topos of practice describes customs, rituals, and techniques of agriculture as (eco)culture enacted by those committed to making a sacred bond with the land and all its inhabitants. Consistent with the other agrarian topoi that we have covered, mythmakers appealing to the topos of practice often collapse the dominant Western individualist dualism of nature and culture, in favor of an ontology of mutuality in which agriculture is culture and nature is a measure of agriculture.[66] Elaborating on how this holistic philosophy guides and is reproduced through agricultural practices, Paul Thompson identifies industrial agriculture as driven by a view of agriculture as a food-production business that becomes "just another sector in the industrial economy." In contrast, agrarians practice "agriculture as performing a social function above and beyond its capacity to produce food and fiber."[67] While the industrial vision in its most rigid forms reduces the relationship between farmer and land to commodity production at the lowest possible price for the greatest number of people, an agrarian vision counters this with a broader spectrum of experience and deeper meanings. Agrarianism promotes practices that foster the well-being of places, people, and communities while adding the value of a permeating sense of connection.[68]

Revisionist agrarian mythmakers have built upon this "broad spectrum" approach to agrarianism by positioning agrarianism as a domain of the "generalist" rather than a "specialist." As a generalist practice, agrarianism is more than an occupation or economic means to an end, as food is more than a consumer product or the provision of sustenance.[69] Here, the experiential and spiritual aspects of the agrarian topos of practice render it especially attractive to urbanites, as it contests the specialization and secularization of urban societies while suggesting remedies to its dehumanizing effects. Similarly, Southern Agrarian Allen Tate wrote that the modern Americans stuck in the "half-religion of work" cannot see the half that completes the "whole horse." Caught up in categories and abstraction, they recognize only "that half which may become a dynamo, or an automobile, or any other horse-powered machine."[70] Generalist and well-rounded agrarian living in small-scale farming communities is instead of a way of living a whole life. It is consistent with but also broader than any singular agrarian manifestation, including cultivation of food using organic and natural methods, as well as other alternative food movements for local food, slow food, food sovereignty, or food justice.

New-agrarian rhetorics of practice in these movements stress quality over, or in balance with, quantity; holistic ecological thinking over specialization and isolation; and the attempt to rethink the meanings of effectiveness and efficiency to account for hidden human and ecological costs. Further, as Wes Jackson states, agrarianism adapts science and technology to normative practices without diminishing their use in the name of modern "technological fundamentalism."[71] For instance, advocates of sustainable agriculture are generally not anti-science and technology, but critical of the alliance between applied science and corporate interests that gave rise to industrial agriculture. For agrarians, good land and model communities begin with a broader appreciation for how they fit within broader ecological systems. As an embodied agricultural ecological sensibility, agrarian practice is sensitive to how agriculture may help to maintain the biodiversity and interdependence on farmlands and in the neighboring ecosystems, such as woods; streams or ponds; wilderness areas; or town, suburban, or urban lands.

Agrarianism presents a holistic viewpoint on agriculture, though it is not the only powerful narrative of farming and, as scholars have shown, captures some, albeit persistent, aspects of how many farmers self-identify. Pressured by the rise of agrarianism in the culture, industrial agricultural interests increasingly deploy a "hybrid" myth that reframes the self-sufficient farmer as a traditional family-oriented caretaker who is also a utilitarian individualist and technology-savvy businessman on an infinite and brave frontier.[72] In her study of modern industrial farmers' narratives on environmental conservation, Tarla Rai Peterson shows how this picture often fails to cohere as farmers' agrarian myth-inspired self-recognition as caretakers conflicts with other aspects of their self-image.[73] While farmers agreed in principle with an environmental caretaker role, they paradoxically viewed it through an industrial model.[74] They viewed themselves as technicians facing relentless external threats who must use technology to implement the frontier imperative to control and exploit wilderness, as well as to fulfill the caretaker role.[75] Peterson found that these industrial agrarian narratives reflect "incompatible values" and mythic permutations in which multiple myths converge to create an implicit "hierarchy of motives."[76]

Similarly, Patrick H. Mooney and Scott A. Hunt find in agricultural discourse emerging during and after industrialization a shift from a stance of agrarian caretaking of place toward a more generalized sense of managing the land. Mooney and Hunt document the rise of what Paul H. Johnstone identified over fifty years earlier as "agrarian fundamentalism."[77] This discourse retains the Jeffersonian idea of farmers as virtuous citizens, while often glossing over significant details regarding farming methods and farm size.[78] This ambiguous appeal to the farmer's virtue may invite blind legitimation of exploitive, industrial-agricultural practices. As these scholars begin to suggest, a new agrarianism faces the threat of ideological distortion. To summarize, this threat includes but is not limited to the following: the unchecked universalization of agrarian discourse, where virtually anyone can claim an agrarian practice; the use of agrarianism to claim inherent virtue without regard to practice on the ground; and the disassociation of the caretaker role from self-identification as technicians conquering nature. Each of these rhetorical moves empties agrarianism of its traditional meanings; it begins to appear as an empty signifier used to suggest that any agricultural experience aligns with the mythic past.

We turn to "So God Made a Farmer" as an engagement with agrarian practice that dramatizes these contradictions. As the analysis below suggests, this rendering of practice pivots on the marketer's ability to simply affirm and offer the brand as the fulfillment of desires consumerism has itself produced. Here, agrarian virtue is open to all of us if we are daring enough to actualize our inner farmer. However, the version of agrarianism we are asked to buy into, while filled with the rhetoric of virtue as only Harvey could convey it, is a cornerstone of the hydrocarbon economy as evidenced by the technical "tools" that are finally the subject of the ad's testimony and its embodiment of place, solidarity, and practice. While consumers may be comforted by agrarian virtues, the farming is increasingly deeded over to the machines.

RefRAMe: Practicing Universal Consent

In a classic study, Edward S. Herman and Noam Chomsky note that the media strategies they call "the manufacture of consent" also included the handling of dissent. The beauty of the system of consent, they write, is that "dissent and inconvenient information are kept within bounds and at the margins." This allows for a simulation of dissent, in which the system is shown to tolerate opposing views, but these are "not large enough to interfere unduly with the operation of the official system."[79] In "So

God Made Farmer," the manufacture of consent functions somewhat differently. Here, the universalizing potential of agrarianism is not cast to the margins but absorbed and reframed at the heart of the hydrocarbon economy from which industrial agriculture derives. By removing the threatening aspects of resistant agrarianism, RAM articulates agrarian practice as a means of consent that reifies the status quo as virtuous, natural, and enduring. Surfacing the past as a nostalgia that looks favorably on the present, RAM prepares for the reign of smart equipment, an agricultural Internet of Things.

Agrarian practice in its workaday forms is a focus of the ad, which offers a generalist view of the farmer's actions through examples of adaptation and "making do" through resourcefulness. The generalist depicted has a variety of hands-on skills that contribute to an agrarian ethic of self-sufficiency and making the most of the resources available to him to complete his caretaking tasks. The tasks described symbolize the agrarian commitment to conservation and to homemade goods rather ceaseless and wasteful capitalist consumption and disposal of mass manufactured products. The farmer is noteworthy not only for his ingenuity and thrift but for his capacity for work even when enduring hardship and pain. Heroically, the farmer's quiet, honest sacrifice benefits all of us. This farmer is both ordinary and extraordinary; he represents the best of working people but also stands out as one willing to resist the easy temptations of consumerism and complacency. Both self-sufficient and community-minded, the rugged farmer is one of us but pitched at a higher level, closer to the Almighty's purpose. As a figure from an agrarian myth of democratic virtue, he is at once common and superlative.[80]

The images that depict this pantheon of practices and qualities visualize, at first, the farmer's character. We see, for instance, a pair of hands, folded, resting against the top of a fence or the back of an old chair. They are large, soiled from work; one thumbnail is chipped, the other blackened from a bruise—the thumb may have been smashed. They are reverent, peaceful, but coiled with power as if the farmer has paused for a brief break, or prayer, before another long stretch of labor begins. Similarly, the reference to the long work week and associated pains ends with the chorus, "So God made a farmer," and the image of a farmer with his tired head leaning against his hand. He is in the darkened house in the evening. His head is hatless, his face is deeply furrowed by lines of care, but his eyes are alive and sharp as he regards the camera. Tired as he is, he is not yet at rest, but awaits the call that may well come of work yet to be done.

As the range and duration of the farmer's work, and the scope of activities on the land, evoke the agrarian perspective on farming and rural life, a subtle narrowing of experience is also underway. Political dimensions of the farmer's work are truncated or omitted, including farm policy concerns that often color farmers' daily lives. Absent, too, is any mention of economics, including global price structures or market demands that may dictate what crops are planted as well as where they are shipped. A more realistic picture of farm life might, for instance, include a narrative or visual reference to the computer, where the farmer tracks prices and may even trade futures contracts to hedge the investment made in seed, fertilizer, and equipment. The ad also tells us little about the about a farmer's practices regarding environmental stewardship or public health, or the economic or cultural life of rural communities. Instead, we are given an example of Johnstone's agrarian fundamentalist discourse that hides the absence of details regarding farming practices with appeals to moral virtue.[81] As Debbie Dougherty rightly argues, the romanticized appeal of the concept of the farmer's work often masks wide differences across farmers' relationships with the soil in their daily lives. In the age of megafarms and niche global markets, the farmer may be more of a desk-bound, strategic business leader than someone who spends a great deal of time engaged in physical labor.[82]

With the image of the farmer cast as limited and tied to the agrarian past, a more vital party emerges, as shown in the images wedged between those of farmers in Harvey's catalog of the ingenuity and difficulty of agrarian work. The reference to the improvised horseshoe is illustrated by the image of a horse, cropped so that its head does not appear. The focus is instead on the body and the saddle holding a wound rope, as if the horse is but a tool for carrying other tools. The next images supplant the horse with more recent technologies, tractors and pickups. In one, a farmer stands up in the open cab on a tractor. The tractor is not moving. The farmer is abruptly upright, like an extension of the machine, his stiffened body not as tall as the tractor's massive back tire. He looks to the distance as if he, too, awaits an order from an external party that does not appear. As Harvey speaks of the farmer's initial forty-hour week, the images shift to a strange scene in which no farmer, or any human, is present. Cattle have moseyed up to a mud-splattered pickup truck. We can make out the bale of hay in the bed, surely this is what they are after, but they seem to be congregating around the gas tank, as if they, too, run on diesel. The farmer may be in the cab, but it is hard to say due to reflection in the glass. Perhaps he is around the back of the truck, barely out of view. We cannot say, but the truck itself seems to have been working hard all day with more work coming quickly. Its front tire is pointed at a sharp angle as if it had no time to properly park itself and is poised for a rapid exit when the cattle are fed.

Starkly different from the farmers shown in the ad, who tend to look weathered and are most often clad in old work clothes or weather-beaten jackets and hats, most farm equipment is new. Even the pickup truck shown being used to feed the cattle, while muddy, shines with a fresh paint job and is free of dents. They are the prizes of considerable investments, not only on the part of individual farmers, but also of a culture. The RAM ad's message, which it trusts is shared across the culture that it addresses, is essential not only to life on the farm, but to our sense of American identity. Further, these machines embody their own set of virtues. Mechanized equipment carries the farmer and his family to town, enables him to plant and harvest the crops and feed the animals, and can be relied upon in the most mundane of circumstances as well as in times of trouble. The farmer's steadiness and attachment to the land passed down through generations enacts agrarian practice as a resilience that cannot be thwarted by weather, setbacks, tragedies, or the prospect of struggle with little financial reward. However, the famed resilience of farmers, RAM will have us see, is based on the utter dependability of the partners who remain not only stoic but silent until such time that they roar powerfully to work in the burning of fossils. In RAM and its brethren, then, we see the true path of endurance, and the special case of the agrarian American, as the life and energy of eons is concentrated in the here and now. It is as if even the farmer's moral virtue is a luxury or conceit made possible by the bedrock of RAM.

Although RAM, of course, is the featured brand of hydrocarbon animation in this ad, it should be noted that the tractors shown are manufactured by Case, a brand also owned by RAM's parent company, Fiat Chrysler. If God produced the image of this mythical farmer, it is multinational conglomerates that have reproduced it to sell industrial farm equipment and reshape farmers' lives. Despite its nostalgic appeals and provocative portrayals of agrarian life, the ad manufactures consent for advancing industrialization and new rounds of mechanization. By etching hydrocarbons and their golems onto a vision of agrarianism, RAM absorbs the resistant energies and reframes them as consent to the status quo. Despite its evocation of farming as a quintessential, almost forgotten, American experience, and the embodiment of our virtues as an exceptional people, "So God Made a Farmer" becomes another part of the Super Bowl experience to be recounted at the office or viewed once more online.

Next Year's Crop

The acclaim given RAM's "So God Made a Farmer" television commercial demonstrates the wide and enduring cultural appeal of American agrarian myth in the twenty-first century.[83] Following the success of this ad, RAM issued a series of other video ads in a similar vein. Some followed the "farmer in all of us" motif showcasing other types of work with a distinctly American quality that illustrated moral attributes such as perseverance or faith. While several videos feature "real people" telling their stories, including why they are loyal to RAM, another series features "storytellers" including the country roots musician Brent Cobb, whose hybrid style of country and blues is often considered an exemplar of the "Americana" movement in music. One episode follows Cobb's attempts to write a song, from the frustrations of failed attempts in his hotel room, to the memory of a brief interaction with a window washer who told Cobb that all he had to get by in the world was his good name. This line comes back in memory and becomes the basis of a tune that celebrates work and honor on life's journey.[84]

This exploration of resilience also revisits farming in a mini "documentary" called "Next Year's Crop," released in 2014.[85] Unlike "So God Made a Farmer," this video is not a nostalgic album of photographs covered by voiceover narration. Rather, the video is a compilation of "beauty shot" video footage of farmers, land, and families, supported by interview clips (presented as voiceover) and a few statistics about farming in the United States. Here, farmers describe the meaning and importance of farming in their lives and in the health of the nation and world. Among the emerging themes are the qualities of the farmer, including the familiar agrarian virtues of caring and stewardship, as well as self-sufficiency, ingenuity, and intelligence. As in "So God Made a Farmer," the work of farming is associated with love for family and a sense of belonging on the land. The overarching value that emerges from the documentary is resilience. The farmers discuss what drives them to persist and their hopes that their children will continue the family tradition and enjoy its many gifts. "No matter how tough it gets," a farmer says, "we'll figure out a way to get into next year. It's always been next-year country, long as I remember, boy, next year, next year." Looking to the future where a profitable crop may be delivered is the dream that holds the farmer to the land despite the hardships or misfortunes of the past and present. "Next year's still coming," the voice of the farmer says with a chuckle, as on the screen we see a farmer in silhouette leaning against a pickup in the first light of dawn.

Here, RAM's ideological rhetoric closes the circle. Despite the passion and fulfillment its adherents experience through farming, rhetorical representation of farmers and farming becomes a means of social consolidation that suppresses systemic change. With the promise of better times to come, this rhetoric reifies the centrality of the farmer's exceptional personal attributes. RAM presents a case of the co-option of agrarian myth that recognizes its mass appeal while dissipating other, socially resistive articulations of it. But even with this attempt to contain and temper agrarianism, the story is not complete, for the creation of "So God Made a Farmer" and RAM's continuing branding also reveals the relevance of agrarian myth among Americans.

As noted, one of the reasons for the popularity of the Super Bowl ad is that many people who have never farmed and were not raised in rural settings have nonetheless been affected, one way or another, by agrarian myth. The American agrarian topoi of place, solidarity, and practice traced in this chapter as providing a loose structure for American agrarian rhetoric today serve as a ready reserve for critical analysis of the (re)production of a nostalgic public imaginary. This chapter has argued that examining configurations of agrarian topoi can help to reveal the nature and organization of linkages between myth and ideology. Examining "So God Made a Farmer" through these topoi, this chapter attests to the thick and ideology-serving ambiguity of a text that embraces agrarianism as an embodiment of virtue, while also limiting the meanings of agrarianism and quelling its subversive functions. While this reading constitutes a criticism of the ad, it is also an explanation of its mass appeal, as RAM projected a sentimental, bounded version of agrarianism that Americans could readily recognize and embrace. Unfortunately, it spoke not only to the farmer in us all but to a captive nation of consumers. By aligning the agrarian past with the industrial present, it assures us that no changes are required even as it heralds an era of agrarianism via machines.

The case of "So God Made a Farmer" brings our broader arguments to a vexing conundrum. RAM's television commercial fulfills the imagistic and discursive commonplaces associated with agrarianism while thwarting the resistant ideologies that have propelled agrarianism as a critique of modernization and industrialism. Here, in the guise of a nostalgic appeal to the past, the critical synergies of agrarianism run aground amid corporate co-option. Even in this case, however, the ambiguous, ironic nature of agrarian myth remains. By identifying Americans with the agrarian, the ad stole the scene of our ultimate self-referential media spectacle and tapped a sacred cultural script buried deep in the national imagination. In a modern testament to agrarianism as a sacrament in this poignant, well-wrought ad, RAM summoned forth a spirit that even a king-size pickup truck could not contain. It found its farmers in the virtual land in the guise of millions of viewers, quieted to near reverence for a few long moments.

CONCLUSION

Agrarians Greet the Apocalypse

One version of an agrarian apocalypse involves a return to the deep agricultural past. Keith Ferrell, the former editor of *Omni* magazine, known for envisioning possible futures of technology and society, has spent the last decade learning to work the land like a relatively self-sufficient, eleventh-century farmer. On his former country retreat in Virginia, now a full-time residence, Ferrell juxtaposes millennia: "My car can take me to neighbours, to stores, to town. Across the meadow, my house, a (mostly) converted barn, contains telephone and internet connecting me to friends, relatives, colleagues, a universe of information and distraction, the modern world. Right between them lies the sliver of land I use to try my hand at agriculture, as it was practised 1,000 years ago."[1]

Living through what he terms a "personal apocalypse" that included career setbacks and his wife's chronic health condition, Ferrell turned to the land for solace and sustenance, taking farming as a vocation even as writing continued to provide income. Ferrell farms in an unusual manner that even a traditionalist agrarian might consider a bit too retro—or painful. Using only hand tools, such as a scythe and a mattock, Ferrell works his land. He attempts to grow much of the food that he and his wife need, while keeping at bay the invasive scrub pine that is so poised to take over neglected fields. His project is an experiment in ancient agricultural practices, but he is drawn to something more: "The transition from hunter-gatherer to farmer has always fascinated me. The ability to plant, cultivate and harvest crops stands alongside the emergence of self-awareness, control of fire, the wheel, and the development of mathematics and written language as one of humanity's transformational events. We became something different once we began to farm."[2] Ferrell tries to recast this historical transformation as a modern person becoming a farmer, in a manner that is at once diluted and obsessively pure. By displacing the abstractions of consumer identity, he attempts to enact the primordial connection of humans to workable, life-giving land. Ferrell is after farming as an essence, at once practical, philosophical, and spiritual; farming as a fulfilling way of being human. With a back-to-the-land spirit, he toils on the earth in a solitary manner reminiscent of a monastic figure whose gift to the collective is testimony of the soul. While Ferrell's apocalypse is personal, he holds it open through writing for others to witness.

As we have seen throughout this book, American agrarian myth always speaks of "something more," placing agricultural practice in rich, sometimes surprising, contexts. In this chapter, we conclude our study by looking to new-agrarian futures, where productive, inspirational permutations of American agrarian myth help to unite the committed individual with collective movements. In our opening salvo, we chronicled how American agrarian myth began with Jefferson and his contemporaries' conception of a nation of farmers, spreading democratic virtues as they laid claim to the land of a vast continent. Our study then moved to World War I, where the myth was modernized for collective memory in a way that propelled the nation toward world leadership through sustenance and abundance. In the reactionary period after the war, Country Life fought for the status of the farmer as the embodiment of democracy and the legitimator of the state. The Southern Agrarians sought a refuge from the ravages of industrialism where the land could again be viewed as sacred. Rodale similarly connected soil to spirit through organic practices that countered a poisoned culture.

The cases in part 2 of this book also evoke agrarian myth as a principle for searching, whether the object is a vernacular community fit for its setting as in the SCFarm, the purity of a new beginning realized through farming "clean" food in Chipotle, or the farmer's enduring spirit as the mark of authenticity for a nation of nostalgic consumers in RAM. We have also seen how the larger forces of modern society, from industrialism to consumerism, have turned agrarianism from the basis of the nation to an ambiguous and exploitable ethic of resistance. We are left wondering whether agrarianism and America agrarian myth are still viable, not only in the present, but in a future that looks increasingly perilous. In this conclusion, we peer toward agrarian futures while realizing that, as Ferrell reminds us, this will necessitate thinking again about the past.

Apocalypse without End

We engage agrarian myth as a bridge from past to future by (re)viewing it as an evolving rhetoric of apocalypse. We use this loaded term with caution, to evoke not only its connection to the end times, as manifest in discourses ranging from Biblical eschatology to science fiction literature and films, but also the original derivation of the term. "Apocalypse" comes from the Greek *apokalypsis*, meaning "to uncover" or "reveal," part of the Greek project of bringing to light or out of the shadows, and a root distinct from *eskhatos*, meaning "last."[3] In this conclusion, we look to emerging trends in farming and food, surveyed against what is admittedly a disturbing world picture, to consider what may come to pass. In doing so, we extend the red thread in this study connecting agrarian myth to the discursive genres that emerge from the Bible, from the gospel of gardening in Charles Lathrop Pack and the search for redemption among the Southern Agrarians, to the jeremiad of Rodale and the Genesis story of Paul Harvey's farmer as caretaker. As Charles Dana Gibson charged his team of artists making the agrarian propaganda of World War I, we aim to show the "spiritual side of the conflict."[4] In apocalypse, the spirit seeks to light the forward path in a darkening that may well forebode doom.

In a classic essay from rhetorical literary criticism, Robert Alter builds on the inventive potentials of apocalypse by defining the "apocalyptic temper" ambiguously, not only as a miscellany of descriptions of doom, but a site of potential transformation born of facing up to even the most difficult problems.[5] He begins with a quotation from Saul Bellow: "We must get it out of our head that this is a doomed time, that we are waiting for the end, and the rest of it. . . . Things are grim enough without these shivery games. . . . We love apocalypses too much."[6] A key text for Alter's theory, however, comes from the theologian Martin Buber, who in "Prophecy, Apocalyptic, and the Historical Hour" distinguishes "apocalypse" and "prophecy": while visions of apocalypse are retreats "from a history that has become unbearable," prophecy is "courageous engagement in even the most threatening history."[7] The prophets, according to Buber, vividly evoked a future not to show a sealed fate, but because they believed in the power of human actions to alter the trajectory that would bring doom into being. The prophet did not predict, Buber argues, but sought a confrontation with decisions. For prophets, the unique quality of humans was to be not the subject of fate but a "center of surprise in creation."[8]

This distinction of apocalypse as sealed fate and prophecy as inspiration of surprise, parallels the theory of William A. Covino that links rhetoric to different manifestations of magic. For Covino, a rhetoric that remains trapped in its limits is an example of an arrested magic that relies, at best, on familiar sleights-of-hand. A generative rhetoric, however, draws on the greater power of magic to shape-shift, making things disappear or come into being, or deeply transform them.[9] As a binding principle, American agrarian myth connects agricultural practices to larger purposes that relate humans to each other and the world. Seeing this myth as prophetic/apocalyptic, however, allows us to hone the message further. From Jeffersonian democracy to community gardens, agrarian myth is about *summoning forth*, allowing the full dimensionality of our connections to land, community, nation, and world to be made manifest. While this magic is often diluted to nostalgia, the promise remains of a discourse that remakes us by regenerating our connections to place, community, and others, including those sowing futures we may not have considered. As we consider emerging trends in food production, we read them as Buber does the prophets. Do they prepare us for difficult decisions or yield to the unveiling of fate?

Post Farms

In 2018, Waymo, the autonomous vehicle subsidiary of Alphabet, parent company of Google, made what *The Atlantic* has termed "the most important self-driving car announcement yet." The firm's announcement that it had purchased 20,000 electric, self-driving cars with distinctive sporty styling from the premium car brand Jaguar had drawn headlines. However, the news with the most impact was buried in the company's press release. With Waymo also purchasing a fleet of autonomous transit vans from Chrysler, the company claimed that its ride-hailing service would have the capacity to deliver millions of driverless trips each day by 2020.[10] Driverless technologies were reaching the mainstream for American consumers.

While these developments may seem remote from the farm, parallel transformations are occurring in agriculture with rapid onset. A 2016 article in *Popular Science* recounts the magazine's long history covering the tractor as a tool and vehicle that revolutionized farming. The tractor allowed for larger-scale operations by greatly increasing the range a farmer could cultivate, while reducing expenses by deploying machines powered by fuel instead of animals fed with grains. The tractor also ushered in a sense of detachment that may reduce a farmer's feel for the land. As we discuss below, the efficiencies of the tractor and the detrimental type of plowing it allowed was a contributing factor to the disaster of the Dust Bowl on the Southern Plains in the 1930s, with the more systemic cause being a lack of appreciation for conditions of land, climate, and soil. *Popular Science* now reports that the next round of the revolution of mechanization and automation has arrived with the driverless tractor. The article features a prototype developed by Case, the subsidiary of Fiat Chrysler that also made the farm-equipment kin of the pickup trucks featured in RAM's "So God Made a Farmer."

The *Popular Science* piece is quick to note that this prototype is but one example in the larger movement of autonomous farm equipment. Companies began developing driverless tractors as early as 2011, while firms such as Autonomous Tractor Corporation have refitted existing tractors to operate autonomously.[11] As noted by Big Ag, a website catering to industrial farmers, autonomous tractors are only part of advanced robotics and intelligent systems that will revolutionize industrial agriculture. John Deere, for instance, has developed integrated systems for more efficient management, including technologies that allow tractors to communicate with combines and other equipment and row sensing and tracking systems to help steer tractors away from growing plants or ensure full coverage of fertilizer applications.[12] The farm, it would seem, is another site for AI, monitoring and adjusting along a node in the Internet of Things.[13]

This specter of the farm as a site not for agrarian democracy, but bloodless systems-management seems another step in a fateful direction that has led to the depopulation of rural regions and the expansion of mega-farms. With resource-intensive models for production and the globalization of markets, American farms' production and profit have been consolidated in large operations requiring fewer farmers or agricultural workers. According to the United States Department of Agriculture (USDA), the number of farms in the United States declined from 6.8 million in 1935 to 2.1 million in 2012, without a parallel decline in the total acreage in production. Farms today are on average three times larger than those in 1935.[14] They have also become more specialized with mono-crop cultivation the norm among larger producers.[15]

This narrative of the displaced farmer, however, does not tell the full story. A large, partly latent, population of small farmers, near-farmers, used-to-be farmers, or would-be farmers remains and is born out in statistics as well. As we noted in discussing the state of American farms in relation to RAM, farms with annual sales of less than $10,000 account for 53 percent of all farms in the United States. However, these smaller farms account for only 1 percent of farm production.[16] This still only tells part of the story of the small operations. The USDA uses the term *Post Farm* to delineate a category of farm with annual sales of less than $1,000, but with crop or livestock production that surpasses $1,000. This may include farmers raising food at small scale for their families or selling at small farmers' markets or from their homes on a cash basis. In 2012, Post Farms accounted for 20.3 percent of American farms.[17] Meanwhile, most American farmers derive income from other sources: 91 percent of American farm families have at least one family member working off the farm. Further, these are generally not "second jobs" for extra income. The percentage of farm operators or their spouses holding management positions at other enterprises is higher than the percentage of Americans not associated with farms who hold such demanding professional positions.[18]

While the production of commodity crops for domestic and international markets has largely consolidated into mega-producers, some Americans proudly persist at the edges of agriculture, producing small quantities of crops in micro-operations and pursuing careers or income off the farm. People with farming roots straddle the farm and city, helping to bind urban society to a rural experience that is still far from defunct. Farming, for these shadow agrarians, is not a means of production as much as a way of life they are wont to surrender. A richness and familiarity, tied to personal and cultural memories, holds Americans to the land. It is this extra share of meaning, we argue, that creates the generative possibilities of agrarian myth, even as the nature of farming and food practices stretch or displace the original figure of the yeoman farmer.

The Agrarian Beyond

"The United States was born in the country and has moved to the city," Richard Hofstadter notes.[19] But Americans "born to the city" are also subject to the pull of the farm and the agrarian imaginary. Surveying the future of farming in the late 1980s, Gene Logsdon predicted that while mega-farms were not about to cease to exist, the leading edge of agriculture was with the small operators whose numbers were likely to rise over time. He points to several reasons, including the need for a network of small farms to service an increasingly dense urban population. He highlights basic economics as well, given that when the costs are tallied, it is cheaper to produce zucchini in a garden than on a large farm. However, the main driver resides at the nexus of entrepreneurship, changing market demands, and spiritual longing. Farming speaks to the whole person, especially if that person is an enterprising American who wants to cut a different path. Logsdon noticed a flood of young people entering agriculture who did not come from the farm, and who recognized their ability to fill market niches with diverse agricultural products while improving their lives by connecting to the land through small-scale farming. Logsdon writes: "Thus a historic shift takes place. The Vergilian ecology and careful husbandry of the traditional yeoman farmer that gave way to the all-consuming dreadnought of agribusiness economics now reappears in the unlikely form of the ex-urban farmer."[20]

This persistence of agriculture at smaller scales is taking new forms as agriculture is adapted to fit the needs of the residents of urban and suburban America. We have analyzed the case of one well-known community garden in Los Angeles that offered a compelling vision of agrarian practice adapted to culturally diverse, urban settings. Despite the struggles over land ownership and use, to which the SCFarm eventually succumbed, emerging scholarship suggests that the broader and growing community-garden movement in the United States is likely to prove more sustainable than earlier manifestations of collective urban gardening that were typically seen as a necessity in the management of social crisis or war. The current movement is broad-based, supported by research, and sponsored by an array of private, public, and community interests.[21]

Community gardens are joined by other forms of urban agricultural practice, including urban farming and community-supported agriculture. Urban farming may be defined as producing food in urban setting, but unlike community gardens that are typically volunteer and non-profit enterprises, urban farming "assumes a level of commerce" in which the grown products are sold, typically to urban audiences that may include consumers or restaurants.[22] Community-supported agriculture further refines urban farming by asking "non-farmers" to share the risks and rewards of farming by buying "shares" in the upcoming harvests of local farms. This model provides farmers with social support while drawing consumers closer not only to the results of farming in locally grown foods but also to the process. For instance, produce is available only when it is harvested, rather than being omnipresent as at supermarket chains.[23]

Despite the ascent of post-human technologies such as the driverless tractor, the "human touch" in agriculture remains and is expanding in myriad directions. From hipsters raising chickens in cities to a growing homestead movement that takes Americans "off the grid," new-agrarian sensibilities form alternatives to vapid consumer identity. Some emerging models for food production that now seem on the periphery may move to the center with surprising speed, as questions of food and farming deepen with climate change, rising populations, political instability, and environmental degradation. Hydroponics, often described as part of the future of agriculture, is one such development. Hydroponic facilities depend upon technological advances; crops are raised under LED lights and fed by engineered solutions of ionic compounds attuned to the variety of plant.[24] These facilities are typically sited in urban spaces such as converted warehouses, as hydroponics has emerged from the laboratory to become an urban practice. Although these methods require electricity to run lights and equipment, the energy savings are considerable: switching to nutrient baths requires no hydrocarbon-based fertilizers or pesticides. Hydroponics offer potential reductions, too, in fuel used for transportation. Crops can be distributed in the cities in which they are grown.[25] Despite the gross overfishing of the oceans, other forms of aquaculture are also on the rise, including symbiotic productive ecosystems where varieties of shellfish reside to be harvested alongside algae.[26]

As surprising as it may seem to some Americans, the cultivation of insects is another development that shares some of these same new-agrarian sensibilities. Societies around the world have long found in ants, crickets, grasshoppers, and beetles a delicious protein fix. Americans, meanwhile, have viewed insects as pests to be poisoned at scale, lest they eat valuable crops. However, as recounted in documentaries such as *Bugs: The Film* and *The Gateway Bug*, the cultivation of insects and their consumption is on the rise in the United States, starting among younger people concerned with sustainability of the species and planet. With up to 45 percent of the land on Earth used to feed cattle, insects that require no land, little feed and small quantities of water are a sane alternative that builds on world food traditions.[27] For the present study, the cultivation of insects prompts intriguing questions. Is the raising of crickets for consumption in a prefabricated metal building an agrarian act, even if it involves no land or soil, and only a minimal commitment to place? Is this a feat of technology, even though not much equipment is required? If enough crickets are raised for market, is there a point at which cricket farming becomes industrial, or will insect farms remain a niche that operates on a local habitat-to-table model with freshness guaranteed? As surprising as it may seem, practices such as hydroponics and insect farming aid our ability to feed ourselves in a sustainable manner, locally and globally. The generative magic of such prophetic practices suggests that the future of food and farming, and what exactly counts as new agrarianism may be hard to anticipate.

Resentment and Adaptation

While the powerful messages of care and connection at the center of American agrarian myth may be adaptable on a broad scale, the status of the farmer remains ambiguous. More than many Americans, farmers are acutely aware of their entwinements in a global economy, which is made apparent whenever they check commodity prices. Decisions about crop varieties are often made to appeal to international markets. At the time of this writing, the American role in international farm trade has been torn asunder due in part to the sudden eruption of trade wars linked to federal-protectionist trade policies. With the massive Chinese market all but closed, farmers in the United States are challenged to cultivate new markets in other countries while also accepting federal aid to keep their operations afloat.[28] Despite these obvious international connections, farmers also resist engaging global issues such as climate change. According to studies, US farmers are widely skeptical of the idea of human-made climate change, while also believing that agriculture has received undue blame for accelerating global warming.[29] These farmers often cite warming trends as manifesting "natural cycles" that are out of the control of humans and as old as the planet.[30] This view is buttressed by the federal policies that pulled the United States out of the Kyoto Treaty and removed references to climate change and the science that tracks it from federal reports and websites. The president has repeatedly called climate change a hoax perpetuated by the Chinese to weaken the US economy.[31] Agrarian myth here becomes an apocalyptic limit, providing, at best, the shelter of the familiar against a global threat.

Farming, however, is often about adaptation. Paradoxically, but perhaps predictably, American farmers who are skeptical about human-made climate change and resent being called out as a cause are adjusting their practices to fit changes in climate on the ground. The *New York Times* reports the example of Doug Palen, a fourth-generation grain farmer in

Kansas who has adapted to drought, warming temperatures, and freak snowstorms without attributing these disruptive conditions to climate change.[32] Palen's methods, however, have changed to reflect these conditions, while helping to conserve water and protect the soil. He practices carbon sequestration, which traps enriching microbes in the soil while reducing released carbons. In his no-till operation, the remains of last year's crop blanket the soil with protective biomass, while his crop rotation of wheat with sunflowers, sorghum, and alfalfa emulates the original diversity of the prairie.[33] The *Times* cites research suggesting that such restorative farming may prove significant in stabilizing the climate.[34] Palen hardly sees himself as an emissary of climate best practices, at least within the political sphere as currently constituted, where climate change is an issue identified with liberals. Instead, he states that like most farmers he wants to be left alone to farm without interference. The politicians and public can exhaust themselves arguing about climate; he has an operation to run and knows its requirements better than anyone.[35]

While this farmer sees his adaptions as a local matter, climate change on a world-wide scale is an issue that most people across the world, including farmers, readily acknowledge. The everyday demands of farming, as well as the agrarian ethos that enshrines connection to a local place as a virtue, may keep farmers from recognizing their position as potential political agents in a globalized world. Farmers, however, may be well-suited to make a difference, not only due to the nature of their work, which can be turned toward carbon solutions, but also due to their global presence. A worldwide lobby of farmers sharing values and pushing politicians would be a powerful force. We envision a Country Life movement at scale, directed toward other ends.

Dust and Doom

The effects of climate change on farms in the United States and around the world will continue to be seen. Researchers suggest that the varieties of crops cultivated in the American Plains do not appear able to withstand the changes in temperature projected to come with climate change.[36] However, the UN projects that North American agriculture may benefit from a longer growing season associated with higher temperatures, while places such as West Africa and India will see profound drops in farming yields.[37] Another round of consolidation and increases in the size of the large operations may also be foreseen if arable lands become increasingly uncommon or changes to crop varieties strengthen the grip of monopolistic corporate practices regarding seeds or chemical fertilizers or pesticides. Farming is also part of a larger picture of economics, politics, environment, and social cohesion, all of which are likely to be transformed, perhaps indelibly, by climate change. While we may imagine, for instance, farmers reaping profits as their faltering crops are highly valued amid food shortages, other outcomes may include the loss of markets or collapse of prices due to social instability or war. It is also possible that farming adapted to take advantage of any opportunities presented by climate changes may exacerbate the environmental degradations perpetuated by industrial practices as well as a warmer Earth.

To recognize how climate, politics, economics, and the limited scope of human perceptions affect one another, we may look to the Dust Bowl in the Southern Plains in the 1930s. As a prophetic warning or moral tale linked to place, the Dust Bowl is as an apocalyptic tale of epic, if not biblical, proportions. It began with a land rush. Real-estate marketing swindles combined with federal government policies sold migrants on the idea that a vast dry prairie could be converted wholesale to the cultivation of wheat. During the "great plow up," as the period leading to the disaster is known, millions of acres—stretches of land totaling the areas of several Eastern states—turned from prairie grassland to farms. This included the section of the panhandle of Oklahoma that was known, perhaps with good reason, as No Man's Land.[38] The cattle ranchers and hands who had populated or passed through the region warned farmers that the area was subject to drought and could scarcely support cattle, much less wheat.[39] The grasses that covered the land needed little water and had evolved over thousands of years to be effective at holding the loose soil in place against the constant winds.[40] As one old-timer in the area put it, the Dust Bowl came when the ground had been turned wrong-side up.[41]

Unfortunately, the migrants whose lives were now staked to the place were swayed by the illusion broadcast by promoters that "rain follows the plow."[42] This was backed by the claims of scientists that a relatively wet period in the Southern Plains in the 1920s signaled a permanent change in climate.[43] With rains and high prices, crops were profitable for several years, leading to what Timothy Egan has called a "classic American bubble."[44] The land was cultivated with efficiency, using new technologies including tractors and shallow plowing, that led to loosened soil.[45] Speculators known in the Plains as "suitcase farmers" came from the city, bought or leased land, and paid others to plant and harvest crops, reaping profits detached from the agrarian mythos of home.[46] As the boom took hold, hired tractor operators worked in shifts, plowing all night, to prepare more land for planting.[47] With the coming of the Great Depression, grain prices fell and farmers in the Southern Plains responded with a frenzy of cultivation, believing their only hope to service their debts was to raise their yields. As prices continued to plummet, however, the market for wheat dried up.[48] With elevators full, surplus wheat was simply piled up, to be sold if priced recovered. The plowed land was susceptible to the drought that began in in the early 1930s and soon began to blow in the near constant wind. The storms of blowing dust begin in 1932 and continued for seven devastating years.[49]

The transition to contoured plowing allowed for some wheat farming to return to the Southern Plains, while the formation of soil conservation drew together parcels of land with different owners and operators for common management.[50] However, as Egan reports, the days of widespread cultivation in the region were over. Prairie has returned, with large parcels now federally protected grasslands. The towns of the Southern Plains remain greatly reduced in size from their peak at the start of the Dust Bowl. Egan sees the aftermath of the Dust Bowl as emblematic of the wider changes in society, including urbanization and the ascent of corporate farming. Farm profits from government subsidies began with Franklin Roosevelt's attempts to halt the frenzy of production that brought on the Dust Bowl.[51] Subsidies are now paid in massive amounts to farmers growing crops that already glut the market, forcing small operators out of business. "The money," Egan notes, "has almost nothing to do with keeping people on the land or feeding the average American."[52] Instead, it incentivizes detachment from the land. Farming is often not about growing crops, but is more about managing capital tied to commodity prices and the perceived value of agricultural lands. When we add to this picture climate change, along with allied developments such as AI for mega-sized operations, monopolistic practices regarding chemicals and seeds, and commodity crops grown for ingredients in processed (non)foods, the future looks arid and empty. Are we careening toward an agriculture devoid not only of humans but of systemic controls? Have we without realizing it been cultivating a larger patch of No Man's Land never made to sustain the likes of us? On the last day, will God let loose a soil storm of such proportions that not even a trusty caretaker can clean up, no matter how practiced his squint? We cannot fully consider this non-human, anti-agrarian apocalypse without updating the weaponization of food and the progress of starvation.

Human Rights/The Right Humans

Surveying American agriculture in the twenty-first century, Paul B. Thompson finds a state of arrest. He describes a "bipolar organizational structure," with one pole consisting of industrial agriculture, including processing and packing companies, and major grocery stores and restaurant chains. The second pole, designated alternative or sustainable agriculture, consists of "a loose network of organic and regional producers, chefs, nongovernmental organizations, and ordinary food consumers."[53] Mindful of the effects of climate change, Thompson hopes to enrich and more deeply unify this alternative network by relating the perspective of American agrarianism as a philosophical ethic of sustainability. Thompson is careful to specify that he is not calling for strict allegiance to a Jeffersonian agrarian perspective or literal devotion to a Jeffersonian agricultural practice. Instead, he seeks to define the dialectical tension between the now-dominant technical perspective arising from the Enlightenment and the agrarian viewpoint that predates Jefferson and goes to back to ancient civilizations. Beginning with these poles, Thompson suggests two hybrids for the future; the first is the marketing of consumer health through organic food, which gains market share but is delivered at scale by the methods of industrial agriculture. In the second, agricultural practices foster sustainability through alternative technologies. Both outcomes short-circuit debate and reduce meanings, which is part of their attraction in a consumerist system; as Thompson notes, "that is, in many respects, the American way."[54]

For Thompson, such reductions push the interesting questions to the margins, while an agrarian perspective tied to environmentalism restores them to the center. More than agricultural practices, an informed, new agrarianism constitutes an experience of connectedness that allows people to develop in relations of care for others and for places.[55] Thompson may, however, consider this ethos on too limited terms. As we prepare to evoke hard decisions, in a manner reminiscent of Buber's prophets, let us restate agrarian myth for a future where visions of Dust Bowls without end may perch on the edge of perception. The problem with climate change is the nature of the problem: it is *everywhere*; diffused through countless effects and relations, it spreads across modern experience and emanates from its industrial hothouse. It is not that it affects us or that we cannot evade it: we are it in ways we have only started to trace. While new agrarianism is not a solution, it may offer what we most need to survive: not technological fixes or carbon markets, but a way of being more attuned to the earth and prepared for a long run.

Throughout this book, we have examined agrarian myth as a rhetoric of critique and invention that resists the modernity of an overdetermined, hegemonic system headed toward doom. We have looked at a variety of cases and harvested a miscellany of rhetorical concepts and practices informing and evolving a new-agrarian myth of resistance. From heterotopias and counterpublics to vernacular discourses, prosthetic memories, and agrarian topoi, we demonstrate the heuristic value of a mythic approach to the discourses and politics of food and agriculture. As shown, one of the important functions of such an approach is that it is able to critically uncover methods for co-opting as well as suppressing the potential of American agrarian myth for fostering social change. Perhaps the most complicated of these methods is not the consumerist branding or watering-down of agrarian actions to Americana or nostalgia, but a potential hyper-focus on the local as the place where time stands still. In this area and others, there is still much to learn about the strategic role of rhetoric for building new-agrarian resistance among diverse groups in the United States and abroad.[56] It is evidently clear, however, that in order to combat issues such as climate change, new agrarians in the United States must ally with their counterparts around the world, whether international movements for micro-farming and gardening, eco-system adaptions to droughts and climate change, or crowd-sourced methods for resisting multinational corporations.[57]

While agrarianism powered by Jeffersonian myth remains a viable force, the vast and entrenched problems we face call for the prophetic, generative experience of pan-agrarian consciousness. A "big tent" approach not interested in giving into doom, whether it comes as dust, machines, or war, may generate the critical mass and variety of experience we need to make, rather than simply refer to, systemic change. We will need to "scale up," enacting the topoi of place, practice, and especially solidarity in a manner that, without crushing difference, forms a united agrarian front able to bring more and more of the world's people into its fold. Forming such a pan-agrarian consciousness is a solution, so far largely overlooked, by which the species might begin to chart a new destiny, an alternative apocalyptic summoning into being. While climate change provides the powerful reason for this mutuality, other issues, also dramatic and pressing, may provide occasions to form a united front of "country life" gone planetary.

This brings us back to an original, modern experience, the weaponization of food. While we have discussed it directly, it remains the ghost that haunts our history. Developments in industrialism often begin as advancements in the efficiency of killing, with crossovers also common; tractors, for instance, served as the original chaises for tanks, beginning in World War I. Farmers in the Country Life movement had mobilized for war, only to be abandoned as the industrial machine no longer needed them. Rodale knew that chemicals used in farming grew of out of the same industrial processes that yielded the poisons of World War I. The South Central area of LA bears the scars of its role in the war machine. Fiat, Daimler, and Chrysler, the lineage of RAM, all had shares in war production. Even the battle of brands for market share, a feature of capitalism that comes to seem natural, relies on strategies and terminologies born of war, from "campaigns" to industrial "espionage." The innocence projected by Chipotle's *Food with Integrity* campaign offered an exit from the machinery for killing so poignantly captured at Crow Foods. When we speak of agrarian resistance, then, we refer, in part, to movements against war.

As the third decade of the twenty-first century begins, the weaponization of food has been reclaimed as a strategy for killing and otherwise diminishing life with women and children the usual targets. In one year, 2016, the number of humans facing extreme hunger rose from 777 million to 825 million, with the increase mainly attributable to the weaponization of food in war zones through means including blockades and the refusal of regimes to allow the distribution of food aid to non-combatant populations.[58] Many of those in danger of starvation or diseases provoked by malnutrition in places including Syria and Yemen have had their homes destroyed and families or communities torn asunder. They lack not only the means for democratic participation, but the safety and security related to place that agrarians hold dear.

From the standpoint of agrarianism, this is unacceptable. From coalitions of agricultural scientists and activists to global movements to protect heirloom seeds, the world is awash with activity to secure food and farming futures, in spite of profound and worsening ecological changes. Across cultures, activities, and mythos, a latent coalition appears of those concerned with food and the future who would seek to reverse modern mistakes with resonant practices from a still-living past. With war exacerbating the effects of climate change, rendering populations more vulnerable and reducing the ability of other communities to respond, ending the weaponization of food is essential and feasible. This latent coalition of those engaged in farming and food, future and past, should it go live, would be difficult to resist. In 2018, the nongovernmental organization Human Appeal, in response to protracted conflicts where aid has been blocked and aid workers met with violence, issued a call to end the weaponization of food. Human Appeal called for the distribution of food by neutral parties, with others respecting this neutrality and not interfering with the distribution of food.[59]

Now we come to mobilization and the confrontation with decision in Buber's theory of prophecy as doom counteraction. Agrarians are precisely the people to end this catastrophic development, the weaponization of food. While American farmers have heroically "done their part" for the country in times of war and have benefited from wartime economies, especially the mobilizations of World Wars I and II, agrarian values and the work of the farm draws, too, from another reality, apart from killing machines. While it may take the form of a business for profit, or become, as Jefferson envisioned, a basis for full democratic participation, farming is at heart the cultivation of life. It is about the propagation of generations and a flourishing of living things. Despite international competition for markets and the other residuals of war in our economics, the weaponization of food for the purposes of killing is diabolical from an agrarian point of view.

Forming a "big tent" agrarianism to end this condition of malevolence requires the mobilization of myth, not to turn gardens to munition plants but something akin to the reverse, if not repurposing munition plants for crickets, reserving a sacred place for the garden no matter the circumstance, even when it is war. Considering how agrarian actions may be distributed widely, community gardens, urban agriculture, local food and the continued development of the organic revolution of Rodale and his contemporaries are parts of the picture. Advancements in food justice and fair trade are also essential. Perhaps a new-agrarian movement can find its broadest common company by articulating secure sustenance, the right to food, and a safe place to eat it—as not only core agrarian virtues, but inalienable human rights. An eighth-generation farmer up late at the computer working on the books, an app developer mixing a perfected blend of nutrients, a consumer choosing a post-Chipotle brand, and a merchant setting up at a local market anywhere in the world can all agree that part of the future of food should be secure access to it. Is this not why God made a farmer? In the time of developments from AI to climate change that seem able to push this species aside, perhaps we may reclaim a common humanity through a retold agrarian myth that begins with the comfort of security in food.

NOTES

Introduction

1. Freyfogle, "Introduction."
2. The global environmental and human health effects of industrial meat production are particularly detrimental. See Gerber et al., *Tackling*.
3. Brewster, "Toward"; Vigden, *Food Literacy*.
4. Alkon and Agyeman, "Introduction"; Broad, *More than Just Food*, 1–15; Fiskio, "Cultivating"; M. White, "D-Town Farm."
5. Thomson, "Big Food," 1–7.
6. Major, *Grounded*, 3.
7. Major, *Grounded*, 19. See also W. K. Kellogg Foundation, *Perceptions of Rural America*, 2002, https://www.wkkf.org/resource-directory/resource/2002/12/perceptions-of-rural-america; Conlogue, *Working*, 3–24. Several late twentieth-century survey studies in rural sociology also attest to the pervasive acceptance of agrarian values and ideology. For a review, see Dalecki and Coughenour, "Agrarianism in American Society."
8. Major, *Grounded*, 19–20.
9. Pollan, "Naturally."
10. Pollan, "Naturally."
11. Bladow, "Milking It," 15–16; Brasier, "Ideology"; Browne et al., *Sacred Cows*, 11–18, 141–43; Goldman and Dickens, "Selling"; Guthman, *Agrarian Dreams*, 1–22, 172–86; Kelsey, "Agrarian Myth."
12. Greer and Bruno, *Greenwash*.
13. We borrow this terminology of nodal points, moments, and discursive articulation from Laclau and Mouffe, *Hegemony*, 93–145.
14. Wirzba, "Introduction," 4
15. On the historical and political emergence of this captive government, or corporate-state governance, in US food and agriculture, see Hooks, "From an Autonomous.
16. Freyfogle, "Introduction"; Major, *Grounded*, 1–61; P. Thompson, *Agrarian Vision*, 1–17; Wirzba, "Introduction," 5–16.
17. Rushing, "Mythic Evolution," 291; Peterson, "Telling," 292; Retzinger, "Cultivating."
18. Buell, *Environmental*, 44.
19. Berry, *What Are People For?*, 145; Berry, *Unsettling*, 39.
20. Berry, "Agrarian Standard," 26.
21. Jackson, *Consulting*, 19–44; Jackson, *Nature*, 93–114.
22. Major, *Grounded*, 13, 26–28.
23. Wirzba, "Introduction," 4.
24. Freyfogle, "Introduction," xiv–xvii.
25. Major, *Grounded*, 149–172; K. Smith, "Wendell Berry's Feminist Agrarianism."

26. For example, see Carlson, *Natural Family*; Freyfogle, *Agrarianism*; Logsdon, *Mother*; Peters, *Wendell Berry*; K. Smith, *Wendell Berry*.

27. Kazin, *Populist Persuasion*; Burkholder, "Kansas Populism."

28. Allan Carlson examines agrarian voices ranging from Liberty Hyde Bailey and back-to-the-land advocate Ralph Borsodi, to the mid-twentieth century advocacy of Roman Catholic priest Luigi Lugutti, and the late twentieth and early twenty-first century's farmer/writer Wendell Berry. See Carlson, *New Agrarian Mind*.

29. See Gordon and Hunt, "Reform."

30. On the urban-cosmopolitan academic, see Major, *Grounded Vision*, xi; and V. Hanson, *Other Greeks*, 178. Also see Zencey, "Rootless Professors."

31. Motter and Singer, "Review Essay," 451.

32. Hughes, *Myths*, 2.

33. H. Smith, *Virgin Land*; Hofstadter, *Age*; Slotkin, *Regeneration*; Slotkin, *Fatal*; Slotkin, *Gunfighter*.

34. Hofstadter, *Age*, 24.

35. Hofstadter, *Age*, 3–59; Hofstadter's conceptualization of agrarian myth derives in part from Eisinger, "Freehold Concept." Also see Inge, *Agrarianism*, xiii-xx. For a more recent but similar description of American agrarian myth, see Heinze, "From Virgil," 6.

36. Crowley, *Toward a Civil Discourse*, 26.

37. Major, *Grounded*, 2–61.

38. V. Hanson, *Other Greeks*, xi–xxiv, 106–08.

39. Jefferson, "Commerce," 818.

40. Jefferson, "Commerce," 818.

41. Thompson, *Agrarian Vision*, 47.

42. Jefferson, "Commerce" 818; Montmarquet, *Idea*, 90.

43. McCoy, *Elusive Republic*, 76–85, 95–104.

44. R. Hanson, *Democratic Imagination*, 81, 129.

45. P. Johnstone, "Old Ideals," 117; Hofstadter, *Age*, 23.

46. Hofstadter, *Age*, 28.

47. Hofstadter, *Age*, 25.

48. Hofstadter, *Age*, 25.

49. Hofstadter, *Age*, 24.

50. Slotkin, *Gunfighter*, 15–16; Slotkin, *Fatal*, 52–68; 81–110.

51. Slotkin, *Fatal*, 52.

52. Slotkin, *Fatal*, 52.

53. On United States farmers, agrarian myth, and the Populist and Progressive eras, see Hofstadter, *Age*, 60–148.

54. Slotkin, *Fatal*, 52.

55. H. Smith, *Virgin*, 123–260.

56. Slotkin, *Fatal*, 52.

57. White, "Frederick Jackson Turner," 11, 27.

58. White, "Frederick Jackson Turner," 11–13.

59. Turner, "Significance," 43–45.

60. White, "Frederick Jackson Turner," 15.

61. White, "Frederick Jackson Turner," 13.

62. Slotkin, *Fatal*, 69–72.

63. H. Smith, *Virgin*, 250–60, 155–64, 211–49.
64. Smith, *Virgin*, 123.
65. Smith, *Virgin*, 250–60.
66. Smith, *Virgin*, 124.
67. Smith, *Virgin*, 135.
68. Hofstadter, *Age*, 38.
69. Hofstadter, *Age*, 24.
70. Hofstadter, *Age*, 30–31.
71. Montmarquet, *Idea*, viii.
72. Browne et al., *Sacred*, 11–18. Other scholars make similar arguments; see Brasier, "Ideology"; and Kelsey, "Agrarian Myth."
73. Browne et al., *Sacred*, 15.
74. Browne et al., *Sacred*, 15.
75. Browne et al., *Sacred*, 15.
76. McGuire, "Mythic Rhetoric," 13.
77. Foucault, "Of Other," 333.
78. Foucault, "Of Other," 331.
79. Foucault, "Of Other," 334.
80. Foucault, "Of Other," 334.
81. See Flores, "Creating."
82. Burke, "Ideology," 200.
83. Burkholder, "Kansas," 304.
84. Browne et al., *Sacred*; Burkholder, "Kansas"; Dorsey, "Frontier Myth"; Harter, "Masculinity(s)"; Mooney and Scott A. Hunt, "Repertoire"; Motter and Singer, "Review Essay"; T. Peterson, "Will"; T. Peterson, "Jefferson's Yeoman"; T. Peterson, "Telling"; Retzinger, "Cultivating"; Short, "'Hello Americans'"; Walter, "'Curious Blend.'"
85. Beus and Dunlap, "Endorsement"; Buttel and Flinn, "Sociopolitical"; Buttel and Flinn, "Sources"; Carlson and McLeod, "Comparison"; Craig and Phillips, "Agrarian Ideology"; Dalecki and Coughenour, "Agrarianism"; Flinn, "Agrarian Values"; Flinn and Johnson, "Agrarianism among Wisconsin Farmers"; Mulnar and Wu, "Agrarianism, Family Farming"; Singer and de Sousa, "Sociopolitical Consequences"; P. Smith, "Agrarian Ideology."
86. Major, *Grounded*, 149–91.
87. Baudrillard, *Simulacra*.

Chapter 1

1. "Women in Bread Riot." The Food Riots are also recounted in Weiss, *Fruits of Victory*, 23–24.
2. "Women in Bread Riot."
3. Marcy, "Food Riots."
4. Marcy, "Food Riots."
5. International News Service, cited in Marcy, "Food Riots."
6. Nestle and Dalton, "Food Aid."
7. Veit, *Modern Food*, 3.
8. Veit, *Modern Food*, 6.

9. Camporesi, *Bread of Dreams*.
10. Trentmann, "Coping," 13.
11. Pack, *War Garden*, 3.
12. Broadberry and Harrison, "Economics," 12–13.
13. Pack, *War Garden*, 5.
14. See Hawkins, *Starvation Blockades*.
15. Davis, *Home Fires*, 180–187.
16. Vincent, *Politics of Hunger*.
17. Grebler, *Cost*. The shortage of food also impacted the German armed services, as, for instance, troops from agricultural backgrounds were given leave for farm work and then accused of hoarding food brought from home upon their return. Erich Maria Remarque's *All Quiet on the Western Front* includes episodes of soldiers risking their lives for food, including attacking French trenches in hopes of acquiring corned beef. See Sass, "WWI Centennial."
18. For a detailed recounting, see Pipes, *Russian Revolution*. Orlando Figes cites the famine of 1891 as another key event leading to revolution, as it showed the inability of the Tsarist administration and Russian economy in feeding the populace; see Figes, *People's Tragedy*.
19. Outrage over the Assad regime in Syria using starvation as a weapon has spurred recounting of starvation in the region during World War I. See, for instance, Ciezadlo, "War."
20. For an account of Persian starvation in World War I, see Majd, *Great Famine*.
21. Wiebe, *Search for Order*, xiii.
22. Wiebe, *Search for Order*, 11–12.
23. In 1903, arrivals surpassed 850,000; in 1905, 1 million. See Lund, "Boundaries."
24. Wells, "Mobilizing."
25. Patten, *New Basis*.
26. Treidler is quoting Wilson. See Treidler, "Hunger."
27. W. Wilson, "Extract."
28. Wilson, "Extract."
29. Wilson, "Extract."
30. Capozzola, *Uncle Sam*, 8.
31. Veit, *Modern Food*, 4.
32. Weiss, *Fruits of Victory*, 23.
33. Conversely, drafting men for agricultural work was also considered. For a detailed examination of these ideas and discussions at the federal and state levels, see, Weiss, *Fruits of Victory*, 24–30.
34. Eighmey, *Food Will Win*, 22.
35. Veit, *Modern Food*, 15.
36. Hoover cited in Mullendore, *History*, 41.
37. Eighmey, *Food Will Win*, 29.
38. Veit, *Modern Food*, 19.
39. Eighmey, *Food Will Win*, 32.
40. Veit, *Modern Food*, 19, 33.
41. Capozzola, *Uncle Sam*, 7.
42. Capozzola, *Uncle Sam*, 7.
43. Hoover, "Introduction," cited in Kingsbury, *For Home*, 37.

44. Knauth, "Farmer's Income."
45. Zeiger, *America's Great War*, 73.
46. Zeiger, *America's Great War*, 30.
47. Zeiger, *America's Great War*, 49.
48. Creel, *How We Advertised*, 12.
49. Creel, *How We Advertised*, 5.
50. Wiebe, *Search for Order*, 296.
51. Axelrod, *Selling*, 66.
52. Bernays provides his viewpoint on the American propaganda of World War I and the work of the CPI in *Propaganda*.
53. Creel, *How We Advertised*, 184–99.
54. Capozzola, *Uncle Sam*, 5.
55. Capozzola, *Uncle Sam*, 5.
56. Capozzola, *Uncle Sam*, 5.
57. Capozzola, *Uncle Sam*, 5.
58. Capozzola, *Uncle Sam*, 5.
59. James, "Reading," 2.
60. Sontag, "Introduction."
61. Landsberg, *Prosthetic Memory*, Kindle location 39.
62. Landsberg, *Prosthetic Memory*, Kindle location 39.
63. Landsberg, *Prosthetic Memory*, Kindle location 48.
64. Landsberg, *Prosthetic Memory*, Kindle location 150.
65. Axelrod, *Selling*, 138.
66. James, "Reading," 4.
67. James, "Reading," 10.
68. Axelrod, *Selling*, 141.
69. Axelrod, *Selling*, 141.
70. Schnapp, "Epilogue," 372.
71. Schnapp, "Epilogue," 373.
72. Leach, *Land of Desire*, 4.
73. Leach, *Land of Desire*, 4.
74. Hunter, "Story Behind."
75. Foner, *Story of American Freedom*, 148.
76. "Indian Queens and Indian Princesses: Allegorical Representations of America." *Native American History at the Clements Library*, web exhibit, University of Michigan, https://clements.umich.edu/public-programs/exhibits/#online-exhibits.
77. Kern, "Embodiment."
78. Dewey, *Art of Ill Will*, 13.
79. Schnapp, "Epilogue," 373.
80. Chambers, "Food Will Win."
81. Lasswell, *Propaganda*, 68.
82. Weschler, "'Destroy.'"
83. While the British propagandists accused Germans of eating nettles, there is some evidence that the Germans used nettles in the production of sandbags and rucksacks during the war, as the British and Americans controlled the cotton trade; see "A Heavy Nettle Hero," *The Independent*, August 3, 2004. https://www.independent.co.uk/news/science/a-heavy-nettle-hero-49944.html.

84. After the war, the story was found to be untrue. Joachim Neander and Randal Marlin have shown that it hinged on a willful mistranslation of German terms referring to the cadavers of horses, not humans. These authors suggest that cynicism left by the debunking of this story contributed to disbelief of the initial accounts of the Nazi death camps. See Neander and Marlin, "Media."
85. Schaffer, *America*, 192.
86. Pack, *War Garden*, 112.
87. Pack, *War Garden*, 112.
88. Pack, *War Garden*, 112.
89. Pack, *War Garden*, 113.
90. Pack, *War Garden*, 112–113.
91. Pack, *War Garden*, 23.
92. Pack, *War Garden*, 23.
93. Pack, *War Garden*, 7.
94. Pack, *War Garden*, 7.
95. Pack, *War Garden*, 8.
96. Pack, *War Garden*, 10.
97. Pack, *War Garden*, 18.
98. Pack, *War Garden*, 35.
99. Pack, *War Garden*, 33.
100. Pack, *War Garden*, 33.
101. Pack, *War Garden*, 24.
102. Barney, "War Gardens Over the Top," in Pack, *War Garden*, 12.
103. Pack, *War Garden*, 9.
104. Pack, *War Garden*, 28.
105. Pack, *War Garden*, 109.
106. Pack, *War Garden*, 46.
107. Pack, *War Garden*, 48.
108. Pack, *War Garden*, 49–50.
109. Pack, *War Garden*, 13.
110. Pack, *War Garden*, 14.
111. Pack, *War Garden*, 53.
112. Pack, *War Garden*, 97.
113. Pack, *War Garden*, 96–97.
114. Pack, *War Garden*, 73–74.
115. Pack, *War Garden*, 73.
116. Pack, *War Garden*, 66.
117. Barney, "War Garden Victorious," in Pack, *War Garden*, 14.
118. Pack, *War Garden*, 52.
119. Clinker and Dwyer, "Don't Waste."
120. Burke, "Definition."
121. Several studies take the development and characterization of this system as a theme, including Wiebe, *Search for Order*; Schaffer, *America*; Capolozza, *Uncle Sam*; and John F. McClymer, *War and Welfare*.
122. Veit, *Modern Food*, 2.
123. Veit, *Modern Food*, 2.
124. Pack, *War Garden*, 150.

125. Pack, *War Garden*, 150.
126. Pack, *War Garden*, 150.
127. Deaths from the influenza virus may be partly attributable to the arrangements of food in modern warfare. Epidemiologists have traced an initial outbreak to a hospital and staging camp near the western front, where the virus jumped from hogs massed for slaughter to humans. See Byerly, *Fever of War*; and "U.S. Military."

Chapter 2

1. Phillips, *This Land*, 36.
2. Sanderson, "People," 5.
3. Taylor, "Interpretation," 6.
4. "Commission on Country Life," *Liberty Hyde Bailey: A Man for All Seasons*. Cornell University Library, http://rmc.library.cornell.edu/bailey/commission/.
5. Phillips, *This Land*, 36.
6. Danbom, *Born*, 187.
7. Hurt, *American Agriculture*, 221.
8. Calhoun, "Introduction," 8.
9. Fraser, "Rethinking," 67.
10. Squires, "Rethinking," 463.
11. Squires, "Rethinking," 463.
12. Habermas, *Communication*, 178, italics in original.
13. Habermas, *Communication*, 178.
14. Cohen, "Procedure," 95.
15. Benhabib, "Toward a Deliberative Model," 68.
16. Payrow Shabani, *Democracy*, 124.
17. Habermas, *Communication*, 178–79.
18. Baldwin, "Professional," 3.
19. Lowden, "Presidential Address," 3.
20. Poe, "Democracy," 7.
21. Poe, "Democracy," 7.
22. Lowden, "Presidential Address," 3–4.
23. Lowden, "Presidential Address," 4.
24. Danbom, *Born*, 189.
25. Culp, "End of the Rural School," 5.
26. Butterfield, *Farmer*, 226.
27. Butterfield, *Farmer*, 225.
28. M. Wilson, "Education," 4.
29. Hurt, *American Agriculture*, 267.
30. Culp, "End of the Rural School," 5.
31. American Country Life Association, "Declaration," 11.
32. American Country Life Association, "Declaration," 11.
33. Lowden, "Presidential Address," 4.
34. Lowden, "Presidential Address," 4.
35. Taber, "Some National Issues," 11.
36. Tigert, "Objectives of Rural Education," 9.
37. Butterfield, *Farmer*, 208.

38. Butterfield, *Farmer*, 208.
39. American Country Life Association, "Objectives," 24.
40. Perelman and Olbrechts-Tyteca, *New Rhetoric*, 190.
41. Perelman and Olbrechts-Tyteca, *New Rhetoric*, 412.
42. Culp, "End of the Rural School," 5.
43. Theiss, "Victory Gardens," 57.
44. American Country Life Association, "Platform," 2.
45. Butterfield, *Farmer*, 55.
46. Holt, "More Hell," 5.
47. Butterfield, *Farmer*, 15–16, italics in original.
48. Ward, "On Our Way," 4.
49. Laclau and Mouffe, *Hegemony*, 105.
50. Ward, "On Our Way," 4.
51. M. Wilson, "Education," 4.
52. M. Wilson, "Education," 4.
53. Frame, "American Country," 6.
54. Butterfield, *Farmer*, 212.
55. Butterfield, *Farmer*, 211.
56. Terpenning, "Toward Greater Security," 7–8.
57. Motter and Singer, "Review Essay," 441.
58. L. Bailey, *Outlook*, 96.
59. Liberty Hyde Bailey as quoted in Frame, "American Country," 7.
60. L. Bailey, *Country-Life*, 204–5.
61. Rutherford, "Types," 11.
62. Brunner, "Our Challenge," 14.
63. L. Bailey, *Outlook*, 92, italics in original.
64. L. Bailey, *Outlook*, 92.
65. American Country Life Association, "Back," 15.
66. Melvin, "Local," 3.
67. M. Wilson, "Rural Discussion," 6.
68. L. Bailey, *Outlook*, 92.
69. Allen, *Talking*, xvi.
70. Allen, *Talking*, 186.
71. Bennett, "Passing," 39.
72. M. Wilson, "Education," 5.
73. M. Wilson, "Education," 5.

Chapter 3

1. This group is also known at the Nashville Agrarians, as several were affiliated with Vanderbilt University.
2. Berry, *Unsettling*, 13.
3. One of us, Stephanie, has previously published on the rhetoric of the Southern Agrarians and its connections to contemporary food movements. See Grey, "Gospel."
4. Singer, "Visualizing."
5. Burke, *Counter-Statement*, 168.

6. Sweet, "Economy."
7. C. Johnstone, "Thoreau."
8. See Twelve Southerners, *I'll Take*; and Agar and Tate, *Who Owns*.
9. Conkin, *Southern Agrarians*, 20–25.
10. Twelve Southerners, "Introduction," xliv.
11. Twelve Southerners, "Introduction," xliv.
12. Moss, "Reconceptualization."
13. For analyses of conversion as a rhetorical form see Griffin, "Rhetoric"; Lynch, "Prepare"; and D. Bailey, "Enacting."
14. Weaver, *Ideas*. Interestingly, Weaver idealized Southern culture for what he perceived as its non-materialistic ideology, a concept that would deeply influence his own rhetorical theories on the relationship between ethics and persuasion. See Weaver, *Southern Tradition*.
15. See Anderson, *Imagined Communities*. The Southern Agrarians would be caught between the nexus of projecting their ideal community as a fiction and the process of recovery. While these agrarians viewed themselves as recovering the Old South, they would gradually be forced to come to terms with forces similar to those described by Anderson, in which they transformed a fragmented culture into a coherent cultural body—a much different rhetorical process.
16. Burke, *Rhetoric*, 142–46.
17. Davidson, "Mirror," 31.
18. Lytle, "Hind," 202.
19. Malvasi, *Unregenerate South*, 35.
20. Ransom, "Reconstructed," 15.
21. Twelve Southerners, "Introduction," xlii.
22. Twelve Southerners, "Introduction," xlviii.
23. Davidson, *Attack*.
24. Twelve Southerners, "Introduction," xlvi.
25. Duck, "Postsouthern."
26. Twelve Southerners, "Introduction," xliv.
27. Davidson, "Mirror," 51.
28. Twelve Southerners, "Introduction," xlv.
29. Davidson, "Mirror," 40.
30. Davidson, "Mirror," 30–31.
31. Grammar, *Pastoral*.
32. P. Murphy, *Rebuke*.
33. It is important to note that only social scientist Henry Nixon addressed the issue of racism in *I'll Take My Stand*. He was not solicited for contribution in later work. The Southern Agrarians initially did not address the issue of slavery as a moral failure, but instead sublimated its rejection into their evolving vision of the authentic Southern heterotopia. It was not until after the 1930s that some of these thinkers acknowledged the abhorrent ethical nature of the slave system and the cultural damage wrought by racist ideology in the South.
34. Karanikas, *Tillers*.
35. See Rushing, "Rhetoric"; and Carpenter, "Frederick Jackson Turner."
36. See Balthrop, "Culture." See also S. Smith, *Myth*, for an analysis of the historical implications of the lost cause mythology.

37. Bingham and Underwood, "Introduction," 1–35.
38. It should be noted that blood-and-soil arguments are frequently linked to fascism and genocide. See Kiernan, *Blood*. The Southern Agrarians' anti-federalist tendencies prevented them from falling into this category; we discuss this in the conclusion to the chapter.
39. Malvasi, *Unregenerate*, 142.
40. Tate, "Remarks," 162.
41. Tate, "Remarks," 166.
42. Burgmann and Milner, "Futures."
43. Bram, "Narrative."
44. Ransom, "Reconstructed," 5.
45. Malvasi, *Unregenerate*, 174.
46. Lytle, "Hind," 209.
47. Hodin, "Mechanisms."
48. Owsley, "Irrepressible," 65.
49. Tate, "Remarks," 175.
50. Tate, *Fathers*.
51. Adams, "'Painfully.'"
52. Ransom, "Reconstructed," 7.
53. Ransom, "Reconstructed," 9.
54. Tate, "Remarks," 172.
55. Lytle, "Hind," 211.
56. Tate, "Remarks," 168.
57. Weaks-Baxter, *Reclaiming*.
58. See Voelz, *Transcendental*. For specific influences on Thoreau's writings, see Newman, "Thoreau's Natural Community." These movements differed from the Southern Agrarians in that they persisted in the realization of utopian ideals, ultimately failing to generate a sustained counter-space.
59. Burke, *Rhetoric*, 22
60. Twelve Southerners, "Introduction," xlvii.
61. Lytle, "Hind," 207.
62. Ransom, "What," 248.
63. Ransom, "What," 249.
64. Ransom, "What," 251–52.
65. Lytle, "Small," 319.
66. Lytle, "Small," 321.
67. Ransom, "What," 245.
68. Romines, *Home Plot*.
69. Mary Shattuck Fisher, "The Emancipated," 413.
70. Allen Tate, "Notes," 111.
71. Frank Lawrence Owsley, "The Foundations," 77.
72. John Crowe Ransom, "Corporate," 95.
73. Ransom, "What," 250.
74. Davidson, "That," 158.
75. For example, see Rushkoff, *Life Inc*.
76. Davidson, "That," 153.
77. Ransom, "What," 243.

78. Kiernan, *Blood*, 23.
79. Zagacki, "Preserving."
80. Nader, *Unstoppable*.
81. For examples, see Winne, *Food Rebels*; and Reed, *Rebels*.

Chapter 4

1. Rodale, "Introduction."
2. Rodale, *Pay*.
3. The publisher's synopsis featured on *Pay Dirt*'s dust jacket asserts that the book is the first to be written and published in the United States with a focus on organic farming and gardening.
4. Dobrow, *Natural*, 18; Gross, *Our Roots*, 69; Rodale, "20 Years," 17.
5. C. Jackson, *J.I.*, 30.
6. C. Jackson, *J.I.*, 114.
7. C. Jackson, *J.I.*, 114.
8. Gross, *Our Roots*, 69.
9. Gross, *Our Roots*, 55–56, 62–63; Obach, *Organic*, 37, 22–38; S. Peters, "Organic," 252–58.
10. Paull, "Biodynamic"; Paull, "Betteshanger"; S. Peters, "Organic," 252–54.
11. Steiner, who saw unity between the spiritual and natural worlds, is also known in part for the philosophy of Anthroposophy, which provides a basis in several fields for the practical application of natural-science methods to spiritual experience. McDermott, *Essential Steiner*, 3–11, 392–95.
12. Biodynamic Association, "Who."
13. Obach, *Organic*, 37.
14. Obach, *Organic*, 37.
15. Some well-known back-to-the-land advocates include Edmund Morris, author of the late nineteenth century and early twentieth century sensation *Ten Acres Enough*; urban community-garden champion Bolton Hall; and later, Lewis Mumford and Ralph Borsodi, well-known voices for urbanites heading back to the country in the years before and during and Great Depression. In *Pay Dirt*, Rodale quotes Borsodi, who experimented with Howard's compost methods and published at least one article in *Organic Farming and Gardening*. Rodale, *Pay Dirt*, 90–91; Brian Obach has noted Borsodi's influence on Rodale. See Obach, *Organic*, 37. On this movement, also see Danbom, "Past," 10–11; Prody, "Tracing."
16. Shprintzen, *Vegetarian*; Carson, *Cornflake*, 15–27; Haydu, "Cultural."
17. Carson, *Cornflake*, 43–60; A. Smith, *Eating*, 30–33; also see Graham, *Lectures*.
18. On these changes and others, see Gates, *Farmer's Age*.
19. Danbom, "Past," 6–8; Rasmussen, "Civil War."
20. Danbom, "Past," 12–13; Rasmussen and Stone, "Toward."
21. C. Jackson, *J.I.*, 25.
22. Gross, *Our Roots*, 33–37, 77.
23. Gross, *Our Roots*, 56–58.
24. Gross, *Our Roots*, 33–34.
25. Rodale Institute, "About Us," accessed June 26, 2016, http://rodaleinstitute.org/about-us/mission-and-history/.

26. Paull, "Betteshanger," 14.
27. Day, "Ehrenfried."
28. Conford, *Origins*, 100. For more on the life and work of Howard and his connection to the organic movement, see Heckman, "History."
29. Gross, *Our Roots*, 55–56.
30. Paull, "Betteshanger," 14; Obach, *Organic*, 22–38; S. Peters, "Organic," 252–58.
31. A. Smith, *Eating*, 193.
32. A. Smith, *Eating*, 193.
33. A. Smith, *Eating*, 193.
34. Danbom, "Past," 12–13; Hooks, "From," 29–43. Something akin to this happened in Great Britain as well. See Conford, *Origins*, 29.
35. Gross, *Our Roots*, 66.
36. S. Peters, "Organic," 257.
37. Perenyi, "Apostle."
38. C. Jackson, *J.I.*, i.
39. C. Jackson, *J.I.*, 29.
40. "J.I. Rodale Dead"; Greene, "Guru."
41. McGrath, "Bizarre."
42. On Rodale's cool temper and his spats with drama critics, see C. Jackson, *J.I.*, 239. Regarding the FTC, see C. Jackson, *J.I.*, 148–90. On the AMA, see C. Jackson, *J.I.*, 140–44; and "J.I. Rodale Dead," 42.
43. "J.I. Rodale Dead," 42; C. Jackson, *J.I.*, 148–90.
44. C. Jackson, *J.I.*, 123.
45. Miller, *Errand*, 6.
46. Bercovitch, *American Jeremiad*.
47. Darsey, *Prophetic*, 28, 16, 18.
48. For example, see Bercovitch, *American Jeremiad*; Howard-Pitney, *Afro-American*; Johannesen, "Ronald"; J. Murphy, "Time."
49. Howard-Pitney, *Afro-American*, 11.
50. Buell, *Environmental*; Check, "Mortification"; Opie and Elliot, "Tracking"; Rosteck and Frentz, "Myth"; Slovic, "Epistemology"; Wolfe, "Ecological."
51. Opie and Elliot, "Tracking," 9–10, 33–35.
52. Wolfe, "Ecological," 11, 12.
53. Dalecki and Coughenour, "Agrarianism"; Eisinger, "Freehold"; V. Hanson, *Other Greeks*; Hofstadter, *Age*, 3–59; Inge, *Agrarianism*, xiii-xx; Peterson, "Telling," 293–97; H. Smith, *Virgin*, 123–260.
54. Eisinger, "Freehold," 42–47; Hofstadter, *Age*, 23–59; Montmarquet, *Idea*, 25–38.
55. Opie and Elliot, "Tracking," 10, 32.
56. On conversion as a rhetorical form, see Griffin, "Rhetoric"; Lynch, "Prepare"; D. Bailey, "Enacting."
57. Walsh, *Scientists*, 5.
58. C. Jackson, *J.I.*, 123–37.
59. In *Pay Dirt*, Rodale relies heavily on Howard, *Agricultural Testament*; Pfieffer, *Biodynamic*; Balfour, *Living*; and Northbourne, *Look*, among others.
60. Rodale, *Pay*, vii.
61. Rodale, *Pay*, 11.
62. Rodale, *Pay*, 100.

63. Rodale, *Pay*, 101.
64. Rodale, *Pay*, 103.
65. Rodale, *Pay*, 66.
66. Rodale, *Pay*, 3.
67. Rodale, *Pay*, 11.
68. Rodale, *Pay*, 25.
69. Rodale, *Pay*, 65.
70. Rodale, *Pay*, 112.
71. Rodale, *Pay*, 113.
72. Rodale, *Pay*, 23–24.
73. Rodale, *Pay*, 181.
74. Rodale, *Pay*, 239.
75. Rodale, *Pay*, 234.
76. Various ancient and modern agrarian writers have written of agriculture as a way of life. Our use of the specific expression "agriculture as culture" comes from Wendell Berry. Among other places, see Berry, *Unsettling*, 39. On agrarianism as happiness, see, for example, Eisinger, "Freehold," 43–44.
77. Rodale, *Pay*, 234.
78. Rodale, *Pay*, 234.
79. Rodale, *Pay*, 234.
80. Dorsey, "Frontier," 7.
81. Rodale, *Pay*, 130.
82. Rodale, *Pay*, 114.
83. Rodale, *Pay*, 13.
84. Burkholder, "Kansas"; Hofstadter, *Age*, 60–93.
85. Rodale, *Pay*, 100.
86. Rodale, *Pay*, 201.
87. Rodale, *Pay*, 204; Eisinger, "Freehold," 43–44.
88. Rodale, *Pay*, 204–5. Further demonstrating his fervent belief that the land is the root of sacred civic virtues that cannot be reached by artificial means, Rodale also critiques other trends toward a landless citizenry such as hydroponic greenhouse gardening.
89. On agrarianism as moral legitimation, see Browne et al., *Sacred*, 13. Hofstadter, *Age*, 3–59; Mooney and Hunt, "Repertoire."
90. Rodale, *Pay*, 158.
91. Rodale, *Pay*, 220, 222.
92. Rodale, *Pay*, 158.
93. Rodale, *Pay*, 171.
94. Rodale, *Pay*, 172, 24.
95. Rodale, *Pay*, 173.
96. Rodale, *Pay*, 171.
97. Rodale, *Pay*, 172.
98. Rodale, *Pay*, 172.
99. Rodale, *Pay*, 171.
100. Rodale, *Pay*, 231.
101. Rodale, *Pay*, 113.
102. Rodale, *Pay*, 51.

103. Rodale, *Pay*, 83.
104. Rodale, *Pay*, 56–57.
105. Rodale, *Pay*, 221.
106. Rodale, *Pay*, 131.
107. Rodale, *Pay*, 180.
108. Rodale, *Pay*, 221.
109. Rodale, *Pay*, 65.
110. Rodale, *Pay*, 68.
111. Rodale, *Pay*, 72.
112. Rodale, *Pay*, 223.
113. Rodale, *Pay*, 224.
114. Rodale, *Pay*, 203.
115. Rodale, *Pay*, 113.
116. Rodale, *Pay*, 232.
117. Rodale, *Pay*, 203.
118. Rodale, *Pay*, 25.
119. Rodale, *Pay*, 231.
120. Rodale, *Pay*, 231.
121. Rodale, *Pay*, 232.
122. Johnstone, "Old Ideals," 117.
123. Hofstadter, *Age*, 63.
124. Opie and Elliot, "Tracking," 9–10, 33–35. This difference between ecological prophets does not, however, entirely explain why Carson's *Silent Spring* jeremiad has become so widely known compared to Rodale's in *Pay Dirt*. Carlton Jackson posits that despite offering some of the same critiques of chemical agriculture's discourse of progress decades before Rachel Carson, Carson's timing, message, and persona may explain why *Silent Spring* received more publicity. Jackson suggests that "Carson's technique is livelier and more learned" than Rodale's, and the public was more receptive in 1962 to a book of this type than it was in 1945, due possibly in part to television as well as Carson having won over many scientists. Also, as highlighted by Jackson, Rodale's image had been that of a guru, and because some perceived him as overly enthusiastic about natural food supplements, he took on a more mystic persona that some audiences distrusted (C. Jackson, *J.I.*, 34–35).
125. Belasco, *Appetite*, 71–73.
126. Greene, "Guru," SM30; C. Jackson, *J.I.*, 27; "J.I. Rodale Dead," 42.
127. S. Peters, "Organic," 108.
128. S. Peters, "Organic," 108. For a slightly different view, see Conford, *Origins*, 102.
129. S. Peters, "Organic," 115.
130. Sayre, "Politics," 38–40.
131. Sayre, "Politics," 38–40.

Chapter 5

1. The first author previously took part in this scholarly conversation about the SCFarmers. See Singer, "Visualizing." Although a general interest in new agrarianism carries from this praxis forum essay into the present chapter, the analysis and arguments in this chapter are entirely original. Other environmental and food

studies on the SCFarm include the following: Broad, "Ritual"; Emmett, "Community"; Enck-Wanzer, "Race"; Foust, "Considering"; Irazábal and Punja, "Cultivating"; LaGreco and Leonard, "Building; Mares and Peña, "Urban"; Mares and Peña, "Environmental"; Retzinger, "Eleven Miles."

2. Irazábal and Punja, "Cultivating," 1.
3. Hoffmann, "History"; "Synopsis: *The Garden*."
4. Footage of some SCFarmers speaking alongside Kennedy at film showings is found in the "Extras" in the DVD version of the film.
5. Ono and Sloop, "Critique," 20.
6. Enck-Wanzer, "Race," 364; Enck-Wanzer now goes by the surname Wanzer-Serrano.
7. Mares and Peña, "Urban," 246.
8. Boyd, "Deep."
9. Ono and Sloop, "Critique," 23.
10. Ono and Sloop, "Critique," 23.
11. Rushing, "Mythic," 291.
12. On this point, also see Enck-Wanzer, "Race," 366.
13. Mooney and Hunt, "Repertoire."
14. Burkholder, "Kansas"; Kazin, *Populist*.
15. T. Peterson, "Telling," 303–05; T. Peterson, "Will," 14, 18–20.
16. Milstein et al., "Communicating," 496–97, 488.
17. Irvine et al., "Community"; Hanna and Oh, "Rethinking."
18. Saldivar-Tanaka and Krasny, "Culturing."
19. Bassett, "Reaping"; Hynes, *Patch*; Warner, *To Dwell*; Irvine et al., "Community"; Hanna and Oh, "Rethinking."
20. Francis et al., *Community*; Schmelzkopf, "Urban."
21. Saldivar-Tanaka and Krasny, "Culturing," 399.
22. Saldivar-Tanaka and Krasny, "Culturing," 399.
23. Saldivar-Tanaka and Krasny, "Culturing," 399.
24. Malakoff, "What Good"; Bicho, "Simple."
25. Larson et al., "Neighborhood"; Sloane et al., "Improving."
26. Lyson, *Civic*, 63.
27. Armstrong, "Survey"; Baker, "Tending," 305; K. Day, "Active"; Kuo and Sullivan, "Environment"; Saldivar-Tanaka and Krasny, "Culturing"; Tranel and Handlin, "Metromorphosis"; Voicu and Benn, "Effect."
28. Cenzatti, "Heterotopias."
29. Emmett, "Community," 68.
30. Broad, "Ritual," 24.
31. Hoffmann, "History."
32. Bullard, "Threat"; Peeples, "Trashing."
33. Hoffmann and Petit, "14 Acres."
34. Irazábal and Punja, "Cultivating," 1.
35. Broad, "Ritual," 24.
36. Broad, "Ritual," 25.
37. Broad, "Ritual," 25.
38. Irazabal and Punja, "Cultivating," 1–2.
39. Mares and Peña, "Environmental," 206.

40. Hecht, "Future."
41. Hoffmann, "L.A. Urban."
42. Hoffmann, "History."
43. Hoffmann, "History"; Hoffmann, "L.A. Urban."
44. Hecht, "Future"; Hoffmann, "History."
45. Hoffmann, "L.A. Urban"; Morain and Chung, "Westly."
46. Caldwell, "Judge."
47. Broad, "Ritual," 26.
48. Archibold, "Hollywood."
49. Broad, "Ritual," 26; Hopkins, "End."
50. "Offer Made to Purchase Farm," *KTLA News*, June 8, 2006, http://ktla.com/category/news/local-news/.
51. Hopkins, "End."
52. Gencer, "Reaping."
53. Jessica Hoffmann, "South Central Farm."
54. Kennedy described his purpose this way in his comments following a screening of the film and academic discussion panel regarding it (in which the first author participated) at the 2010 National Communication Association Convention in San Francisco, CA.
55. "Synopsis: *The Garden*."
56. "Synopsis: *The Garden*."
57. Nichols, *Introduction*, 172–79.
58. Nichols, *Introduction*, 31.
59. Nichols, *Introduction*, 31, 154–57, 167–71.
60. Nichols, *Introduction*, 156–57.
61. Ono and Sloop, "Critique," 25.
62. Burkholder, "Kansas," 294–95; also see Mooney and Hunt, "Repertoire," 184–88.
63. Mooney and Hunt, "Repertoire," 183–88. Mooney and Hunt document how agrarian uprisings going back to the colonial era rely on this appeal, or "producer frame," usually within broader discourses of anti-monopolistic economic fairness.
64. Hofstadter, *Age*, 24–25.
65. Berry, *Citizenship*, 64.
66. Berry, *Citizenship Papers*, 64.
67. Hoffmann, "History."
68. Laffey, *South Central Farm*.
69. On commodified discourse of space-making, see Ritzer, *McDonaldization*, 1–22, 97–112; on space-making discourse as xenophobic and racist territorialization, and national security state surveillance, see Shome, "Space."
70. Mares and Peña, "Urban," 246.
71. Mares and Peña, "Environmental," 207.
72. Hoffmann and Petit, "14 Acres."
73. On widely distributed private property "freeholding" as a key foundation of American agrarian myth, see Eisinger, "Freehold."
74. Lyson, *Civic*, 63.
75. Mignolo, "Delinking," 459.
76. Mignolo, "Delinking," 453–56.

77. Mignolo, "Delinking," 453.
78. Mares and Peña, "Urban"; Enck-Wanzer, "Race."
79. Ono and Sloop, "Critique," 23.
80. Broad, "Ritual," 20–21.
81. Eisinger, "Freehold," 42–44. Agrarianism as "good" land-use reflects in the agrarian myth's notion of the farmer as God's caretaker of the beneficent earth. See Hofstadter, *Age*, 24–25. Environmental stewardship has become a particularly prominent theme among new-agrarian mythmakers ranging from Liberty Hyde Bailey to Wendell Berry. Berry often uses the expression "kindly use."
82. Enck-Wanzer, "Race," 364.
83. Broad, "Ritual," 21.
84. Oravec, "Ideological," 383–84.
85. Oravec, "Ideological," 390.
86. Irazábal and Punja, "Cultivating," 2.

Chapter 6

1. Liquified Squid, "CNN."
2. CDC, "Multistate."
3. Houck, "43 Chipotle."
4. CDC, "Multistate."
5. Schumaker, "Chipotle's Food."
6. Schumaker, "Chipotle's Food."
7. Ferdman and Bhattarai, "There's a Crisis."
8. Berfield, "Inside."
9. Berfield, "Inside."
10. Abrams, "Chipotle Is Subpoenaed."
11. Abrams, "Chipotle Is Subpoenaed."
12. Abrams, "Chipotle Is Subpoenaed."
13. Berfield, "Inside."
14. CDC, "Burden."
15. Roberts, *End of Food*, 182.
16. As Deleuze and Guattari suggest in *A Thousand Plateaus*, subjectivity can be understood as a product of the shifting symbolic systems of late capitalism, with consumer desire now constituting the primary means through which the citizen is projected across the discursive field. See Deleuze and Guattari, *Thousand*.
17. Dickinson, "Joe's Rhetoric," 7–10.
18. Banet-Weiser, *AuthenticTM*, 9.
19. Johnston and Baumann, *Foodies*, 60.
20. Johnston and Baumann, *Foodies*, 61; italics in original.
21. Wickstrom, *Performing*.
22. Jenkins, *Special Affects*.
23. Jenkins, *Special Affects*, 189.
24. Dickinson, "Joe's Rhetoric," 5–27; Muniz and O'Guinn, "Brand"; Thompson and Arsel, "Starbucks."
25. Chipotle Mexican Grill, "Back."
26. Fitzgerald, "Sincerity."

27. Banet-Weiser, *AuthenticTM*, 10.
28. Laclau and Mouffe, *Hegemony*.
29. Camerini and Diviani, "Activism."
30. Ritzer, *McDonaldization*, 6.
31. *Behind the Counter*.
32. Foust, "Considering."
33. Stock and Wong, "Chipotle."
34. Chipotle Mexican Grill," Scarecrow."
35. Weber, "Fiona Apple."
36. Hodin, "Mechanisms"; Montmarquet, *Idea*.
37. Jenkins, *Special Affects*, 13.
38. Jenkins, *Special Affects*, 7.
39. Milstein, "'Somethin' Tells Me.'"
40. Debord, *Society*.
41. Baudrillard, *Simulacra*, 13.
42. Douglas, *Purity*, 35.
43. Ragas and Roberts, "Agenda."
44. Piper, *Farmed*.
45. Lockie, "Responsibility," 193.
46. Hillis, "Warning"; Zelinkova and Wenzl, "Occurrence."
47. H. Peterson, "Mystery."
48. Peterson, "Mystery."
49. Berfield, "Inside Chipotle."
50. Berfield, "Inside Chipotle."
51. Berfield, "Inside Chipotle."
52. Banet-Weiser, *AuthenticTM*, 217.
53. "Chipotle Revs Up."
54. Weber, "Fiona Apple."
55. "Honest Scarecrow."
56. Carducci, "Culture Jamming."
57. Baertlein and Herbst-Bayliss, "Chipotle Founder."
58. Rubino, "Chipotle Hiring."
59. Stewart, "Chipotle's New Mantra."
60. Chipotle Mexican Grill, "Love Story."
61. Kell, "Panera."

Chapter 7

1. "Super Bowl Just Shy"; "Official RAM Trucks." The 23 million YouTube views noted is the approximate number as of June 15, 2018, and this number does not include over 1 million additional views of a longer version of the ad (40 seconds longer) on YouTube.
2. Shepardson, "Chrysler's."
3. Seidl, "Wow."
4. See Harvey, "What It Is"; T. Smith, Letter to the Editor; and "Dirt Farmers."
5. Carden, "Remembering/Engendering"; Goldman and Dickens, "Selling"; Retzinger, "Cultivating"; Short, "'Hello Americans'"; Thompson, "Agriculture and Working-Class."

6. T. Peterson, "Telling," 292.
7. For a more recent account of industrial agriculture's distorted, unethical use of American agrarian myth, see Singer, "Agrarian."
8. Motter and Singer, "Review Essay."
9. Banet-Weiser, *AuthenticTM*, 10.
10. See Motter and Singer, "Review Essay."
11. Motter and Singer, "Review Essay," 440.
12. Thompson, *Agrarian Vision*, loc. 738–743, Kindle.
13. Montmarquet, *Idea*, 88–89.
14. Jefferson, "Commerce," 818.
15. Inge, "Introduction," xiii–xiv.
16. On Jefferson's agrarian contemporaries, see Montmarquet, *Idea*, 45–57, 86–93. For a relatively concise survey and synopsis of key voices in the evolution of American agrarianism since Jefferson's time, see Montmarquet, *Idea*, 221–50.
17. Freyfogle, "Introduction," xiii.
18. Inge, "American Arcadia," 1.
19. Nash, *Wilderness*.
20. H. Smith, *Virgin Land*.
21. Marx, *Machine*.
22. Berry, *Gift*, 215.
23. Thompson, *Agrarian Vision*, loc. 105, Kindle.
24. On various skeptical and optimistic theories of globalization from above, below, and everything in between, see Giddens, *Runaway*, 6–19; also see Ritzer, *Globalization*, 6–58.
25. Giddens, *Runaway*, 6–19; on glocal hybridity, see Ritzer, *Globalization*, 15–30.
26. On theoretical distinctions between place and space, see Stewart and Dickinson, "Enunciating," 280–87. On how these distinctions apply to food and agriculture institutions, and how these institutions have used technologies of mastery, exploitation, and control to become model space-makers, see Dickinson, "Joe's Rhetoric"; Knight, "Supersizing"; Ritzer, *McDonaldization*; Shiva, *Stolen Harvest*; Singer, "Corporate."
27. Tippett, "Agriculture."
28. Govan, "Agrarian."
29. Freyfogle, "Introduction," xxv, xxvi.
30. Berry, *Gift*, 220.
31. Also see Freyfogle, "Private"; and Motter, "Yeoman."
32. Eisinger, "Freehold."
33. Eisinger, "Freehold," 43.
34. Eisinger, "Freehold," 44.
35. Eisinger, "Freehold," 44.
36. Eisinger, "Freehold," 44–45.
37. Short, "'Hello, Americans'" 43.
38. Franke-Ruta, "Paul Harvey's 1978."
39. Retzinger, "Cultivating," 52–53.
40. Goldman and Dickens, "Selling," 585, italics in original.
41. Brasier, "Ideology"; Browne et al., *Sacred Cows*, 15; Kelsey, "Agrarian"; Mooney and Hunt, "Repertoire," 183.
42. Barthes, *Mythologies*, 130–31, 142–43.

43. Lakoff, "Why It Matters," 71, 74–75.
44. Danbom, "Romantic"; Montmarquet, *Idea*, 183–216.
45. Retzinger, "Cultivating," 46, 57; Singer, "Agrarian."
46. Leopold, *Sand*, 201–26.
47. Leopold, *Sand*, 204. For a glimpse of Leopold's impact on new-agrarian mythmaking, see Freyfogle, *Bounded*. We should note here that, as important as Leopold's ideas have been, the much earlier writings of Liberty Hyde Bailey should be credited for initially instilling this strong biocentric sensibility in the new agrarianism. New agrarians such as Wendell Berry and Wes Jackson Bailey credit Bailey's ideas.
48. Leopold, *Sand*, 222.
49. Leopold, "Farmer." Leopold presented "The Farmer as Conservationist" during the University of Wisconsin-Madison's observance of Farm and Home Week in 1939. The same year, the address was published in *American Forests*.
50. Berry, *Unsettling*, 46.
51. Berry, *Unsettling*, 4, 7.
52. Berry, *Unsettling*, 7, 14.
53. Berry, *Home*.
54. Logsdon, "Green."
55. Berry, "People," 193–94.
56. Cauley, "Merits," 28–36, 32–33.
57. Berry, *Gift*, 221.
58. Donahue, "Resettling," 46–50.
59. Mooney and Hunt, "Repertoire," 182.
60. Burkholder, "Kansas."
61. Mooney and Hunt, "Repertoire," 183.
62. Retzinger, "Cultivating," 57.
63. Wirzba, "Introduction," 6–7.
64. Freyfogle, "Introduction," xv.
65. Donahue, "Resettling," 46–50; Thompson, *Agrarian Vision*, 39; Wirzba, "Introduction."
66. Hilde and Thompson, "Introduction," 5; W. Jackson, *Nature*, 93–114.
67. Thompson, *Agrarian Vision*, 30.
68. We want to reiterate that on actually-existing farms, these strict and dichotomous categorizations of agrarian and industrialism break down through the blending of (contradicting) practices. Take, for example, the industrial-scale, petroleum input-intensive farm that specializes in organic black beans, the Concentrated Animal Feeding Operation that creates horrific environmental pollution through humane, cage-free chicken eggs, or the small-scale, local family-owned orchard that regularly uses pesticides and relies on migrant workers who work for much less than the legal minimum wage.
69. Berry, "Whole," 236–48.
70. Tate, "Remarks, 169. For a modern engagement with and extension to these ideas, see Berry, "Whole," 236–48.
71. Jackson, *Consulting*, 85, 85–91.
72. T. Peterson, "Telling," 296–302.
73. T. Peterson, "Telling," 296–302.

74. T. Peterson, "Telling," 294–95.
75. T. Peterson, "Telling," 299–300.
76. T. Peterson, "Telling," 290, 292.
77. Mooney and Hunt, "Repertoire," 183; Johnstone, "Old," 117.
78. Mooney and Hunt, "Repertoire," 183; Johnstone, "Old," 117.
79. Herman and Chomsky, *Manufacturing*, loc. 78–80, Kindle.
80. On these mythic qualities, see T. Peterson, "Telling," 297–98.
81. Johnstone, "Old," 117.
82. Dougherty, *Reluctant*, 201.
83. Most post–Super Bowl media responses stressed the ad's brilliance and emotional appeal, with some describing it as momentarily silencing Super Bowl parties and captivating audiences. A few responses critiqued the ad's romanticization of industrial agriculture. Some of these responses took the form of cynical parody videos posted on YouTube.
84. "Long Live the Storytellers/Brent Cobb/RAM 1500/Episode 3," RAM, https://www.youtube.com/watch?v=jNoA6kycHws.
85. "Next Year's Crop," *RAM Zone*, April 25, 2014, http://blog.ramtrucks.com/features/next-crop-project/.

Conclusion

1. Ferrell, "Farming."
2. Ferrell, "Farming."
3. The root of apocalypse is from the *Oxford Living Dictionaries*, "Apocalypse," *English: Oxford Living Dictionaries*, https://en.oxforddictionaries.com/definition/apocalypse. *Eskhatos* is provided as a root for the term eschatology at the same source, "Eschatology," *English: Oxford Living Dictionaries*, https://en.oxforddictionaries.com/definition/eschatology.
4. Axelrod, *Selling*, 141.
5. Alter, "Apocalyptic."
6. Bellow, *Herzog*, quoted in Alter, "Apocalyptic."
7. Buber, "Prophecy," quoted in Alter, "Apocalyptic."
8. Buber, "Prophecy," quoted in Alter "Apocalyptic."
9. Covino, *Rhetoric*.
10. Madrigal, "Most Important."
11. Atherton, "Autonomous"; "Autonomous Tractors."
12. "Autonomous Tractors."
13. While such advancement may change the work life of the farm owner or operator, it also threatens the livelihoods of some farm workers, already among the most exploited and vulnerable of workers in the Unites States. Such developments may be related to a spike in suicides of farm workers, as reported by *The New Food Economy*, June 21, 2018, https://newfoodeconomy.org/farmer-suicide-crisis-cdc-study/.
14. Tippett, "Agriculture."
15. Dimitri et al., "20th Century."
16. Tippett, "Agriculture."
17. USDA National Agricultural Statistics Service, "Iowa."

18. Brown and Weber, "Off-Farm."
19. Hofstadter, *Age*.
20. Logsdon, *Living*, 108.
21. Birkey, "Modern."
22. "What Is Urban Farming?"
23. "Community Supported."
24. Hydroponics, too, has historical antecedents offering lower-tech forms that may provide adaptable to different conditions. The Aztecs developed floating farms and Marco Polo recounted floating gardens during his travels in China. See "Hydroponics and the Future."
25. "Hydroponics and the Future."
26. B. Smith, "Coming."
27. Johnsen, *Bugs*; Kelly, *Gateway Bug*.
28. As an example of the discourse around these issues, see this report of the farm subsidies designed to mitigate the damage of trade wars for farmers: Bryan, "Trump."
29. Arbuckle et al., "Climate."
30. Doll, "Skeptical."
31. For a characteristic example of Donald Trump's climate-change denial and its impacts, see Landler and Davenport, "Dire."
32. Tabuchi, "In America's."
33. Tabuchi, "In America's."
34. Stockmann et al., "Knowns," cited in Tabuchi, "In America's."
35. Tabuchi, "In America's."
36. See Schaurbeger et al., "Consistent." For a broader view of impacts in the United States, see U.S. EPA "Climate Impacts."
37. The UN forecasts drops in farming yields of 2.9 percent in West Africa and 2.6 percent in India by 2050. See "State of Agricultural Commodity Markets."
38. Our readings of the events around the Dust Bowl and their causes follows that of Timothy Egan in *The Worst Hard Time*. We also draw on the Ken Burns documentary on the Dust Bowl, to which Egan was a key contributor and source: Burns, *Dust Bowl*.
39. Egan, *Worst*, 25.
40. Egan, *Worst*, 101.
41. Egan, *Worst*, 114.
42. Egan, *Worst*, 24.
43. "Episode One: The Great Plow Up."
44. Egan, interviewed in "Episode One: The Great Plow Up."
45. "Episode One: The Great Plow Up."
46. Egan, *Worst*, 50.
47. "Episode One: The Great Plow Up."
48. According to Egan, in 1930, the price of wheat was $0.41 a bushel, 1/8 its price in 1920. Egan, *Worst*, 76.
49. Egan, *Worst*, 113.
50. "Episode Two: Reaping the Whirlwind"; Egan, *Worst*, 311.
51. Egan, *Worst*, 310.
52. Egan, *Worst*, 310.

53. P. Thompson, *Agrarian Vision*, loc. 585–88, Kindle.
54. P. Thompson, *Agrarian Vision*, loc. 825–833.
55. P. Thompson, *Agrarian Vision*, loc. 95–99.
56. African Americans have a long history of agrarian resistance we have not reviewed here but can be engaged in sources such as this "backgrounder": Davy et al., "Black Agrarianism." For international work in resistance, see "Food Is Power."
57. Regarding micro-farming, see Frederick, "Gardens." As an example of ecosystem-based adaptation, see Munang and Andrews, "Despite." Finally, for a crowd-sources protest, see the "Occupy Monsanto" movement: "Recall Roundup."
58. Curtin, "Using."
59. "Hunger as a Weapon of War." See also the full Human Appeal report, https://reliefweb.int/sites/reliefweb.int/files/resources/advocacy-report-v013-final.pdf.

BIBLIOGRAPHY

Abrams, Rachel. "Chipotle Is Subpoenaed in Criminal Inquiry over Norovirus Outbreak." *New York Times*, last modified January 6, 2016. https://nyti.ms/22KKH K.
Adams, Amanda. "'Painfully Southern': *Gone with the Wind*, the Agrarians, and the Battle for the New South." *Southern Literary Journal* 40, no. 1 (2007): 58–75.
Agar, Herbert, and Allen Tate, eds. *Who Owns America? The New Declaration of Independence*. New York: Houghton Mifflin, 1936.
Alkon, Allison Hope, and Julian Agyeman. "Introduction: The Food Movement as Polyculture." In *Cultivating Food Justice: Race, Class, and Sustainability*, edited by Allison Hope Alkon and Julian Agyeman, 1-20. Cambridge: MIT Press, 2011.
Allen, Danielle S. *Talking to Strangers: Anxieties of Citizenship Since Brown v. Board of Education*. Chicago: University of Chicago Press, 2004.
Alter, Robert. "The Apocalyptic Temper." *Commentary*, June 1966, https://bit.ly/2OgUrk.
American Country Life Association. "Back to the Cracker Box?" *Rural America* 5 (April 1927): 15.
———. "A Declaration by Farm Women." *Rural America* 6 (October 1928): 11.
———. "The Objectives of Country Life: A Summary Statement of the Conclusions of the Committees of the First National Country Life Conference." In *Country Life Reconstruction: Proceedings of the First National Country Life Conference*, 15–24. Chicago: University of Chicago, 1919.
———. "Platform for Rural Democracy." *Rural America* 9 (December 1931): 2.
Anderson, Benedict. *Imagined Communities: Reflections on the Origin and Spread of Nationalism*. London: Verso, 2010.
"Apocalypse," *English: Oxford Living Dictionaries*, https://bit.ly/2zpy1Sj.
Arbuckle, J. Gordon, Jr., Linda Stalker Prokopy, Tonya Haigh, Jon Hobbs, and Tricia Knoot. "Climate Change Beliefs, Concerns, and Attitudes Toward Adaptation and Mitigation Among Farmers in the Midwestern United States." *Climatic Change* 117 (2013): 943–50. https://bit.ly/2CSPN3F.
Archibold, Randal C. "Hollywood Stars Shine Down on Protest to Preserve and Urban Farm." *New York Times*, June 9, 2006. https://nyti.ms/2Rqblsi.
Armstrong, Donna. "A Survey of Community Gardens in Upstate New York: Implications for Health Promotion and Community Development." *Health and Place* 6, no. 4 (2001): 319–27.
Atherton, Kelsey D. "Autonomous Tractor Concept Takes the Farmers Out of Farming." *Popular Science*, September 1, 2016. https://bit.ly/2ABztCU.
Atkinson, Joshua. "Thumbing their Noses at 'The Man': An Analysis of Resistance Narratives about Multinational Corporations." *Popular Communication* 1, no. 3 (2003): 163–80.
"Autonomous Tractors: The Future of Farming?" *Big Ag*, January 18, 2018. http://www.bigag.com/topics/equipment/autonomous-tractors-future-farming/.

Axelrod, Alan. *Selling the Great War: The Making of American Propaganda.*
   New York: Palgrave MacMillan, 2009.
Baertlein, Lisa, and Svea Herbst-Bayliss. "Chipotle Founder Out as CEO as Investor
   Patience Expires." *Reuters*, November 29, 2016. https://www.reuters.com/
   article/us-chipotle-move-ceo/chipotle-founder-out-as-ceo-as-investor
   -patience-expires-idUS KBN1DT1UI.
Bailey, David. "Enacting Transformation: George W. Bush and the Pauline
   Conversion Narrative in A Charge to Keep." *Rhetoric and Public Affairs* 11,
   no. 2 (2008): 215–42.
Bailey, Liberty Hyde. *The Country-Life Movement in the United States.* Norwood,
   MA: Norwood Press, 1911.
———. *The Holy Earth.* New York: Charles Scribner's Sons, 1916.
———. *The Outlook to Nature.* New York: Macmillan, 1915.
Baker, Lauren E. "Tending Cultural Landscapes and Food Citizenship in Toronto's
   Community Gardens." *Geographical Review* 94, no. 3 (2004): 305–25.
Baldwin, Robert D. "Professional Leadership in Rural Education." *Rural America*
   (May 1936): 3–6.
Balfour, Eve B. *The Living Soil: Evidence of the Importance to Human Health of Soil
   Vitality, with Special Reference to Post-War Planning.* London: Faber and
   Faber, 1943.
Balthrop, V. William. "Culture, Myth, and Ideology as Public Argument: An
   Interpretation of the Ascent and Demise of Southern Culture."
   Communication Monographs 51, no. 4 (1984): 339–52.
Banet-Weiser, Sarah. *AuthenticTM: The Politics of Ambivalence in a Brand Culture.*
   New York: New York University Press, 2012.
Barney, Maginel Wright. "War Gardens Over the Top." Poster. National War Garden
   Commission, 1919. Library of Congress POS-US .B383.
———. "War Gardens Victorious." Poster. National War Garden Commission, 1919.
   Library of Congress POS-US .B383.
Barthes, Roland. *Mythologies.* Translated by Annette Lavers. New York: Farrar,
   Strauss and Giroux, 1972.
Bassett, Thomas J. "Reaping on the Margins: A Century of Community Gardening
   in America." *Landscape* 25, no. 2 (1981): 1–8.
Baudrillard, Jean. *Simulacra and Simulation.* Translated by Sheila Glaser. Ann
   Arbor: University of Michigan Press, 1981.
Behind the Counter: Inside Chipotle Mexican Grill. Produced by Bloomberg
   Television. Video, July 9, 2013, 22:46. https://bloom.bg/2yJfctB.
Belasco, Warren J. *Appetite for Change: How the Counterculture Took on the Food
   Industry, 1966–1988.* New York: Pantheon Books, 1989.
Bellow, Saul. *Herzog.* New York: Viking, 1964.
Benhabib, Seyla. "Toward a Deliberative Model of Democratic Legitimacy." In
   Democracy and Difference: Contesting the Boundaries of the Political, edited
   by Seyla Benhabib, 67–94. Princeton: Princeton University Press, 1996.
Bennett, Jeffrey A. "Passing, Protesting, and the Arts of Resistance: Infiltrating the
   Ritual Space of Blood Donation." *Quarterly Journal of Speech* 94, no. 1 (2008):
   23–43.
Bercovitch, Sacvan. *The American Jeremiad.* Madison: University of Wisconsin
   Press, 2012.

———. *The Puritan Origins of the American Self*. New Haven: Yale University Press, 1975.
Berfield, Susan. "Inside Chipotle's Contamination Crisis: Smugness and Happy Talk About Sustainability Aren't Working Anymore." *Bloomberg*, December 22, 2015. https://www.bloomberg.com/features/2015-chipotle-food-safety-crisis/.
Bernays, Edward. *Propaganda*. Introduction by Mark Crispin Miller. Brooklyn: IG Publishing, 2004.
Berry, Wendell. "The Agrarian Standard." In Wirzba, *Essential Agrarian Reader*, 23–33.
———. *Citizenship Papers*. Washington, DC: Shoemaker & Hoard, 2003.
———. *The Gift of Good Land: Further Essays Cultural and Agricultural*. San Francisco: North Point Press, 1981.
———. *Home Economics*. New York: North Point Press, 1987.
———. "People, Land, and Community." In Wirzba, *Art of the Commonplace*, 182–204.
———. *The Unsettling of America: Culture and Agriculture*. San Francisco: Sierra Club Books, 1977.
———. *What Are People For?* San Francisco: North Point, 1990.
———. "The Whole Horse." In Wirza, *Art of the Commonplace*, 236–48.
Beus, Curtis E., and Riley E. Dunlap. "Endorsement of Agrarian Ideology and Adherence to Agricultural Paradigms." *Rural Sociology* 59, no. 3 (1994): 462–84.
Bicho, Ariane N. "The Simple Power of Multicultural Community Gardening." *Community Greening Review* 5, no. 1 (1996): 2–11.
Bingham, Emily, and Thomas Underwood. "Introduction." In *The Southern Agrarians and the New Deal*, edited by Emily Bingham and Thomas Underwood, 1–35. Charlottesville: University of Virginia Press, 2001.
Biodynamic Association. "Who Was Rudolf Steiner?" https://www.biodynamics.com/steiner.html.
Birkey, Joshua. "The Modern Community Garden Movement in the United States: Its Roots, Its Current Conditions, and Its Prospects for the Future." Master's thesis, Department of Geography, University of South Florida, 2009.
Bladow, Kyle. "Milking It: The Pastoral Imaginary of California's (Non)Dairy Farming."
Gastronomica: The Journal of Critical Food Studies 15, no. 3 (2015): 9-17.
Boyd, Todd. "Deep in the Shed: The Discourse of African American Cinema." *Iowa Journal of Literary Studies* 11, no. 1 (1991): 99–104.
Bram, Shahar. "The Narrative Facet of the Epic Tradition: Imagining the Past as Utopian Future." *Journal of the History of Ideas* 4, no. 1 (2006): 1–19.
Brasier, Kathryn J. "Ideology and Discourse: Characterizations of the 1996 Farm Bill by Agricultural Interest Groups." *Agriculture and Human Values* 19, no. 3 (2002): 239–53.
Brewster, Cori. "Toward a Critical Agricultural Literacy." In *Reclaiming the Rural: Essays on Literacy, Rhetoric, and Pedagogy*, edited by Kim Donehower, Charlotte Hogg, and Eileen E. Schnell, 34–51. Carbondale: Southern Illinois University Press, 2012.
Broad, Garrett. *More than Just Food: Food Justice and Community Change*. Berkeley: University of California Press, 2016.

———. "Ritual Communication and Use Value: The South Central Farm and the Political Economy of Place," *Communication, Culture & Critique* 6, no. 1 (2013): 20–40.
Broadberry, Stephen, and Mark Harrison. "The Economics of World War I: A Comparative Quantitative Analysis," August 2, 2005. https://docs.google.com/viewer?url=http%3A%2F%2Fwww2.warwick.ac.uk%2Ffac%2Fsoc%2Feconomics%2Fstaff%2Facademic%2Fharrison%2Fpapers%2Fww1toronto2.pdf.
Brown, Dona. *Back to the Land: The Enduring Dream of Self-Sufficiency in Modern America*. Madison: University of Wisconsin Press, 2010.
Brown, Jason P., and Jeremy G. Weber. "The Off-Farm Occupations of U.S. Farm Operators and Their Spouses: A Report Summary from the Economic Research Service." USDA, September 13, 2013. https://www.ers.usda.gov/webdocs/publications/43789/40008_eib-117_summary.pdf?v=41523.
Browne, William J., Jerry R. Skees, Louis E. Swanson, Paul B. Thompson, and Laurian J. Unnevehr. *Sacred Cows and Hot Potatoes: Agrarian Myths in Agricultural Policy*. Boulder: Westview Press, 1992.
Brunner, Edmund S. "Our Challenge to Ourselves." *Rural America*, February 1935.
Bryan, Bob. "The Trump Administration Is Throwing $4.7 Billion at US Farmers to Try and Make Up for the Trade War Pain." *Business Insider*, August 27, 2018. https://read.bi/2Rq3c73.
Buber, Martin. "Prophecy, Apocalyptic, and the Historical Hour." In *Pointing the Way: Collected Essays*, edited and translated by Maurice Friedman. New York: Harper & Row, 1957.
Buell, Lawrence. *The Environmental Imagination: Thoreau, Nature Writing, and the Formation of American Culture*. Cambridge: Harvard University Press, 1995.
Bullard, Robert D. "The Threat of Environmental Racism." *Natural Resources and Environment* 7, no. 3 (1993): 23–26.
Burgmann, Verity, and David Milner. "Futures without Financial Crisis: Utopian Literature in the 1890s and 1930s." *Continuum* 23, no. 6 (2009): 839–53.
Burke, Kenneth. *Counter-Statement*. Berkeley: University of California Press, 1965.
———. "Definition of Man." In *Language as Symbolic Action: Essays on Life, Literature and Method*, 3–24. Berkeley: University of California Press, 1968.
———. "Ideology and Myth," *Accent* 7 (1947): 195–205.
———. *A Rhetoric of Motives*. Berkeley: University of California Press, 1969.
Burkholder, Thomas R. "Kansas Populism, Woman Suffrage, and the Agrarian Myth: A Case Study in the Limits of Mythic Transcendence." *Communication Studies* 40, no. 4 (1989): 292–307.
Burns, Ken, dir. *The Dust Bowl: A Film by Ken Burns*. Arlington: Florentine Films; PBS Distribution, 2012.
Buttel, Frank H., and William L. Flinn. "Sociopolitical Consequences of Agrarianism." *Rural Sociology* 41, no. 4 (1976): 473–83.
———. "Sources and Consequences of Agrarian Values in American Society." *Rural Sociology* 40, no. 2 (1975): 134–51.
Butterfield, Kenyon. *The Farmer and the New Day*. New York: Macmillan, 1920.
Byerly, Carol R. *Fever of War: The Influenza Epidemic in the U.S. Army During World War I*. New York: New York University Press, 2005.

———. "The U.S. Military and the Influenza Epidemic of 1918–19." *Public Health Reports* 125, Suppl. 3 (2010): 82–91.
Caldwell, Tanya. "Judge Rules for Farmers." *Los Angeles Times*, July 13, 2006. http://articles.latimes.com/2006/jul/13/local/me-farm13.
Calhoun, Craig. "Introduction." In *Habermas and the Public Sphere*, edited by Craig Calhoun, 1–50. Cambridge, MA: MIT Press, 1992.
Camerini, Luca, and Nicola Diviani. "Activism, Health, and the Net: Are New Media Shaping Our Perception of Uncertainty?" *Interactions* 3, no. 3 (2012): 335–43.
Camporesi, Piero. *Bread of Dreams: Food and Fantasy in Early Modern Europe*. Translated by David Gentilcore. Reprint ed. Chicago: University of Chicago Press, 1994.
Capozzola, Christopher. *Uncle Sam Wants You!: World War I and the Making of the Modern American Citizen*. New York: Oxford University Press, 2010.
Carden, Mary Paniccia. "Remembering/Engendering the Heartland: Sexed Language, Embodied Space, and America's Foundational Fictions in Jane Smiley's 'A Thousand Acres.'" *Frontiers: A Journal of Women's Studies* 18, no. 2 (1997): 181–202.
Carducci, Vince. "Culture Jamming: A Sociological Perspective." *Journal of Consumer Culture* 6, no. 1 (2006): 116–38.
Carlson, Allan C. *The Natural Family Where It Belongs: New Agrarian Essays*. New Brunswick, NJ: Transaction Publishers, 2014.
———. *The New Agrarian Mind: The Movement Toward Decentralist Thought in Twentieth-Century America*. New Brunswick, NJ: Transaction Publishers, 2004.
Carlson, John E., and Maurice E. McLeod. "A Comparison of Agrarianism in Washington, Idaho, and Wisconsin." *Rural Sociology* 43, no. 4 (1978): 17–30.
Carpenter, Ronald H. "Frederick Jackson Turner and the Rhetorical Impact of the Frontier Thesis." *Quarterly Journal of Speech* 63, no. 2 (1977): 117–29.
Carson, Gerald. *Cornflake Crusade*. New York: Rinehart and Company, 1957.
Cauley, Troy J. "The Merits of Agrarianism." In Inge, *Agrarianism in American Literature*, 28–36.
Centers for Disease Control and Prevention. "Burden of Foodborne Illness: Findings." Last modified July 15, 2006. https://bit.ly/2OgTTEK.
———. "Multistate Outbreaks of Shiga Toxin Producing *Escherichia Coli* O26 Infections Linked to Chipotle Mexican Grill Restaurants (Final Update)." February 1, 2016. https://www.cdc.gov/ecoli/2015/o26-11-15/index.html.
Cenzatti, Marco. "Heterotopias of Difference." In *Heterotopia and the City: Public Space in a Postcivil Society*, edited by Michiel Dehaene and Lieven De Cauter, Kindle locations 1678–1889. New York: Routledge, 2008.
Chambers, Charles E. "Food Will Win the War." Poster for the United States Food Administration, 1917. National Archives and Records Administration Identifier No. 512499.
Check, Terrence. "Mortification and Moral Equivalents: Jimmy Carter's Energy Jeremiad and the Limits of Civic Sacrifice." In *Green Voices: Defending Nature and the Environment in American Civic Discourse*, edited by Richard D. Besel and Bernard K. Duffy, 175–97. Albany: State University of New York Press.

Chipotle Mexican Grill. "A Love Story." YouTube, July 7, 2016. Video, 04:15. http://www.youtube.com/watch?v=DDG-kumi6Pk.

———. "Back to the Start." Produced and directed by Nexus Studios. *Vimeo.* May 23, 2012. Video, 02:15. https://vimeo.com/42713491.

———. "Food with Integrity." http://www.chipotle.com/en-US/fwi/fwi.aspx.

———. "The Scarecrow." Produced and directed by Moonbot Studioes. *Vimeo,* September 11, 2013. Video, 03:22. https://vimeo.com/98800121.

"Chipotle Revs Up More Anti-Ag Propaganda." *Drover's,* January 27, 2014. https://www.drovers.com/article/chipotle-revs-more-anti-ag-propaganda.

Ciezadlo, Annia. "The War on Bread: How the Syrian Regime is Using Food as a Weapon." *New Republic,* February 14, 2014. http://www.newrepublic.com/article/116615/syrian-war-crimes-regime-bombs-bakeries-uses-starvation-weapon.

Clinker, L.C., and M.J. Dwyer. "Don't Waste Food While Others Starve." Poster. United States Food Administration, 1917. Temple University Special Collection, PT 020126.

"Community Supported Agriculture," *Local Harvest,* accessed June 17, 2018. https://localharvest.org/csa/.

Cohen, Joshua. "Procedure and Substance in Deliberative Democracy." In *Democracy and Difference: Contesting the Boundaries of the Political,* edited by Seyla Benhabib, 95–119. Princeton: Princeton University Press, 1996.

Conford, Philip. *The Origins of the Organic Movement.* Edinburgh: Floris Books, 2001.

Conkin, Paul. *The Southern Agrarians.* Knoxville: University of Tennessee Press, 1988.

Conlogue, William. *Working the Garden: American Writers and the Industrialization of Agriculture.* Chapel Hill: University of North Carolina Press, 2001.

Covino, William A. *Rhetoric, Magic and Literacy: An Eccentric History of the Composing Imagination.* Albany: State University of New York Press, 1994.

Craig, R.A., and K.J. Phillips. "Agrarian Ideology in Australia and the United States." *Rural Sociology* 48, no. 3 (1983): 409–20.

Creel, George. *How We Advertised America: The First Telling of the Amazing Story of the Committee on Public Information that Carried the Gospel of Americanism to Every Corner of the Globe.* London: Forgotten Books, 2012.

Crowley, Sharon. *Toward a Civil Discourse: Rhetoric and Fundamentalism.* Pittsburgh: University of Pittsburgh Press, 2006.

Culp, V.H. "The End of the Rural School." *Rural America* 10 (November 1932): 5–6.

Curtin, Michael. "Using Food as a Weapon of War." *International Policy Digest,* Nov. 27, 2017. https://bit.ly/2BroHdx.

Dalecki, Michael G., and C. Milton Coughenour. "Agrarianism in American Society." *Rural Sociology* 57, no. 1 (1992): 48–64.

Danbom, David B. *Born in the Country.* Baltimore: Johns Hopkins University Press, 2006.

———. "Past Visions of American Agriculture." In *Visions of America,* edited by William Lockeretz, 3–16. Ames: Iowa State University, 1997.

———. "Romantic Agrarianism in the Twentieth Century." *Agricultural History* 65, no. 4 (1991): 1–12.

Darsey, James. *The Prophetic Tradition and Radical Rhetoric in America*. New York: New York University Press, 1997.

Davidson, Donald. *Attack on Leviathan: Regionalism and Nationalism in the United States*. Chapel Hill: University of North Carolina Press, 1938.

———. "Mirror for Artists." In Twelve Southerners, *I'll Take My Stand*, 28–60.

———. "That this Nation May Endure—The Need for Political Regionalism." In Agar and Tate, *Who Owns America?*, 149–76.

Davis, Belinda J. *Home Fires Burning: Food, Politics, and Everyday Life in World War I Berlin*. Chapel Hill: University of North Carolina Press, 2000.

Davy, Dānia C., Edward "Jerry" Pennick, Savonala Horne, and Tracy Lloyd McCurty. "Black Agrarianism: Resistance." *Food First* 6 (2016). https://foodfirst.org/wp-content/uploads/2016/11/DR6_final-3.pdf.

Day, Bill. "Ehrenfried Pfeiffer, the Threefold Community, and the Birth of Biodynamics in America." *Biodynamics*, Fall 2008. https://www.biodynamics.com/threefold-day.

Day, Kristen. "Active Living and Social Justice: Planning for Physical Activity in Low-Income, Black, and Latino Communities." *Journal of the American Planning Association* 72, no 1. (2006): 88–99.

Debord, Guy. *The Society of the Spectacle*. Translated by Donald Nicholson-Smith. Detroit: Black & Red, 1970.

Deleuze, Gilles, and Félix Guattari. *A Thousand Plateaus: Capitalism and Schizophrenia*. Translated by Brian Massumi. London: Continuum, 1980.

Dewey, Donald. *The Art of Ill Will: The Story of American Political Cartoons*. New York: New York University Press, 2007.

Dickinson, Greg. "Joe's Rhetoric: Finding Authenticity at Starbucks." *Rhetoric Society Quarterly* 32, no. 4 (2002): 5–27.

Dimitri, Carolyn, Anne Effland, and Neilson Conklin. "The 20th Century Transformation of U.S. Agriculture and Farm Policy." USDA Economic Research Service, June 2005. https://ageconsearch.umn.edu/bitstream/59390/2/eib3.pdf.

"Dirt Farmers Gave All For Someone Else's Land." *The Oklahoman*, November 14, 1982. https://oklahoman.com/article/2002835/dirt-farmers-gave-all-for-someone-elses-land.

Dobrow, Joe. *Natural Prophets: From Health Foods to Whole Foods—How the Pioneers of the Industry Changed the Way We Eat and Reshaped American Business*. New York: Rodale Press, 2014.

Doll, Julie E. "Skeptical but Adapting: What Midwestern Farmers Say about Climate Change." *Weather, Climate and Society* 10 (Oct. 2017). https://bit.ly/2yFEs3J.

Donahue, Brian. "The Resettling of America." In Wirzba, *Essential Agrarian Reader*, 34–51.

Dorsey, Leroy G. "The Frontier Myth in Presidential Rhetoric: Theodore Roosevelt's Campaign for Conservation." *Western Journal of Communication* 59, no. 1 (1995): 1–19.

Dougherty, Debbie. *The Reluctant Farmer: An Exploration of Work, Social Class, and the Production of Food*. Leicester, UK: Troubadour Publishing, 2011.

Douglas, Mary. *Purity and Danger: An Analysis of Pollution and Taboo*. New York: Routledge, 1966.
Duck, Leigh Ann. "Postsouthern and (Increasingly) Post-Agrarian." *Contemporary Literature* 47, no. 2 (2006): 299–303.
Egan, Timothy. *The Worst Hard Time: The Untold Story of Those Who Survived the Great American Dust Bowl*. Boston: Houghton Mifflin, 2006.
Eighmey, Rae Katherine. *Food Will Win the War: Minnesota Crops, Cooks, and Conservation During World War I*. St. Paul: Minnesota Historical Society Press, 2010.
Eisinger, Chester E. "The Freehold Concept in Eighteenth Century American Letters." *William and Mary Quarterly* 4, no. 1 (1947): 42–59.
Emmett, Robert. "Community Gardens, Ghetto Pastoral, and Environmental Justice." *Interdisciplinary Studies in Literature and Environment* 18, no. 1 (2011): 67–86.
Enck-Wanzer, Darrel. "Race, Coloniality, and Geo-Body Politics: The Garden as Latin@ Vernacular Discourse." *Environmental Communication* 5, no. 3 (2011): 363–71.
"Eschatology." *English: Oxford Living Dictionaries*. Accessed September 29, 2018. https://bit.ly/2ETP84p.
Farré, Jordi, Jordi Padres, and Jan Gonzalo. "The Mediatization of the Food Chain in Networked Times." *Catalan Journal of Communication and Cultural Studies* 5, no. 2 (2013): 163–75.
Ferdman, Roberto A., and Abha Bhattarai. "There's a Crisis at Chipotle: Food Poisonings and Other Challenges Are Threatening the Darling of Fast Food's Reputation." *Washington Post*, December 9, 2015. https://wapo.st/2DdRf1m.
Ferrell, Keith. "Farming the Apocalypse." *Aeon*, July 8, 2014. https://bit.ly/2qk6BZV.
Figes, Orlando. *A People's Tragedy: The Russian Revolution: 1891–1924*. New York: Penguin, 1998.
Fisher, Mary Shattuck. "The Emancipated Woman." In Agar and Tate, *Who Owns America*, 401–16.
Fiskio, Janet. "Cultivating Community: Black Agrarianism in Cleveland, OH." *Gastronomica: The Journal of Critical Food Studies* 16, no. 2 (2016): 18–30.
Fitzgerald, Jonathan D. "Sincerity, Not Irony, Is Our Age's Ethos." *The Atlantic*, November 12, 2012. https://www.theatlantic.com/entertainment/archive/2012/1/sincerity-not-irony-is-our-ages-ethos/265466/.
Flagg, James Montgomery. "Sow the Seeds of Victory." Poster. National War Garden Commission, 1917. National Archives Identifier 512498.
Flinn, William. "Agrarian Values, Farm Structure, and Community Welfare." *Review of Policy Research* 2, no. 1 (1982): 26–32.
Flinn, William L., and Donald E..Johnson. "Agrarianism among Wisconsin Farmers." *Rural Sociology* 39, no. 2 (1974): 187–204.
Flores, Lisa. "Creating Discursive Space through a Rhetoric of Difference: Chicano Feminists Craft a Homeland." *Quarterly Journal of Speech* 82, no. 2 (1996): 142–56.
Foner, Eric. *The Story of American Freedom*. New York: Norton, 1998.
Food and Agriculture Organization of the United Nations. "2018: The State of Food Security and Nutrition in the World." Accessed October 2, 2018. http://www.fao.org/state-of-food-security-nutrition/en/.

"Food Is Power." Food Empowerment Project. Accessed 11/1/2018. http://www.food ispower.org/.

Food Marketing Institute. "Shopping for Health 2011." Accessed June 20, 2017. https://www.fmi.fmi.org/docs/health-wellnesswellness-research-downloads/shoppingforhealth2011.pdf?sfvrsn=2.

Foucault, Michel. "Of Other Spaces: Utopias and Heterotopias." In *Rethinking Architecture: A Reader in Cultural Theory*, edited by Neil Leach, 330–36. New York: Routledge, 1997.

Foust, Christina R. "Considering the Prospects of Immediate Resistance in Food Politics: Reflections on *The Garden*." *Environmental Communication* 5, no. 3 (2011): 350–55.

Fox, Josh, dir. "The Sky Is Pink." *Vimeo*, June 20, 2012. Video, 18:23. https://vimeo.com/44367635.

Frame, Nat T. "American Country Life Planning: The Presidential Address." *Rural America* 12 (November 1934): 3–8.

Francis, Mark L., Lisa Cashdan, and Lynne Paxson. *Community Open Spaces: Greening Neighborhoods through Community Action and Land Conservation*. Washington, DC: Island Press, 1984.

Franke-Ruta, Garance. "Paul Harvey's 1978 'So God Made a Farmer' Speech." *The Atlantic*, February 3, 2013. https://bit.ly/2SAyIAU.

Fraser, Nancy. "Rethinking the Public Sphere: A Contribution to the Critique of Actually Existing Democracy." *Social Text* 25/26 (1990): 56–80.

Frederick, Brian. "Gardens Are Emblems of Resistance: Interview with Slow Food International Vice President." *Foodtank: The Think Tank for Food*, December 2017. https://foodtank.com/news/2017/12/edie-mulkiibi-interview/.

Freyfogle, Eric T. *Agrarianism and the Good Society: Land, Culture, Conflict, and Hope* Lexington: University Press of Kentucky, 2007.

———. *Bounded People, Boundless Lands: Envisioning a New Land Ethic*. Washington, DC: Island Press/Shearwater, 1998.

———. "Introduction: A Durable Scale." In *New Agrarianism: Land, Culture, and the Community of Life*, edited by Eric Freyfogle, xiii-xli. Washington, DC: Island Press, 2001.

———. "Private Property Rights in Land: An Agrarian View." In Wirzba, *Essential Agrarian Reader*, 237–58.

Gates, Paul W. *The Farmer's Age: Agriculture, 1815–1860*. New York: Holt, Rinehart and Winston, 1962.

Gencer, Arin. "Reaping Harvest of Anger at Urban Farm." *Los Angeles Times*, July 6, 2006. https://lat.ms/2EPEJXG.

Gerber, Pierre J., Henning Steinfeld, Benjamin Henderson, Anne Mottet, Carolyn Opio, Jeroen Dijkman, Alessandra Falcucci, and Giuseppe Tempio. *Tackling Climate Change through Livestock—A Global Assessment of Emissions and Mitigation Opportunities*. Rome: Food and Agriculture Organization of the United Nations, 2013. http://www.fao.org/3/i3437e/i3437e00.htm.

Giddens, Anthony. *Runaway World: How Globalization Is Reshaping Our Lives*. New York: Routledge, 2003.

Goldman, Robert, and David R. Dickens. "The Selling of Rural America." *Rural Sociology* 48, no. 4 (1983): 585–606.

Golodryga, Bianna, and Michael Milberger. "Secrets of America's Favorite

Restaurants: Chipotle Elevates Fast Food." *ABC News*, September 15, 2010. https://abcnews.go.com/GMA/Consumer/chipotle-chain-restaurant-bucks-fast-food-image-local/story?id=11634188.

Gordon, Constance, and Kathleen Hunt. "Reform, Justice, and Sovereignty: A Food Systems Agenda for Environmental Communication." *Environmental Communication* 13, no. 1 (2019): 9–22.

Govan, Thomas P. "Agrarian and Agrarianism: A Study in the Use and Abuse of Words." *Journal of Southern History* 30, no. 1 (1964): 35–47.

Graham, Sylvester. *Lectures on the Science of Human Life*. London: Horsell and Aldine Chambers, 1849.

Grammar, John. *Pastoral Politics in the Old South*. Baton Rouge: Louisiana State University Press, 1997.

Grebler, Leo. *The Cost of the World War to Germany and Austria-Hungary*. New Haven: Yale University Press, 1940.

Greene, Wade. "Guru of the Organic Food Cult." *New York Times*, June 6, 1971, SM30.

Greer, Jed and Kenny Bruno. *Greenwash: The Reality Behind Corporate Environmentalism*. New York: Apex Press, 1996.

Grey, Stephanie Houston. "The Gospel of the Soil: Southern Agrarian Resistance and the Productive Future of Food." *Southern Communication Journal* 79, no. 5 (2014): 387–406.

———. "A Growing Appetite: The Emerging Rhetoric of Food Politics." *Rhetoric and Public Affairs* 19, no. 2 (2016): 307–20.

Griffin, Charles. "The Rhetoric of Form in Conversion Narratives." *Quarterly Journal of Speech* 76, no. 2 (1990): 152–63.

Gross, Daniel. *Our Roots Grow Deep: The Story of Rodale*. Emmaus, PA: Rodale Press, 2008.

Guthman, Julie. *Agrarian Dreams: The Paradox of Organic Farming in California*. Berkeley: University of California Press, 2014.

Habermas, Jürgen. *Communication and the Evolution of Society*. Translated by Thomas McCarthy. Cambridge, UK: Polity Press, 1991.

Hanna, Autumn K., and Pikai Oh. "Rethinking Urban Poverty: A Look at Community Gardens." *Bulletin of Science, Technology, and Society* 20, no. 3 (2000): 207–16.

Hanson, Russell L. *The Democratic Imagination in America*. Princeton: Princeton University Press, 1985.

Hanson, Victor Davis. *The Other Greeks: The Family Farm and the Agrarian Roots of Western Civilization*. Los Angeles: University of California Press, 1999.

Harter, Lynn M. "Masculinity(s), the Agrarian Frontier Myth, and Cooperative Ways of Organizing: Contradictions and Tensions in the Experience and Enactment of Democracy." *Journal of Applied Communication Research* 32, no. 2 (2004): 89–118.

Harvey, Paul. "What It Is to Be a Farmer." *Gadsden Times*, August 26, 1975. https://news.google.com/newspapers?id=x6cfAAAAIBAJ&sjid=VtYEAAAAIBAJ&pg=940,4348824.

Hawkins, Nigel. *The Starvation Blockades: Naval Blockades of World War I*. Barnsley, South Yorkshire: Leo Cooper, 2002.

Haydu, Jeffrey. "Cultural Modeling in Two Eras of U.S. Food Protest: Grahamites

(1830s) and Organic Advocates (1960s-70s)." *Social Problems* 58, no. 3 (2011): 461–87.

Hecht, Jamey. "The Future at War with the Past." March 22, 2006. http://www.fromthewilderness.com/free/ww3/032206_war_past.shtml.

Heckman, Joseph. "A History of Organic Farming: Transitions from Sir Albert Howard's War in the Soil to the USDA National Organic Program." Weston A. Price Foundation, July 21, 2007. https://www.westonaprice.org/health-topics/farm-ranch/a-history-of-organic-farming-transitions-from-sir-albert-howards-war-in-the-soil-to-the-usda-national-organic-program/.

Heinze, Kirk L. "From Virgil to Virginia: The Wellsprings of the American Agrarian Mythology." In *The American West on Film*, edited by Michael V. Doyle, 6–25. Dubuque, IA: Kendall-Hunt, 1990.

Henion, Karl E., II, and Thomas C. Kinnear. *Ecological Marketing*. Chicago: American Marketing Association, 1976.

Herman, Edward S., and Noam Chomsky. *Manufacturing Consent: The Political Economy of the Mass Media*. New York: Pantheon Books, 2002. Kindle.

Hilde, Thomas C. and Paul B. Thompson. "Introduction: Agrarianism and Pragmatism." In *The Agrarian Roots of Pragmatism*, edited by Paul B. Thompson and Thomas C. Hilde, 1–21. Nashville: Vanderbilt University Press, 2000.

Hillis, Krista. "Warning: You Are Probably Eating Crude Oil in These Foods." *The Alternative Daily*, February 23, 2017. https://bit.ly/2Jr6K6o.

Hodin, Stephen B. "The Mechanisms of Monticello: Saving Labor in Jefferson's America." *Journal of the Early Republic*, 26, no. 3 (2006): 377–418.

Hoffmann, Jessica. "History of the South Central Farm: How the Community Has Used the Land since 1985." *New Standard*, April 5, 2006, http://newstandardnews.net/content/index.cfm/items/3028.

———. "L.A. Urban Farmers Fight for Community Garden." *New Standard*, April 5, 2006. http://newstandardnews.net/content/index.cfm/items/3027.

———. "South Central Farm Timeline." *The South Central Farm*, last modified Fall 2006. http://www.walzandrew.com/omeka/items/show/12.

Hoffmann, Jessica, and Christine Petit. "14 Acres: Conversations across Chasms in South Central Los Angeles." *Clamor Magazine*, Spring 2006. http://archive.clamormagazine.org/issues/36/people.php.

Hofstadter, Richard. *The Age of Reform: From Bryan to F.D.R.* New York: Vintage, 1955.

Holt, Arthur E. "More Hell and Less Hogs." *Rural America*, December 1931, 5.

"The Honest Scarecrow." *Funny or Die*, September 18, 2013. Video, 2:10. https://bit.ly/1mE14a5.

Hooks, Greg. "From an Autonomous to a Captured State Agency: The Decline of the New Deal in Agriculture." *American Sociological Review* 55, no. 1 (1990): 29–43.

Hoover, Herbert. "Introduction." *Food and the War*. Washington, DC: United States Food Administration, 1918.

Hopkins, Brent. "The End for South Central Farm?" *Los Angeles Daily News*, June 14, 2006. http://www.dailynews.com/article/zz/20060614/NEWS/606149881.

Houck, Brenna. "43 Chipotle Locations Closed Due to E. Coli Outbreak in

Washington and Oregon." *Eater*, November 2, 2015. https://www.eater.com
/2015/11/1/9653980/chipotle-e-coli-outbreak-washington-oregon-closes
-restaurants.

Howard, Albert. *An Agricultural Testament*. London: Oxford University Press, 1956.

Howard-Pitney, David. *The Afro-American Jeremiad: Appeals for Justice in America*. Philadelphia: Temple University Press, 1990.

Hughes, Richard T. *Myths Americans Live By*. Champaign: University of Illinois Press, 2003.

Human Appeal. "Hunger as a Weapon of War: How Food Insecurity Has Been Exacerbated in Syria and Yemen" *reliefweb*, March 26, 2018. https://bit.ly/2GdUSSI.

"Hunger as a Weapon of War: How Food Insecurity Has Been Exacerbated in Syria and Yemen." reliefweb, March 26, 2018. https://reliefweb.int/report/syrian
-arab-republic/hunger-weapon-war-how-food-insecurity-has-been
-exacerbated-syria-and.

Hunter, Walt. "The Story Behind the Poem on the Statue of Liberty." *The Atlantic*, Jan. 16, 2018.

Hurt, R. Douglas. *American Agriculture: A Brief History*. West Lafayette, IN: Purdue University Press, 2002.

"Hydroponics and the Future of Farming." *Big Picture*. https://bigpictureeducation
.com/hydroponics-and-future-farming.

Hynes, H. Patricia. *Patch of Eden: America's Inner City Gardeners*. White River Junction, VT: Chelsea Green, 1996.

Inge, M. Thomas, ed. *Agrarianism in American Literature*. New York: Odyssey Press, 1969.

———. "An American Arcadia: Theoretical Perspectives." In Inge, *Agrarianism in American Literature*, 1–4.

Irazábal, Clara, and Anita Punja. "Cultivating Just Planning and Legal Institutions: A Critical Assessment of the South Central Farm Struggle in Los Angeles." *Journal of Urban Affairs* 31, no. 1 (2009): 1–23.

Irvine, Seana, Lorraine Johnson, and Kim Peters. "Community Gardens and Sustainable Land Use Planning: A Case-Study of the Alex Wilson Community Garden." *Local Environment* 4, no. 1 (1999): 33–46.

Jackson, Carlton. *J.I. Rodale: Apostle of Nonconformity*. New York: Pyramid, 1974.

Jackson, Wes. *Consulting the Genius of the Place: An Ecological Approach to a New Agriculture*. Berkeley: University of California Press, 2010.

———. *Nature as Measure: The Selected Essays of Wes Jackson*. Berkeley: Counterpoint, 2011.

———. *New Roots for Agriculture*. Lincoln: University of Nebraska Press, 1980.

James, Pearl, ed. *Picture This: World War I Posters and Visual Culture*. Lincoln: University of Nebraska Press, 2009.

James, Pearl. "Reading World War I Posters." In James, *Picture This*, 1–36.

Jefferson, Thomas. "Commerce and Sea Power: To John Jay, August 23, 1785." In *Jefferson: Writings*, edited by Merrill D. Peterson, 818–19. New York: Library of America, 1984.

Jenkins, Eric. *Special Affects: Cinema, Animation, and the Translation of Consumer Culture*. Edinburgh: Edinburgh University Press, 2014.

"J.I. Rodale Dead; Organic Farmer: Espoused the Avoidance of Chemical Fertilizers." *New York Times*, June 8, 1971, 42.

Johannesen, Richard. "Ronald Reagan's Economic Jeremiad." *Central States Speech Journal* 37, no. 2 (1986): 79–89.
Johnsen, Andreas, dir. *Bugs: The Film*. Copenhagen: Rosforth, Danish Documentary Productions, 2017.
Johnston, Josée, and Shyon Baumann. *Foodies: Democracy and Distinction in the Gourmet Foodscape*. New York: Routledge, 2015.
Johnstone, Christopher Lyle. "Thoreau and Civil Disobedience: A Rhetorical Paradox." *Quarterly Journal of Speech* 60, no. 3 (1974): 313–22.
Johnstone, Paul H. "Old Ideals versus New Ideas in Farm Life." In *Farmers in a Changing World: The Yearbook of Agriculture*, edited by Gove Hambridge, 111–70. Washington, DC: United States Government Printing Office, 1940.
Karanikas, Alexander. *Tillers of a Myth: Southern Agrarians as Social and Literary Critics*. Milwaukee: University of Wisconsin Press, 1969.
Kazin, Michael. *The Populist Persuasion: An American History*. Revised ed. New York: Basic Books, 1998.
Kell, John. "Panera Bread Says It Hit $1 Billion Digital Sales Target." *Forbes*, June 14, 2017. https://for.tn/2zdDETn.
Kelly, Johanna B., dir. *The Gateway Bug*. USA: Gateway Bug, LLC; Gravitas Ventures Distribution, 2017. Video. https://thegatewaybug.com/.
Kelsey, Timothy. "The Agrarian Myth and Policy Responses to Farm Safety." *American Journal of Public Health* 84, no. 7 (1994): 1171–77.
Kennedy, Scott Hamilton, dir. *The Garden*. Los Angeles: Black Valley Films, 2008. DVD.
Kern, Louis J. "The Embodiment of a Nation: The Iconicity of Uncle Sam and the Construction of a Conflicted National Identity." In *ConFiguring America: Iconic Figures, Visuality and the American Identity*, edited by Klaus Reiser, Michael Fuchs, and Michael Phillips, 176–179. Fishponds, UK: Intellect Books, 2013.
Kiernan, Ben. *Blood and Soil: A World History of Genocide and Extermination from Sparta to Darfur*. New Haven: Yale University Press, 2007.
Kingsbury, Cecilia M. *For Home and Country: World War I Propaganda on the Home Front*. Lincoln: University of Nebraska Press, 2010.
Knauth, Oswald W. "Farmer's Income." In *Income in the United States, Its Amount and Distribution, 1909–19*, vol. 2, 298–313. Edited by Wesley Clair Mitchell. Washington, DC: National Bureau of Economic Research, 1922.
Knight, Andrew J. "Supersizing Farms: The McDonaldization of Agriculture." In *McDonaldization: The Reader*, 2nd ed., edited by George Ritzer, 183–95. Thousand Oaks, CA: Pine Forge Press, 2004.
Kuo, Frances E., and William C. Sullivan. "Environment and Crime in the Inner City: Does Vegetation Reduce Crime?" *Environment and Behavior* 33, no. 3 (2001): 343–67.
Laclau, Ernesto, and Chantal Mouffe. *Hegemony and Socialist Strategy: Towards a Radical Democratic Politics*. New York: Verso, 2001.
Laffey, Shelia A., dir. *South Central Farm: Oasis in a Concrete Desert*. Santa Monica, CA: Echo Mountain Productions, 2007. DVD.
LaGreco, Marianne, and Dawn Leonard. "Building Sustainable Community-Based Food Programs: Cautionary Tales from *The Garden*." *Environmental Communication* 5, no. 3 (2011): 356–62.

Lakoff, George. "Why It Matters How We Frame the Environment." *Environmental Communication* 4, no. 1 (2010): 70–81.

Landler, Mark, and Carol Davenport. "Dire Climate Warning Lands with a Thud on Trump's Desk." *New York Times*, Oct. 8, 2018. https://nyti.ms/2OgKS38.

Landsberg, Alison. *Prosthetic Memory: The Transformation of American Remembrance in the Age of Mass Culture*. New York: Columbia University Press, 2004.

Larson, Nicole I., Mary T. Story, and Melissa C. Nelson. "Neighborhood Environments: Disparities in Access to Healthy Foods in the U.S." *American Journal of Preventive Medicine* 36, no. 1 (2008): 74–78.

Lasswell, Harold D. *Propaganda Technique in World War I*. Cambridge, MA: MIT Press, 1971.

Leach, William. *Land of Desire: Merchants, Power, and the Rise of a New American Culture*. New York: Vintage, 1994.

Leopold, Aldo. "The Farmer as Conservationist." In *The River of the Mother of God and Other Essays by Aldo Leopold*, edited by Susan L. Flader and J. Baird Callicott, 255–65. Madison: University of Wisconsin Press, 1992.

———. *A Sand County Almanac and Sketches Here and There*. New York: Oxford University Press, 1949.

Liquified Squid. "CNN Republican Presidential Town Hall—Kasich, Bush, Trump—Columbia, SC (February 18, 2016)." YouTube video, February 19, 2016. 2:11:43. http://www.youtube.com/watch?v=heHbaCCUFlE.

Lockie, Stewart. "Responsibility and Agency Within Alternative Food Networks: Assembling the 'Citizen Consumer.'" *Agriculture and Human Values* 26, no. 3 (2009): 193–201.

Logsdon, Gene. *Living at Nature's Pace: Farming and the American Dream*. White River Junction, VT: Chelsea Green, 2000.

———. *The Mother of All Arts: Agrarianism and the Creative Impulse*. Lexington: University Press of Kentucky, 2007.

Lowden, Frank O. "Rural Organization." *Rural America* 7 (November 1929): 3–7.

Lund, John. "Boundaries of Restriction: The Dillingham Commission." *University of Vermont History Review* 6 (1994). http://www.uvm.edu/~hag/histreview/vol6/lund.html.

Lynch, John. "Prepare to Believe: The Creation Museum as Embodied Conversion Narrative." *Rhetoric and Public Affairs* 16, no. 1 (2013): 1–23.

Lyson, Thomas. *Civic Agriculture: Reconnecting Farm, Food, and Community*. Lebanon, NH: University Press of New England, 2004.

Lytle, Andrew Nelson. "The Hind Tit." In Twelve Southerners, *I'll Take My Stand*, 201–245.

———. "The Small Farm Secures the State." In Agar and Tate, *Who Owns America?*, 309–26.

Madrigal, Alexis C. "The Most Important Self-Driving Car Announcement Yet." *The Atlantic*, March 28, 2018. https://bit.ly/2P2kOJG.

Majd, Mohammed Gholi. *The Great Famine and Genocide in Persia, 1917–19*. Lanham, MD: University Press of America, 2003.

Major, William H. *Grounded Vision: New Agrarianism and the Academy*. Tuscaloosa: University of Alabama Press, 2011.

Malakoff, David. "What Good Is Community Greening?" *Community Greening Review* 5 (1995): 4–11.
Malvasi, Mark G. *The Unregenerate South: The Agrarian Thought of John Crowe Ransom, Allen Tate, and Donald Davidson*. Baton Rouge: Louisiana State University Press, 1997.
Marcy, Leslie. "Food Riots in America." *Hellraiser's Journal*, April 9, 2017. http://www.weneverforget.org/hellraisers-journal-war-profits-and-starvation-international-socialist-review-on-food-riots-in-america/.
Mares, Teresa M., and Devon G. Peña. "Environmental and Food Justice: Toward Local, Slow, and Deep Food Systems." In *Cultivating Food Justice: Race, Class, and Sustainability*, edited by Alison Hope Alkon and Julian Agyeman, 197–219. Cambridge, MA: MIT Press, 2011.
———. "Urban Agriculture in the Making of Insurgent Spaces in Los Angeles and Seattle." In *Insurgent Public Space: Guerrilla Urbanism and the Remaking of Contemporary Cities*, edited by Jeffrey Hou, 241–54. New York: Routledge, 2010.
Marx, Leo. *The Machine in the Garden: Technology and the Pastoral Ideal in America*. New York: Oxford University Press, 2000.
McCoy, Drew R. *The Elusive Republic: Political Economy in Jeffersonian America*. Chapel Hill: University of North Carolina Press, 1980.
McClymer, John F. *War and Welfare: Social Engineering in America, 1890–1925*. Westport, CT: Greenwood Press, 1980.
McDermott, Robert A., ed. *The Essential Steiner*. San Francisco: Harper and Row, 1984.
McGrath, Maria. "The Bizarre Life (and Death) of Mr. Organic." *New Republic*, August 8, 2014. http://www.newrepublic.com/article/119007/ji-rodale-paranoid-publisher-who-became-godfather-organics.
McGuire, Michael. "Mythic Rhetoric in *Mein Kampf*: A Structuralist Critique." *Quarterly Journal of Speech* 63, no. 1 (1977): 1–13.
Melvin, Bruce L. "Local Rural Social Planning." *Rural America* 13 (December 1935): 3–6.
Meyer, Russ. "A History of Green Brands in the 1960s and 1970s: Doing the Groundwork." *Fast Company*, May 12, 2010. https://bit.ly/2CNwjoh.
Mignolo, Walter D. "Delinking: The Rhetoric of Modernity, the Logic of Coloniality, and the Grammar of Decoloniality." *Cultural Studies* 21, no. 2 (2007): 449–514.
Miller, Perry. *Errand into the Wilderness*. Cambridge: Harvard University Press, 1956.
Milstein, Tema. "'Somethin' Tells Me It's All Happening at the Zoo': Discourse, Power, and Conservationism." *Environmental Communication* 3, no. 1 (2009): 25–48.
Milstein, Tema, Claudia Anguiano, Jennifer Sandoval, Yea-Wen Chen, and Elizabeth Dickinson. "Communicating a 'New' Environmental Vernacular: A Sense of Relations-in-Place." *Communication Monographs* 78, no. 4 (2011): 486–510.
Montmarquet, James A. *The Idea of Agrarianism: From Hunter-Gatherer to Agrarian Radical in Western Culture*. Moscow: University of Idaho Press, 1989.
Mooney, Patrick H., and Scott A. Hunt. "A Repertoire of Interpretations: Master Frames and Ideological Continuity in U.S. Agrarian Mobilization." *Sociological Quarterly* 37, no. 1 (1996): 177–97.

Morain, Dan, and Jia-Rui Chung. "Westly Wants $15.5 Million in State Grants Repaid." *Los Angeles Times*, September 15, 2004. https://lat.ms/2OgXFhs.
Moss, Christina. "The Reconceptualization of Southern Rhetoric: A Meta-Critical Perspective." PhD diss., Louisiana State University, 2005.
Motter, Jeff. "Yeoman Citizens: The Country Life Association and the Reinvention of Democratic Legitimacy." *Argumentation and Advocacy* 51, no. 1 (2014): 1–16.
Motter, Jeff, and Ross Singer. "Review Essay: Cultivating a Rhetoric of Agrarianism." *Quarterly Journal of Speech* 98, no. 4 (2012): 439–54.
Mullendore, William C. *History of the United States Food Administration, 1917–1919*. Stanford, CA: Stanford University Press, 1941.
Mulnar, Joseph J., and Litchi S. Wu. "Agrarianism, Family Farming, and Support for State Intervention in Agriculture." *Rural Sociology* 54, no. 2 (1989): 227–45.
Munang, Richard, and Jesica Andrews. "Despite Climate Change, Africa Can Feed Africa." *AfricaRenewal*, Special Edition of Agriculture 2014. https://www.un.org/africarenewal/magazine/special-edition-agriculture-2014/despite-climate-change-africa-can-feed-africa.
Muniz, Albert M., Jr., and Thomas C. O'Guinn. "Brand Communities." *Journal of Consumer Research* 27, no. 4 (2001): 412–32.
Murray, Sarah. *Moveable Feasts: From Ancient Rome to the 21st Century, the Incredible Journeys of the Foods We Eat*. New York: St. Martin Press, 2007.
Murphy, John M. "A Time of Shame and Sorrow: Robert F. Kennedy and the American Jeremiad." *Quarterly Journal of Speech* 76, no. 4 (1990): 401–14.
Murphy, Paul. *The Rebuke of History: The Southern Agrarians and American Conservative Thought*. Chapel Hill: University of North Carolina Press, 2001.
Nader, Ralph. *Unstoppable: The Emerging Left-Right Alliance to Dismantle the Corporate State*. New York: Nation Books, 2014.
Nash, Roderick Frazier. *Wilderness and the American Mind*. 4th ed. New Haven: Yale University Press, 2001.
Neander, Joachim, and Randal Marlin. "Media and Propaganda: The Northcliffe Press and the Corpse Factory Story of World War I." *Global Media Journal*, Canadian ed. 3, no. 2 (2010): 67–82.
Nestle, Marion, and Sharron Dalton. "Food Aid and International Hunger Crises: The United States in Somalia." *Agriculture and Human Values* 11, no. 4 (1994): 19–27.
Newman, Lance. "Thoreau's Natural Community and Utopian Socialism." *American Literature* 75, no. 3 (2003): 515–44.
Nichols, Bill. *Introduction to Documentary*. 2nd ed. Bloomington: Indiana University Press, 2010.
Northbourne, Lord. *Look to the Land*. San Rafael: Angelico Press/Sophia Perennis, 2011.
Obach, Brian K. *Organic Struggle: The Movement for Sustainable Agriculture in the United States*. Cambridge, MA: MIT Press, 2015.
"Official RAM Trucks Super Bowl Commercial 'FZarmer.'" YouTube video, February 3, 2013. http://www.youtube.com/watch?v=AMpZoTGjbWE.
O'Keefe, Daniel Lawrence. *Stolen Lightning: The Social Theory of Magic*. New York: Continuum, 1982.
Ono, Kent A., and John M. Sloop. "The Critique of Vernacular Discourse." *Communication Monographs* 62, no. 1 (1995): 19–46.

Opie, John, and Norbert Elliot. "Tracking the Elusive Jeremiad: The Rhetorical Character of American Environmental Discourse." In *The Symbolic Earth: Discourse and Our Creation of the Environment*, edited by James G. Cantrill and Christine L. Oravec, 9–37. Lexington: University Press of Kentucky, 1996.

Oravec, Christine. "The Ideological Significance of Discursive Form: A Response to Solomon and Perkins." *Communication Studies* 42, no. 4 (1991): 383–91.

Owsley, Frank Lawrence. "The Irrepressible Conflict." In Twelve Southerners, *I'll Take My Stand*, 61–91.

———. "The Foundations of Democracy." In Agar and Tate, *Who Owns America?*, 73–92.

Pack, Charles Lathrop. *The War Garden Victorious*. Philadelphia: J.P. Lippincott, 1919.

Patten, Simon N. *The New Basis of Civilization*. Edited by Daniel M. Fox. Cambridge, MA: Harvard University Press/Belknap Press, 1968.

Paull, John. "The Betteshanger Summer School: Missing Link between Biodynamic Agriculture and Organic Farming." *Journal of Organic Systems* 6, no. 2 (2011): 13–26.

———. "Biodynamic Agriculture: The Journey from Koberwitz to the World, 1924–1938." *Journal of Organic Systems* 6, no. 1 (2011): 27–41.

Payrow Shabani, O.A. *Democracy, Power, and Legitimacy: The Critical Theory of Jürgen Habermas*. Toronto: University of Toronto Press, 2003.

Peeples, Jennifer A. "Trashing South-Central: Place and Identity in a Community-Level Environmental Justice Dispute." *Southern Communication Journal* 69, no. 1 (2003): 82–95.

Perelman, Chaim, and Lucie Olbrechts-Tyteca. *The New Rhetoric: A Treatise on Argumentation*. Translated by John Wilkinson and Purcell Weaver. Notre Dame: University of Notre Dame Press, 1971.

Perenyi, Eleanor. "Apostle of the Compost Heap." *Saturday Evening Post*, July 16, 1966, 30–33.

Peters, Jason, ed. *Wendell Berry: Life and Work*. Lexington: University Press of Kentucky, 2007.

Peters, Suzanne. "Organic Farmers Celebrate Organic Research: A Sociology of Popular Science." In *Counter-Movements in the Sciences*, edited by Helga Nowotny and Hilary Rose, 251–75. Dordrecht: De Reidel Publishing, 1979.

Peterson, Hayley. "The Mystery of Chipotle's E. Coli Outbreak Is Stumping Scientists and Fueling Conspiracy Talk." *Business Insider*, January 7, 2016. http://www.businessinsider.com/chipotles-e-coli-outbreak-is-stumping -scientists-2016-1.

Peterson, Tarla Rai. "Jefferson's Yeoman Farmer as Frontier Hero: A Self-Defeating Mythic Structure." *Agriculture and Human Values* 7, no. 1 (1990): 9–19.

———. "Telling the Farmers' Story: Competing Responses to Soil Conservation Rhetoric." *Quarterly Journal of Speech* 77, no. 3 (1991): 289–308.

———. "The Will to Conservation: A Burkeian Analysis of Dust Bowl Rhetoric and American Farming Motives." *Southern Speech Communication Journal* 52, no. 1 (1986): 1–21.

Pezzullo, Phaedra C. "Contextualizing Boycotts and Buycotts: The Impure Politics of Consumer-Based Advocacy in an Age of Global Ecological Crises." *Communication and Critical/Cultural Studies* 8, no. 2 (2011): 124–45.

Pfieffer, Ehrenfried. *Biodynamic Farming and Gardening: Soil Fertility, Renewal, and Preservation*. Translated by Frederick Heckel. New York: Anthroposophic Press, 1938.

Phillips, Sarah T. *This Land, This Nation: Conservation, Rural America, and the New Deal*. New York: Cambridge University Press, 2007.

Piper, Timothy David, dir. *Farmed and Dangerous*. Piro Productions, 2013. Netflix, http://www.netflix.com/title/50502l6.

Pipes, Richard. *The Russian Revolution*. New York: Knopf Doubleday, 1990.

Poe, Clarence. "Democracy in Cooperative Marketing." *Rural America* 3 (February 1925): 7–9.

Pollan, Michael. "Naturally." *New York Times Magazine*, May 13, 2001. http://www.nytimes.com/2001/05/13/magazine/naturally.html.

Prody, Jessica M. "Tracing the 'Back to the Land' Trope: Self-Sufficiency, Counterculture, and Community." In *The Political Language of Food*, edited by Samuel Boerboom, 1–26. Lanham, MD: Lexington Books, 2015.

Ragas, Matthew W., and Marilyn S. Roberts. "Agenda Setting and Agenda Melding in an Age of Horizontal and Vertical Media: A New Theoretical Lens for Virtual Brand Communities." *Journalism and Mass Communication Quarterly* 86, no. 1 (2009): 45–64.

RAM Trucks. "Official RAM Trucks Super Bowl Commercial 'Farmer.'" YouTube video, February 3, 2013. Video, 2:02. https://bit.ly/2ABenEm.

Ransom, John Crowe. "Corporate and Private Persons." In Agar and Tate, *Who Owns America?*, 93–108.

———. "Reconstructed but Unregenerate." In Twelve Southerners, *I'll Take My Stand*, 1–27.

———. "What Does the South Want." In Agar and Tate, *Who Owns America?*, 233–52.

Rasmussen, Wayne D. "The Civil War: A Catalyst of Agricultural Revolution." *Agricultural History* 39, no. 4 (1965): 187–95.

Rasmussen, Wayne D., and Paul Steven Stone. "Toward a Third Agricultural Revolution." *Proceedings of the Academy of Political Science* 34, no. 3 (1982): 174–85.

"Recall Roundup." Occupy Monsanto. Accessed November 1, 2018. http://occupy-monsanto.com/.

Reed, Matthew. *Rebels for the Soil: The Rise of the Global Organic Food and Farming Movement*. London: Earthscan, 2010.

Retzinger, Jean P. "Cultivating the Agrarian Myth in Hollywood Films." In *Enviropop: Studies of Environmental Rhetoric and Popular Culture*, edited by Mark Meister and Phyllis M. Japp, 45–62. Westport, CT: Praeger, 2002.

———. "Eleven Miles South of Hollywood: Analyzing Narrative Strategies in *The Garden*." *Environmental Communication* 5, no. 3 (2011): 337–43.

Ritzer, George. *The Globalization of Nothing 2*. Thousand Oaks, CA: Sage, 2007.

———. *The McDonaldization of Society 6*. Los Angeles, CA: Pine Forge Press, 2011.

Roberts, Paul. *The End of Food*. London: Bloomsbury, 2008.

Rodale Institute. "About Us." Accessed June 26, 2016. https://bit.ly/2Sve3hr.

Rodale, J.I. "20 Years of Organic Gardening." *Organic Farming and Gardening* 9, no. 6 (June 1962): 17–19.

———. "Introduction to Organic Farming." *Organic Farming and Gardening* 1, no. 1 (May 1942). https://bit.ly/2F59mZl.

———. *Pay Dirt: Farming and Gardening with Composts*. Old Greenwich, CT: Devin-Adair, 1945.

———. *The Organic Front*. Emmaus, PA: Rodale Press.

Romines, Ann. *The Home Plot: Women, Writing, and Domestic Ritual*. Amherst: University of Massachusetts Press, 1992.

Rosteck, Thomas, and Thomas Frentz. "Myth and Multiple Readings in Environmental Rhetoric: The Case of *An Inconvenient Truth*." *Quarterly Journal of Speech* 95, no. 1 (2009): 1–19.

Rubino, Joe. "Chipotle Hiring Former Taco Bell CEO to Replace Founder Steve Ells in Top Spot." *Denver Post*, November 29, 2018. https://dpo.st/2JtTw8X.

Rushing, Janice Hocker. "Mythic Evolution of 'The New Frontier' in Mass Mediated Rhetoric." *Critical Studies in Mass Communication* 3, no. 3 (1986): 265–96.

———. "The Rhetoric of the American Western Myth." *Communication Monographs* 50, no. 1 (1983): 14–32.

Rushkoff, Douglas. *Life Inc: How Corporatism Conquered the World and How We Can Take It Back*. New York: Random House, 2011.

Rutherford, G.W. "Types of Local Government—II." *Rural America* 8 (February 1930): 10–11.

Saldivar-Tanaka, Laura, and Marianne E. Krasny. "Culturing Community Development, Neighborhood Open Space, and Civic Agriculture: The Case of Latino Community Gardens in New York City." *Agriculture and Human Values* 21, no. 4 (2004): 399–412.

Sanderson, Dwight. "The People on the Land." *Rural America* 16 (January 1938): 3–6.

Sass, Eric. "WWI Centennial: Starvation Stalks Europe." *Mental Floss*, April 24, 2016. http://mentalfloss.com/article/85224/wwi-centennial-starvation-stalks-europe.

Sayre, Laura. "The Politics of Organic Farming: Populists, Evangelicals, and the Agriculture of the Middle." *Gastronomica: The Journal of Critical Food Studies* 11, no. 2 (2011): 38–47.

Schaffer, Ronald. *America in the Great War: The Rise of the Welfare State*. New York: Oxford University Press, 1991.

Schauberger, Bernard, Sotirios Archontoulis, Almut Arneth, Juraj Balkovic, Philippe Ciais, Delphine Deryng, et al. "Consistent Negative Response of US Crops to High Temperatures in Observations and Crop Models." *Nature Communications* 8 (Jan. 19, 2017). https://go.nature.com/2Rp15jZ.

Schmelzkopf, Karen. "Urban Community Gardens as Contested Spaces." *Geographical Review* 85, no. 3 (1995): 364–81.

Schnapp, Jeffrey T. "Epilogue." In James, *Picture This*, 369–75.

Schumaker, Erin. "Chipotle's Food Poisoning Issue Is Nationwide. Here's A Map." *Huffington Post*, December 10, 2015. https://www.huffingtonpost.com/entry/chipotle-food-poisoning-outbreak_us_56687740e4b009377b235bod.

Seidl, Jonathon M. "Wow Did That Dodge Ad with Paul Harvey Talking About Farmers Rock the Super Bowl." *The Blaze*, February 3, 2013. https://www.theblaze.com/stories/2013/02/03/paul-harvey-talking-about-farmers-in-dodge-ad-wow-was-that-amazing.

Sellers, Sean. "Chipotle Challenge: Time to Back Up 'Food with Integrity.'" *Grist*, December 11, 2009. https://grist.org/article/steve-ells-will-you-accept-the-chipotle-challenge/.

Seltzer, Walter, dir. *Soylent Green*. Burbank: Warner Entertainment, 1973. Film.
Shepardson, David. "Chrysler's Super Bowl Ad Seen, Shared More Than 10 Million Times." *Detroit News*, February 7, 2013. http://superbowl-ads.com/chryslers-super-bowl-ad-seen-shared-more-than-10-million-times/.
Shiva, Vandana. *Stolen Harvest: The Hijacking of the Global Food Supply*. Cambridge, MA: South End Press, 2000.
Shome, Raka. "Space Matters: The Power and Practice of Space." *Communication Theory* 13, no. 1 (2003): 39–56.
Short, Brant. "'Hello Americans': Paul Harvey and the Rhetorical Construction of Modern Agrarianism." *Journal of Radio Studies* 1, no. 1 (1992): 43–54.
Shprintzen, Adam D. *The Vegetarian Crusade: The Rise of an American Reform Movement, 1817–1921*. Chapel Hill: University of North Carolina Press, 2013.
Singer, Gerald, and Ivan Sergio Freire de Sousa. "The Sociopolitical Consequences of Agrarianism Reconsidered." *Rural Sociology* 48, no. 2 (1983): 291–307.
Singer, Ross. "Agrarian Myth, Public Memory, and the Industrial Food Narrative of American Family Farming at Iowa's Living History Farms Open-Air Museum." In *Foodscapes: Food, Space, and Place in a Global Society*, edited by Carlnita P. Greene, 167–97. New York: Peter Lang Publishing, 2019.
———. "The Corporate Colonization of Communication about Global Hunger: Development, Biotechnology, and Discursive Closure in the Monsanto Pledge." In *Food as Communication/Communication as Food*, edited by Janet M. Cramer, Carlnita P. Greene, and Lynn M. Walters, 405–27. New York: Peter Lang Publishing, 2011.
———. "Neoliberal Style, the American Re-Generation, and Ecological Jeremiad in Thomas Friedman's 'Code Green.'" *Environmental Communication* 4, no. 2 (2010): 135–51.
———. "Visualizing Agrarian Myth and Place-Based Resistance in South Central Los Angeles." *Environmental Communication* 5, no. 3 (2011): 344–49.
Sloane, David C., Allison L. Diamonte, LaVonna B. Lewis, Antronette K. Yancy, Gwendolyn Flynn, Lori Miller Nascimento, et al. "Improving the Nutritional Resource Environment for Healthy Living through Community-Based Participatory Research." *Journal of General Internal Medicine* 18, no. 7 (2003): 568–75.
Slotkin, Richard. *The Fatal Environment: The Myth of the Frontier in the Age of Industrialization, 1800–1890*. New York: Atheneum, 1985.
———. *Gunfighter Nation: The Myth of the Frontier in Twentieth-Century America*. New York: Atheneum, 1992.
———. *Regeneration through Violence: The Myth of the American Frontier, 1600–1860*. Norman: University of Oklahoma Press, 1973.
Slovic, Scott. "Epistemology and Politics in American Nature Writing." In *Green Culture: Environmental Rhetoric in Contemporary America*, edited by Carl G. Herndl and Stuart C. Brown, 82–110. Madison: University of Wisconsin Press, 1996.
Smith, Andrew F. *Eating History: Thirty Turning Points in the Making of American Cuisine*. New York: Columbia University Press, 2009.
Smith, Brendan. "The Coming Green Wave: Ocean Farming to Fight Climate Change." *The Atlantic*, November 23, 2011. https://bit.ly/2SxmxEM.

Smith, Henry Nash. *Virgin Land: The American West as Symbol and Myth.* Cambridge: Harvard University Press, 1950.
Smith, Kimberly. *Wendell Berry and the Agrarian Tradition: A Common Grace.* Lawrence: University Press of Kansas, 2003.
———. "Wendell Berry's Feminist Agrarianism." *Women's Studies* 30, no. 5 (2001): 623–46.
Smith, Pat J. "Agrarian Ideology and Region: The Persistence of Two Variants." *Rural Sociologist* 2, no. 5 (1982): 282–94.
Smith, Stephen. *Myth, Media, and the Southern Mind.* Fayetteville: University of Arkansas Press, 1985.
Smith, Tex. Letter to the Editor. *Ellensburg (WA) Daily Record,* Sept. 10, 1949. https://news.google.com/newspapers?id=uGYKAAAAIBAJ&sjid=-EoDAAAAIBAJ&pg=6888%2C805953.
Sontag, Susan. "Introduction." In *The Art of Revolution: Castro's Cuba, 1959–1970,* edited by Dugald Stermer, not paginated. New York: McGraw Hill, 1970.
"Soylent Meal Replacement Drink, Cafe Vanilla, 14 oz Bottles, 12 Pack." Amazon.com. https://amzn.to/2JpI2Dn.
Squires, Catherine. "Rethinking the Black Public Sphere: An Alternative Vocabulary for Multiple Public Spheres." *Communication Theory* 12, no. 4 (2002): 446–68.
"The State of Agricultural Commodity Markets: Agricultural Trade, Climate Change and Food Security." Food and Agricultural Organization of the United Nations, 2018. https://bit.ly/2xp3FhI.
Stewart, James B. "Chipotle's New Mantra: Safe Food, Not Just Fresh." *New York Times,* January 14, 2016. https://nyti.ms/2Sv8I9X.
Stewart, Jessie, and Greg Dickinson. "Enunciating the Locality in Postmodern Suburbs: FlatIron Crossing and the Colorado Lifestyle." *Western Journal of Communication* 72, no. 3 (2008): 280–307.
Stock, Kyle, and Venessa Wong. "Chipotle: The Definitive Oral History." *Bloomberg,* February 2, 2015. https://www.bloomberg.com/graphics/2015-chipotle-oral-history/.
Stockmann, Uta, Mark A. Adams, John W. Crawford, Damien J. Field, Nilusha Henakaarchchi, Meaghan Jenkins, et al. "The Knowns, Known Unknowns and Unknowns of Sequestration of Soil Organic Carbon." *Agriculture, Ecosystems and Environment* 164 (Jan. 2013): 80–99.
"Super Bowl Just Shy of TV Record." *ESPN* online, last modified February 4, 2013. https://es.pn/2OggfpK.
Sweet, Timothy. "Economy, Ecology, and Utopia in Early Colonial Promotional Literature." *American Literature* 71, no. 3 (1999): 399–427.
"Synopsis: *The Garden*: A Film by Scott Hamilton Kennedy." Black Valley Films. Accessed May 1, 2016. https://bit.ly/2qjUT16.
Taber, L. J. "Some National Issues." *Rural America* 12 (February 1934): 9–11.
Tabuchi, Hiroko. "In America's Heartland, Discussing Climate Change Without Saying 'Climate Change.'" *New York Times,* January 28, 2017. https://nyti.ms/2kE5YoO.
Tate, Allen. "Remarks on the Southern Religion." In Twelve Southerners, *I'll Take My Stand,* 155–75.
———. *The Fathers.* New York: Swallow, 1959.

Taylor, Carl C. "An Interpretation of the Conference." *Rural America* 8 (November 1930): 4–7.
Terpenning, Walter A. "Toward Greater Security." *Rural America*, November 1935.
Theiss, Lewis Edwin. "Victory Gardens." *New Country Life*, March 1919.
Thompson, Craig J., and Zeynep Arsel. "The Starbucks Brandscape and Consumers' (Anticorporate) Experiences of Glocalization." *Journal of Consumer Research* 31, no. 3 (2004): 631–42.
Thompson, Paul B. "Agriculture and Working-Class Political Culture: A Lesson from *The Grapes of Wrath*." *Agriculture and Human Values* 24, no. 2 (2007): 165–77.
———. *The Agrarian Vision: Sustainability and Environmental Ethics*. Lexington: University Press of Kentucky, 2010.
Thomson, Deborah Morrison. "Big Food and the Body Politics of Personal Responsibility," *Southern Communication Journal* 74, no. 1 (2009): 2–17.
Tigert, John J. "Objectives of Rural Education." *Rural America*, December 1926.
Tippett, Rebecca. "Agriculture and Food Statistics: USDA Charts the Essentials." Carolina Demography, Carolina Population Center, University of North Carolina, May 11, 2015. https://unc.live/2SyNTKF.
Tranel, Marl, and Larry B. Handlin Jr. "Metromorphosis: Documenting Change." *Journal of Urban Affairs* 28, no. 2 (2006): 151–67.
Treidler, Adolph. "Hunger." Poster for the United States Food Administration. Cincinnati, OH: Strobridge Lithographing Company, 1918. https://digital collections.hoover.org/objects/35615/the-president-says-hunger-does-not -breed-reform-it-breeds;jsessionid=A79E12AD04CBAEABF4C6ED0BC90 C9990?ctx=616d7c98-8b13-48b0-999b-8d3ea7dc32e1.
Trentmann, Frank. "Coping with Shortage: The Problem of Food Security and Global Visions of Coordination, c. 1890s–1950." In *Food and Conflict in Europe in the Age of the Two World Wars*, edited by Frank Trentmann and Fleming Just, 13–48. Houndsmills, UK: Palgrave Macmillan, 2006.
Turner, Frederick Jackson. "The Significance of the Frontier in American History." In *Rereading Frederick Jackson Turner: "The Significance of the Frontier in American History" and Other Essays*, edited by John Mack Faragher, 31–60. New Haven: Yale University Press, 1994.
Twelve Southerners. *I'll Take My Stand: The South and the Agrarian Tradition*. 1930. Repr.; Baton Rouge: Louisiana State University Press, 1977.
———. "Introduction: A Statement of Principles." In Twelve Southerners, *I'll Take My Stand*, xxxvii–xlviii.
USDA National Agricultural Statistics Service. "Iowa Ag News: Farms and Land in Farms," February 16, 2018. https://bit.ly/2qlwqsq.
U.S. Environmental Protection Agency. "Climate Impacts on Agriculture and Food Supply." Accessed October 2, 2018. https://bit.ly/2SApEft.
Veit, Helen Zoe. *Modern Food, Moral Food: Self-Control, Science, and the Rise of Modern Eating in the Early Twentieth Century*. Chapel Hill: University of North Carolina Press, 2013.
Vigden, Helen, ed. *Food Literacy: Key Concepts for Health and Education*. New York: Routledge, 2016.
Vincent, C. Paul. *The Politics of Hunger: The Allied Blockage of Germany, 1915–1919*. Athens: Ohio University Press, 1985.

Voelz, Johannes. *Transcendental Resistance: The New Americanists and Emerson's Challenge*. Lebanon, NH: University Press of New England, 2010.

Voicu, Ioan, and Vicki Benn. "The Effect of Community Gardens on Neighboring Property Values." *Real Estate Economics* 36, no. 2 (2008): 241–83.

Walsh, Lynda. *Scientists as Prophets: A Rhetorical Genealogy*. New York: Oxford University Press, 2013.

Walter, Gerry. "A 'Curious Blend': The Successful Farmer in American Farm Magazines, 1984–1991." *Agriculture and Human Values* 12, no. 3 (1995): 55–68.

Ward, Gordon H. "On Our Way—Where?" *Rural America* 11 (June 1933): 3–6.

Warner, Sam Bass. *To Dwell Is to Garden: A History of Boston's Community Gardens*. Boston: Northeastern University Press, 1987.

Weaks-Baxter, Mary. *Reclaiming the American Farmer: The Reinvention of a Regional Mythology in Twentieth Century Southern Writing*. Baton Rouge: Louisiana State University Press, 2006.

Weaver, Richard. *Ideas Have Consequences*. Chicago: University of Chicago Press, 1948.

———. *The Southern Tradition at Bay: A History of Post-Antebellum Thought*. New Rochelle, NY: Arlington House, 1968.

Weber, Pete. "Fiona Apple and Chipotle Channel Willy Wonka to Slam Factory Farming." *The Week*, September 13, 2013. https://bit.ly/2Pvoq6b.

Weiss, Elaine F. *The Fruits of Victory: The Woman's Land Army of America in the Great War* Washington, DC: Potomac Books, 2008.

Weiss, Elizabeth. "What Does 'The Scarecrow' Tell Us About Chipotle? *The New Yorker*, September 23, 2013. https://www.newyorker.com/business/currency/what-does-the-scarecrow-tell-us-about-chipotle.

Wells, Robert A. "Mobilizing Public Support for War: An Analysis of Propaganda During World War I." International Studies Association Conference, New Orleans, LA, March 2002. http://isanet.ccit.arizona.edu/noarchive/robert wells.html.

Weschler, Lawrence. "'Destroy this Mad Brute': The African Roots of World War I." *The Believer*, September, 2014. http://lawrenceweschler.com/library/article/destroy-this-mad-brute-the-african-roots-of-world-war-i.

"What Is Urban Farming?" Greensgrow: Growers of Food, Flowers, and Neighborhoods. Accessed June 17, 2018. http://www.greensgrow.org/urban-farm/what-is-urban-farming/.

White, Monica M. "D-Town Farm: African American Resistance to Food Insecurity and the Transformation of Detroit." *Environmental Practice* 13, no. 4 (2011): 406–17.

White, Richard. "Frederick Jackson Turner and Buffalo Bill." In *The Frontier in American Culture: An Exhibition at the Newberry Library, August 26, 1994-January 7, 1995*, edited by James R. Grossman, 6–65. Berkeley: University of California Press, 1994.

Wickstrom, Maurya. *Performing Consumers: Global Capital and its Theatrical Seduction*. New York: Routledge, 2006.

Wiebe, Robert H. *The Search for Order, 1877–1920*. New York: Hill and Wang, 1966.

Wilson, M.L. "Education for Democracy." *Rural America* 14 (September 1936): 3–6.

———. "Rural Discussion and National Democracy." *Rural America* 15 (May 1937): 6–7.
Wilson, Woodrow. "Extract from Woodrow Wilson's Jefferson Day Address to the Common Counsel Club, April 13, 1916." *Jefferson Quotes and Family Letters*, Monticello. http://tjrs.monticello.org/letter/1865.
Winne, Mark. *Food Rebels, Guerrilla Gardeners, and Smart-Cookin Mamas: Fighting Back in an Age of Industrial Agriculture*. Boston: Beacon, 2010.
Wirzba, Norman, ed. *The Art of the Commonplace: Agrarian Essays of Wendell Berry*. Washington, DC: Shoemaker & Hoard, 2002.
———. *The Essential Agrarian Reader: The Future of Culture, Community, and the Land*. Lexington: University Press of Kentucky, 2003.
Wirzba, Norman. "Introduction: Why Agrarianism Matters—Even to Urbanites." In Wirzba, *Essential Agrarian Reader*, 1–22.
Wolfe, Dylan. "The Ecological Jeremiad, the American Myth, and the Vivid Force of Color in Dr. Seuss's *The Lorax*." *Environmental Communication* 2, no. 1 (2008): 3–24.
"Women in Bread Riot at Doors of City Hall." *New York Times*, Feb. 21, 2017. https://timesmachine.nytimes.com/timesmachine/1917/02/21/98240700.pdf.
Zagacki, Kenneth. "Preserving Nature and Heritage during the *War on Terrorism*: The North Carolina Outlying Landing Field (OLF) Controversy." *Southern Communication Journal*, 73, no. 4 (2008): 261–79.
Zeiger, Robert H. *America's Great War: World War I and the American Experience*. Lanham, MD: Rowman & Littlefield, 2000.
Zelinkova, Zuzana, and Thomas Wenzl. "The Occurrence of 16 EPA PAHs in Food—A Review." *Polycyclic Aromatic Compounds* 35 (2–4): 248–84.
Zencey, Eric. "The Rootless Professors." In *Rooted in the Land*, edited by William Vitek and Wes Jackson, 15–24. New Haven: Yale University Press, 1996.

INDEX

A

abstracted universal theme, 102
Ackman, William, 184
adulterated food, 112
African Americans: community gardens, 141, 144, 145; Country Life movement, 66; farm workers, 199; plantation system, 106
agrarianism: agrarian practices, 192, 210–17; alternative food networks, 179, 186, 233; apocalyptic perspective, 221–24, 229, 231–33; authenticity, 192–93; branding strategies, 30, 31–32; climate change impacts, 229–35, 236; connectedness, 4–7, 14; democratic values, 197; fundamentalist perspectives, 213; holistic philosophy, 211, 212; Jeffersonian perspective, 18–21, 23, 25, 38, 192–93, 196, 197, 233; land-as-sacred trope, 83, 86, 194–95, 211; land ownership, 159, 197–98, 201; moral and ethical principles, 90, 112, 121, 153, 191; mythic qualities, 3–17, 151, 194; pan-agrarian consciousness, 234–35; political impact, 9–10, 197; property rights, 159, 197–98; rhetorical functions, 233–34; romanticization, 7–9, 202–3, 206–9; solidarity tropes, 203–9; structural characteristics, 233; traditional vs. new agrarianism, 11–12, 14–15, 149, 166, 191, 204–5; transformative potential, 5, 7, 10; universalization, 213–14; weaponization of food, 37–38, 40–42, 53–54, 235–36. *See also* virtue/virtuousness

Agrarianism in American Literature (Inge), 194

agrarian topoi, 192–93, 198, 211, 218, 234. *See also* place-making tropes; practices, agrarian; solidarity tropes

agrarian virtues: Chipotle Mexican Grill, 181; community gardens, 149; corporate co-option, 203; Country Life movement, 67; ecological jeremiad, 120, 122, 126, 130; emerging trends, 236; industrialism, 129; RAM truck Super Bowl commercial, 203, 213, 217; vernacular narratives, 140

agribusiness, 114, 122–23, 179–81, 227
agricultural extension services, 65–66
agricultural industrialization. *See* industrial agriculture
agricultural literacy, 5
aid blockades, 235–36
Alameda-Barbara Investment Company, 144
Allen, Danielle S., 85
"A Love Story" (animated video), 184–86
Alphabet, 224
alternative food networks, 179, 186, 233
Alter, Robert, 223
American agrarian myth: adaption and reconfiguration strategies, 17, 25–26; agrarian ideologies, 219; agrarian practices, 192, 209–17; apocalyptic perspective, 221–24, 229, 231–33; "Back to the Start" (animated video), 169–72, 173, 174, 177, 181; characteristics, 17, 25–27, 38; Chipotle Mexican Grill, 31–32, 133–34, 164–87; civic virtues, 18–21, 25, 26, 87, 154, 179, 199; contradictory worldviews, 11, 74, 90, 213; cooperative attitudes, 83–86, 154; Cooperian perspective, 21; corporate co-option, 24–25, 108, 191, 202–3; Country Life movement, 65–68, 71–81, 86–88, 133; creation stories, 198; dominant narratives, 22,

25–27; ecological jeremiad, 108, 110, 116–30; economic factors, 21, 24, 32, 66–67, 71–80, 87, 215, 229; emerging trends, 32–33, 223–37; environmental connectedness, 152, 153–54, 212, 214; *Farmed and Dangerous* (documentary-style satire), 179–81; food production and conservation, 44–48; fragmented imagery, 27; frontier hypothesis, 22–23; heterotopian space, 91–92, 101–8, 143, 151, 155, 158; historical perspective, 132–33; holistic philosophy, 211, 212; ideological perspectives, 17–28, 190–91; interconnectedness, 211; Jeffersonian perspective, 18–21, 23, 25, 38, 156, 192–93, 233; land-as-sacred trope, 83, 86, 194–95, 211; land ownership, 159, 197–98, 201; Liberty/Columbia imagery, 52, 53–54, 57, 61; moral virtues, 10, 18–21, 23–24, 31, 133, 149, 152, 154, 158; mythical caretaking farmer, 110, 116, 125, 194, 200, 204–5, 206–8, 212–13, 214; mythic connectedness, 3–17, 151, 194; naturalization rhetoric, 202–3; nostalgic appeal, 166, 174, 202–3, 206–9; organic farming movement, 110–33; pan-agrarian consciousness, 234–35; place-making tropes, 192, 194–98, 201–3, 206–7; policymaking impacts, 24–25, 215; political perspectives, 23–24, 157–59, 230; post-modern perspective, 32; post-World War I era, 64, 66–67, 71–81; primary threats, 21; propaganda campaigns, 39; prosthetic memories, 49, 51, 61; RAM truck Super Bowl commercial, 32, 134, 187, 189–92, 198–203, 206–9, 213–19; redemption and second chances, 26, 88, 99, 121, 165, 184–86; research perspective, 3–17, 28–33; rhetorical functions, 233–34; sacred agrarian tradition, 129–30, 152–54; "The Scarecrow" (animated video), 174–79, 181, 182–83; SCFarm/SCFarmers, 133, 137–61; science-technical progress connection, 132; simplicity of institutions, 82–85, 149; social constructs, 179; socially resistive delinking, 157; sociocultural impacts, 90, 93–97, 157–59; solidarity tropes, 19, 149, 191, 192, 203–9; Southern Agrarian perspective, 89–96, 98–108, 133, 195; spirituality-nature connection, 91–95, 97, 100–106, 108, 152–54, 199; structural characteristics, 233; urban agrarianism, 31, 137–61, 210–17, 226–27; virtuous farmers, 18–21, 25, 116–17, 149, 151; war gardens, 30, 39–40, 50–51, 55–61, 114; weaponization of food, 37–38, 40–42, 53–54, 235–36; White-Indian conflict, 21–22, 133; working myths, 25. *See also* branding strategies; Chipotle Mexican Grill; citizen-farmers; community gardens; farmers/farming; marketing and advertising; virtue/virtuousness; yeoman farmers

American Country Life Association: agrarian resistance, 64, 68, 77, 79, 86–87, 133; critique of communism, 78; cultural diversity, 66; historical perspective, 65–68; sense of community, 81–86; state legitimacy concerns, 67, 69–81, 86–88

American Friends of German Democracy, 48

American Medical Association (AMA), 115

American Review, 97

ancient organic agricultural practices, 108, 119, 122, 129–30, 132, 233

Anguiano, Claudia, 140

animal husbandry, 124, 210

animal manures. *See* manure

animal suffering, 173, 175, 177

Animal Veterinarian Medical Association, 124

animistic mimesis, 168, 169, 172, 176

Annenberg Foundation, 146

anthropomorphic vegetables, 57
antibiotics use, 124
anti-corporate populist movements, 13, 22, 31, 101–8
anti-federalist perspective, 96, 107, 108
anti-loafing laws, 59
ants, 228
apocalypse, definition of, 32
apocalyptic perspective, 221–24, 229, 231–33
Apple, Fiona, 175–76
aquaculture, 228
Armour and Company, 44
Armour, J. Ogden, 44
art and literature, 95–96
artificial fertilizers, 125–26, 127. *See also* chemical agriculture
artificial intelligence (AI), 225, 232
artificiality, 166, 167, 180, 185
asceticism, 118
The Atlantic, 224
Attack on Leviathan (1938), 94
Austria: food shortages, 55; postwar food assistance, 63
authenticity: agrarian mythmaking, 168, 173, 174, 186, 192–93; branding strategies, 166–87, 191; corporate branding, 186, 191; food production, 166–87; heterosexual monogamy, 184–85; lifestyles, 167–68; moral and ethical principles, 167; satirical critiques, 182–83
authentic local theme, 102
authoritarian regimes, 18, 38, 76, 77, 79, 81, 87
Autonomous Tractor Corporation, 225
autonomous vehicles, 32, 224–25
Axelrod, Alan, 47

B

Backstreet Boys, 184–85
back-to-the-land movement, 112, 129
"Back to the Start" (animated video), 169–72, 173, 174, 177, 181
bacterial illnesses, 163–65, 173, 181–82, 185
Bailey, Liberty Hyde, 12, 65, 82, 83, 84, 85

Baldwin, Robert D., 71
Balfour, Lady Eve, 114, 118
Banet-Weiser, Sarah, 167, 171, 182
BarfBlog, 182
Barthes, Roland, 203
Baudrillard, Jean, 32, 178
Baumann, Shyon, 167
Baum, L. Frank, 175
beetles, 228
Behind the Counter: Inside Chipotle Mexican Grill (documentary film), 173
Bellow, Saul, 223
beneficial microbes, 119, 122, 128
Benhabib, Seyla, 70
Bennett, Jeffrey A., 85
Bernays, Edward, 47
Berry, Wendell, 11, 13, 89, 107, 154, 195, 204, 205, 206
Big Ag (website), 225
biodiversity, 153, 154–55, 195, 212
biodynamic agriculture, 111–12, 113
Biodynamic Farming and Gardening (Pfieffer), 113
birds, 125
Black Valley Films, 147–48
Blackwood, Boston B., 190
blight, city, 144
"blood and soil" political philosophy, 97
Bloomberg Television, 173, 182
Bolshevik Revolution, 78
Boone, Daniel, 22
Borgmann, Albert, 198
Boxer, Barbara, 146
Boyd, Todd, 139
Bradley, Tom, 144
branding strategies: authenticity campaigns, 166–87, 191; brandscapes, 168, 169, 171, 176, 178–79, 182, 185–87; Chipotle Mexican Grill, 31–32, 164–87, 235; corporate branding, 186, 191; fast-food industry, 184, 186. *See also* RAM truck Super Bowl commercial
Broad, Garrett, 158–59
Browne, William J., 24, 25
Brunner, Edmund S., 84

Buber, Martin, 223, 233, 236
Bugs: The Film (documentary film), 228
Burger King, 163
Burke, Kenneth, 27, 62, 90, 93, 101–2, 183
Burkholder, Thomas R., 27, 152
Butterfield, Kenyon, 73, 76, 78, 80–81

C

Calhoun, Craig, 68
Camporesi, Piero, 38
cannibalism, 54
capacity-building strategies, 30, 158, 160–61
capitalist systems, 90, 93, 95, 98–99, 154, 172, 196, 205. *See also* corporatism
Capozzola, Christopher, 45, 48–49
carbon sequestration, 230
Cardenas, Tony, 152
caretaking ethos, 110, 116, 125, 194, 200–201, 204–5, 207–8, 212–13, 214
Carlson, Allan, 13
Carson, Rachel, 110, 116, 118, 130, 131
cartel capitalism, 46
Case (tractor manufacturer), 216, 225
Center for Food Safety (University of Georgia), 182
Centers for Disease Control and Prevention (CDC), 165
Central American immigrants, 144–45
Cenzatti, Marco, 143
chemical agriculture, 30, 109–13, 117–23, 125–30, 140, 235
Chen Yea-Wen, 140
Chicago (Illinois) food riots, 37
chickens, 8, 124, 127, 128, 174, 228
China: compost-based agricultural methods, 118; trade wars, 229
Chipotle Mexican Grill: agrarian myth-making, 133–34, 164–87; authenticity campaigns, 166–87; branding strategies, 31–32, 164–87, 235; contradictory worldviews, 178–79; food activism, 179–81; foodborne illness outbreaks, 163–65, 173, 181–82, 185; historical perspective, 173; innocence and purity campaigns, 166, 168–71, 173, 174–87; legal challenges, 165; McDonaldization, 172–73; redemption and second chances, 170–71, 173, 184–87; stock value collapse, 164–65, 184; target market, 172
Chomsky, Noam, 213
Chrysler, 224, 235
citizen-farmers: caretaking ethos, 110, 116, 125; democratic values, 156, 171, 176; economic factors, 156; Jeffersonian perspective, 18–19, 132, 156, 158, 193, 196, 197; marginalized communities, 133; as mythic heroes, 20, 23, 206–8, 214; as noble victims, 140, 158, 167; organic farming movement, 110, 116–17, 121, 125, 128, 129–30; real food production, 167; sacred agrarian traditions, 121; Southern Agrarian perspective, 98; urban community gardens, 148–49; virtue/virtuousness, 17, 125, 129, 140, 149, 151, 171, 213; westward migration movement, 133
citizen legitimacy, 70, 71–81
Citizens of South Central, 150, 156
civic agriculture, 142, 156. *See also* community gardens
civic virtues, 18–21, 25, 26, 87, 154, 179, 199
class conflict, 21–22, 133, 140
climate change, 33, 229–35, 236
Cobb, Brent, 217
Cody, Buffalo Bill, 22
coercive voluntarism, 44
Cohen, Joshua, 69–70
Coldplay, 169
Collins, Seward, 97
colonial resistance, 139, 154, 156–60
Committee on Public Information (CPI), 39, 46–48, 50, 53, 54
commodity crops, 226, 229
commodity culture, 90, 94–96, 98–99
common good, 83, 86
communication networks, 47–48, 51
Communism, 78
communities of color, 142, 145
community capacity-building

strategies, 30, 158, 160–61. *See also* community gardens
community gardens: benefits, 141–43; cultural diversity, 141, 144–61; democratic land-use procedures, 155–56; emerging trends, 227; ethnic agricultural practices, 149, 154, 155; food production and conservation, 55–61; historical perspective, 141–42; internal governance structure, 155–56; place-based resistance, 137, 148, 154, 155, 158; political dynamics, 143; social impacts, 57–63; South Central Farm (SCFarm), 31, 137–61; urban citizen participation, 40; vacant lots, 59, 141, 143–46; volunteerism, 59–60, 227
community, sense of, 81–86, 140, 153, 197, 201, 204, 206–7
community-supported agriculture, 3, 227
compost-based agricultural methods, 118, 125–30
concentrated animal feeding operations (CAFOs), 124
Concerned Citizens of South Central Los Angeles, 144, 145, 149–50
connectedness: agrarianism, 4–7, 14; agrarian mythmaking, 81–82, 140, 151–54, 157–58, 223–24, 229; citizen-farmers, 132–33; Country Life movement, 82; democratic participation, 21, 77, 83–84; environmental connectedness, 152, 153–54, 212, 214; familial connectedness, 151–52, 154, 156; food production, 29, 39, 166–67, 177; gardens, 60, 68; Jeffersonian perspective, 192, 194, 196, 199; marketing and advertising, 203; nostalgic appeal, 23; organic farming movement, 132–33, 222; place-making tropes, 194–96, 199–201, 204–6, 209–11, 230, 233; prosthetic memory, 49, 61; sacred agrarian tradition, 26, 83, 92, 129, 153; science-technical progress, 93–96; Southern

Agrarian perspective, 101, 106, 108; spirituality-nature connection, 151–54, 169, 176, 177, 199, 222; urban citizens, 109, 138; yeoman farmers, 74. *See also* RAM truck Super Bowl commercial
consent, manufacture of, 213–14
constitutional freedoms, 157
consumer-citizens, 166–67
consumerism: agrarian mythmaking, 174, 186–87; as agrarian threat, 13, 21, 90; brandscapes, 168, 169, 171; female emancipation, 103–4; food activism, 179–81; Liberty/Columbia imagery, 52; McDonaldization, 172–73; moral and ethical principles, 177; neoliberal perspective, 159; organic farming movement, 132; place-making vs. space-making tropes, 195–96; social media campaigns, 166–69; Southern Agrarian perspective, 92, 94–96, 101–8; as utopian ideal, 51, 174. *See also* branding strategies; marketing and advertising
contaminated food, 163–65, 168, 173, 174, 179–81. *See also* foodborne illness outbreaks
contested land, 143–46, 149–51
contoured plowing, 232
convenience-food industry. *See* fast-food industry
Coolidge, Calvin, 24
cooperative attitudes, 83–86, 154
Cooper, James Fenimore, 21
corner convenience-store grocery shopping, 142
corporate farming, 232
corporatism: branding strategies, 186, 191; food production, 170–71, 173; mega-farms, 225, 226, 232; place-making vs. space-making tropes, 195–96; RAM truck Super Bowl commercial, 134; Southern Agrarian perspective, 96, 101–8
corruption vs. honesty, 149–50, 158, 170–71, 177–78

counterpublics, 26, 68–69, 71–74, 76–77, 88, 234
Country Life Commission, 65–66
Country Life movement: accomplishments, 86–88; agrarian resistance, 64, 68, 77, 79, 86–87, 133; agrarian solidarity, 205; cultural diversity, 66; economic factors, 71–80; farmers/farming, 64, 65–66, 67, 71–81, 86–88, 133, 171; historical perspective, 65–68; sense of community, 81–86; simplicity of institutions, 82–85; state legitimacy concerns, 67, 69–81, 86–88
court-house meetings, 85
cracker box/cracker barrel politics, 84–85
creation stories, 198
Creel, George, 39, 46–47, 49–50
cricket farming, 228–29
crop failures, 41
crop subsidies, 86
Crowley, Sharon, 17
Culp, V. H., 73–74, 77
cultural contamination, 168, 178
cultural diversity: community gardens, 141, 144–61; farm workers, 199; new agrarianism, 31
cultural identity, 17, 137
cultural syncretism, 139, 158
culture-jamming, 183
culture of convenience, 6

D

Daily Record (Ellensburg, Washington), 190
Daimler, 235
Darsey, James, 116
Davidson, Donald, 91, 92, 93, 94, 95–96, 105
Debord, Guy, 178
decentralization movement, 13
decolonial resistance, 139–40
defense tropes, 89–90, 204
degenerate industrialism, 93, 94, 96–97
democratic strangers, 85–86
democratic values: agrarianism, 197;

articulatory functions, 79–80; characteristics, 81; citizen-farmers, 156, 171, 176; community gardens, 56–57, 60, 155–56; connectedness, 21, 77, 83–84; constitutional freedoms, 156–57; cooperative attitudes, 83–86, 154; Country Life movement, 67–81; dissociation strategies, 76–81; dynamic principles, 80–81; fast-food industry, 172, 174; frontier hypothesis, 80; Jeffersonian perspective, 193; local communities, 83–87; modernized future, 56; organic farming movement, 132, 172; political participation, 71–81; publics/public spheres, 68–74, 76–80; resilience, 39; rural life, 65, 71–81, 84–85, 87, 156, 171, 236; sacred agrarian traditions, 83; sense of community, 81–85, 194–98, 201; simplicity, 82–85, 149; state legitimacy and power, 67, 69–81, 86–88; urban citizen participation, 85–86; virtue/virtuousness, 81, 132, 171, 172, 174, 194, 214, 222
demonization portrayals, 54
Denver Post, 184
Department of Agriculture, 112, 225, 226
deserts, food, 142
desiring machines, 168
Dickens, David R., 202
Dickinson, Elizabeth, 140
Dickinson, Greg, 167
digital marketing, 168–69, 176, 179, 181, 186
dishonest food-marketing "greenwash", 8
Disneyland, 178
displaced farmers, 225–26
dissent, 213
dissociation strategies, 76–81
ditch meetings, 85
documentary films, 173, 179–81, 217–18, 228. *See also The Garden* (documentary)
documentary newsreels, 51

dominant cultures, 139, 141, 155, 158–60
Doughboys, 54, 63
Dougherty, Debbie, 215
Douglas, Mary, 178
Doyle, Michael, 182
driverless vehicles, 224–25
drought conditions, 230, 231, 232, 234
Duck, Leigh Anne, 95
due process rights, 157
Dust Bowl, 225, 231–32
dystopian vision, 92, 93

E

earthworms, 122
E. coli bacteria, 163–64, 165, 181
ecological jeremiad, 108, 110, 116–30, 195
ecologically responsible agricultural practices, 210, 212
economics: agrarian resistance, 133; agrarian solidarity, 205–6, 210; agribusiness, 114, 122–23, 179–81, 227; citizen-farmers, 156; climate change impacts, 231; community gardens, 141; consumerism, 42, 172; Country Life movement, 66, 67, 71–80; economic development, 145, 158–59, 196; exploitation, 30; farmers/farming, 71–80; food resources, 190–91; generalist agrarian practices, 211; heterotopian space, 155; household management, 204; Jeffersonian perspective, 18, 21, 193, 197; land value, 145, 152, 155, 158–59, 196; local communities, 206, 215; moral contracts, 152; neoliberalism, 137, 140, 149; new agrarianism potential, 9, 12, 15, 29; policymaking impacts, 215; post-World War I era, 67; RAM truck Super Bowl commercial, 32, 201; rural life model, 21, 71–80, 192–93, 215; SCFarm/SCFarmers, 137, 145; self-determination, 25; small-scale farms, 177, 225–27; Southern Agrarian perspective, 91, 93–95, 98–102, 105; utopian

countermovements, 107; World War I, 39, 66–67. *See also* branding strategies; marketing and advertising
eco-vernacular discourse, 140
Eddie, Don, 152
Egan, Timothy, 231, 232
eggs, 8, 127, 128
Eisinger, Chester, 123, 197
ejido system, 156
electric self-driving cars, 224
electrification, rural, 86
Ellensburg, Washington, 190
Elliot, Norbert, 116, 117, 130
Ells, Steve, 173, 184
eminent domain, 144, 145
Emmaus, Pennsylvania, 113
Emmett, Robert, 143
environmental connectedness, 152, 153–54, 212, 214
environmental justice movement, 144, 158, 161
environmental movements: eco-vernacular discourse, 140; land-as-sacred trope, 194–95; moral and ethical principles, 195; organic farming, 116, 131; place-making tropes, 194–95; SCFarmers, 137–61
environmental stewardship, 159, 160, 215
environmental sustainability, 3, 149, 177, 193, 195, 197
erosion, 125, 127
ethical behaviors, 8, 191
ethics: agrarian principles, 90, 112, 121, 153, 191; agrarian resistance, 6, 12, 21, 64, 133, 222; authenticity, 167; caretaking ethos, 196; consumerism, 177; environmentalism, 195; hospitality, 205; human relations, 81, 83, 90; interconnectedness, 210; Jeffersonian perspective, 18; land ethic, 204; marketing and advertising, 191; planning processes, 161; political solidarity, 203; redemption and second chances, 121; self-sufficiency, 112, 214; simplicity,

84, 171, 177; social justice, 161; sustainability, 233; work ethic, 199, 200–202, 207, 208
ethnic agricultural practices, 149, 154, 155
ethnic and racial oppression, 140
Europe: colonial perspectives, 196; food shortages, 37, 38, 39, 41–42, 44, 55, 62; food trade systems, 41; postwar food assistance, 63; provincial communities, 98; state-sponsored propaganda campaigns, 52, 53–54; U.S. food acquisition and distribution systems, 44–46
exotic foods, 167, 168
exploitation: agrarian mythmaking, 25; corrupt urban elites, 149; exotic foods, 167; farmers/farming, 74, 78; female labor, 103; food production, 62; labor relations, 3, 10, 74, 103–4; land, 82, 105–6, 155; plantation system, 98–99; rural life, 4, 149; SCFarm/SCFarmers, 155; slavery, 97, 99–100; socioeconomic exploitation, 30, 93; solidarity, 206; Southern Agrarian perspective, 89, 93–94, 97–99, 103; world trade system, 40–41. See also industrialism
expository filming techniques, 147–48
extension service, 65–66

F

factory gardens, 60–61
fair-trade practices, 33
fake food, 184–85
false promises, 78
familial connectedness, 151–52, 154, 156
family solidarity, 205, 207, 208–9
Farmed and Dangerous (documentary-style satire), 179–81
farm equipment, 32–33, 201–2, 216, 224–25, 231–32. See also mechanized agriculture
farmers/farming: agrarian practices, 192, 209–17; alternative food networks, 179, 186, 233; "Back to the Start" (animated video), 169–72, 173,

174, 177, 181; beneficial legislation, 86–88; caretaking ethos, 110, 116, 125, 194, 200–201, 204–5, 207–8, 212–13, 214; chemical agriculture, 30, 109–13, 117–23, 125–30, 140, 235; climate change impacts, 229–35, 236; compost-based agricultural methods, 118, 125–30; cooperative attitudes, 83–86, 154; corporatization, 134, 189–92; Country Life movement, 64, 65–66, 67, 71–81, 86–88, 133, 171; creation stories, 198; cultural diversity, 199–200; democratic participation, 65, 71–81, 84–85, 87, 156, 171, 236; demographic changes, 196–97; displaced farmers, 225–26; economic factors, 21, 24, 32, 66–67, 71–80, 87, 215, 229; educational opportunities, 75; emerging trends, 32–33, 223–37; farm and acreage declines, 225; general design, 127; government subsidies, 232; health-related consequences, 120, 121, 131; heroic portrayals, 20, 23, 206–8, 214; home-front support efforts, 52; Jeffersonian perspective, 18–21, 23, 25, 38, 192–93, 196, 197, 233; labor shortages, 56; land-as-sacred trope, 83, 86, 194–95, 211; land ownership, 159, 197–98, 201; Liberty/Columbia imagery, 53–54, 61; local communities, 83–87, 91–92, 100–108; management practices, 204, 215, 225, 232; marginalization, 73, 76–77, 79, 87, 133; mechanized agriculture, 32–33, 94, 112, 127–28, 131, 216, 224–25, 231–32; megafarms, 225, 226, 232; militant pastoralism, 101–5, 108; modernization impacts, 112–13, 129; mono-crop cultivation, 225, 232; moral contracts, 152; moral virtues, 4, 26, 67, 89, 160, 171, 191, 200–202, 215–17; mythical caretaking farmer, 110, 116, 125, 194, 200, 204–5, 206–8, 212–13, 214; mythic connectedness, 151, 171; mythic vernacular narratives,

154; negative representations, 170, 174–79; nostalgic appeal, 30–31, 32, 171, 189–92, 196–203, 206–9, 214–19; pan-agrarian consciousness, 234–35; pesticide residues, 120; place-making vs. space-making tropes, 196; political perspectives, 24, 215, 230; post-World War I impacts, 29, 55–57, 64, 66–67, 71–81; post-World War II impacts, 196; prototypical farmer, 200; questionable agricultural practices, 123–25, 129, 130; RAM truck Super Bowl commercial, 187, 189–92, 198–203; redemption and second chances, 26, 121; reform movements, 112–13; resilience, 216, 217–18; restorative farming methods, 230; rhetorical functions, 233–34; sacred agrarian tradition, 129–30, 152–54, 171, 194–95; "The Scarecrow" (animated video), 174–79, 181, 182–83; science-technical progress connection, 132; sense of community, 81–85, 140, 153, 194–98, 201, 206–7; simplicity of institutions, 82–85, 149; small-scale farms, 177, 225–27; soil-food connection, 117–20; solidarity tropes, 203–9; Southern Agrarian perspective, 89–99, 101–8; spirituality-nature connection, 91–95, 97, 100–106, 108, 152–54, 171, 199; structural characteristics, 233; tractor use, 127–28, 131; urban farming, 88, 227; virtue/virtuousness, 24, 124, 149, 151, 171, 213–14, 217; virtuous farmers, 18–21, 25, 81, 140, 148, 201–2, 206–9, 213, 214; volunteerism, 56; weaponization of food, 37–38, 40–42, 53–54, 235–36; women workers, 56; work ethic, 199, 200–202, 207, 208; working myths, 25. *See also* citizen-farmers; industrial agriculture; organic farming movement
farmers' markets, 3
Farmers of Forty Centuries (King), 118

Farmer-Stockman (Hartshorne, Oklahoma), 190
farm-to-table restaurants, 3
fast-food industry: alternative food networks, 179, 186; animal suffering, 173, 175; branding strategies, 184, 186; democratic values, 172, 174; foodborne illness outbreaks, 163–65, 173, 181–82, 185; high-fat American diet, 10, 142; McDonaldization, 172–73; redemption and second chances, 170–71, 173, 185–86; satirical critiques, 182–83; social media campaigns, 31–32, 164, 166; as utopian ideal, 174, 186. *See also* Chipotle Mexican Grill
The Fathers (Tate), 99
Federal Trade Commission (FTC), 115
female emancipation, 103
Ferrell, Keith, 221–22
fertile land, 18
fertilizer shortages, 41
Fiat Chrysler, 216, 225, 235
Fifteenth Amendment (U.S. Constitution), 104
Fisher, Mary Shattuck, 103–4
fishing harvests, 228
Flagg, James Montgomery, 48, 50, 51, 52
Foner, Eric, 52
food activism: *Farmed and Dangerous* (documentary-style satire), 179–81; urban agrarianism, 137–61
foodborne illness outbreaks, 163–65, 173, 181–82, 185
food deserts, 142
foodie culture, 167
food insecurity, 10, 33, 131, 137–38, 142
food literacy, 8, 142
food production: agrarian practices, 192, 210–17; alternative food networks, 179, 186, 233; authenticity campaigns, 166–87; "Back to the Start" (animated video), 169–72, 173, 174, 177, 181; commodity crops, 226, 229; community-supported agriculture, 227; connectedness, 29, 39, 166–67, 177; corporate co-option,

131, 219; culture of convenience, 6; ecologically responsible agricultural practices, 210, 212; emerging trends, 32–33, 223–37; ethnic foodways, 155; *Farmed and Dangerous* (documentary-style satire), 179–81; federal legislation impacts, 112; home-front support efforts, 38–40, 44–48, 53–54; mega-farms, 225, 226, 232; mono-crop cultivation, 232; nostalgic appeal, 30–31, 32, 166, 169, 171, 174–79, 181, 202; pan-agrarian consciousness, 234–35; pesticide residues, 120; post-World War I era, 55–61; processed foods, 141, 142, 155; real food, 167; reform movements, 112–13; rhetorical functions, 233–34; safety concerns, 163–65, 173, 181–82, 184; satirical critiques, 182–83; science-technical progress connection, 132; small-scale farms, 225–27; social media campaigns, 166; "So God Made a Farmer" (television commercial), 189–92, 198–203; soil-food connection, 117–20; "The Scarecrow" (animated video), 174–79; urban community gardens, 55–61, 137–61; urban farming, 88, 227; World War I era, 38–40, 44–48, 53–54, 62. *See also* Chipotle Mexican Grill

food resources: acquisition and distribution systems, 46; ecologically responsible agricultural practices, 210–17; ethnic foodways, 154, 155; international trade systems, 40–42; postwar food policies, 54, 55–57, 63; social reorganization impact, 62–64; state-sponsored propaganda campaigns, 47–54, 62; weaponization, 40–42, 53–54, 235–36. *See also* community gardens

food riots, 37, 39, 41

food safety. *See* safety, food

food security, 10, 39, 131, 137–38, 142

food shortages, 37, 38, 39, 41–42, 44, 55, 62

food studies: agrarian mythmaking, 190–91; criticisms, 14; non-food social actors, 191; social change movement, 13–14

Food with Integrity media campaign (Chipotle), 31, 164, 166–67, 169–74, 182, 235

Fort Dix, New Jersey, 58

Forty-First and Alameda, Los Angeles, 143–46

Foucault, Michel, 26

Four-Minute Men, 47

Fourteenth Amendment (U.S. Constitution), 104

Frame, Nat T., 80, 82–83

France, 119

Fraser, Nancy, 69

freedom of speech and assembly, 157

freehold land ownership, 19, 71, 123, 128, 197–98

Freud, Sigmund, 47

Freyfogle, Eric T., 197, 204

frontier hypothesis, 22–23, 25, 80, 129, 194, 212

frontier-romance literary genre, 21–22

fundamentalism, agrarian, 213

fungi, 119, 122, 128

Funny or Die (comedy group), 182–83

Future Farmers of America, 190

G

Gadsden, Alabama, 190

Gaia hypothesis, 195

The Garden (documentary), 28–33, 138–41, 143, 146–58, 169–70

"Garden of the West" myth, 23–24, 25

gardens: communication networks, 57–58; compost-based agricultural methods, 118, 125–27; cooperative programs, 59–60; Country Life movement, 64; cultural diversity, 141, 144–61; ecologically responsible agricultural practices, 210; emerging trends, 227; ethnic agricultural practices, 149, 154, 155; kitchen gardens, 155; manuals, 56; military bases, 58–59; place-based resistance, 137, 148, 154, 155, 158; post-World War I era, 55–61, 64; propaganda poster campaigns, 50–51, 57, 61; as sacred spaces, 26; social impacts, 57–63; South Central Farm (SCFarm), 31, 137–57; Victory Garden movement, 30, 40; volunteerism, 59–60, 227; war gardens, 30, 39–40, 50–51, 55–61, 114. *See also* community gardens

The Gateway Bug (documentary film), 228

generalist agrarian practices, 211, 214

genetically modified foods, 155

genocide, 106, 194

German immigrants, 47–48

Germany: demonization portrayals, 54; food shortages, 41, 55; postwar food assistance, 63; state-sponsored propaganda campaigns, 52, 53

Gibson, Charles Dana, 50, 223

"Gibson Girl" illustrations, 50

Giddens, Anthony, 195

The Gift of Good Land (Berry), 195

globalization impacts, 195–96, 205

global warming, 33, 229–35, 236

GMO foods, 180, 182, 183

Goldman, Robert, 202

Gone with the Wind (Mitchell), 99–100

good vs. evil dualism: industrial consumerism, 91, 93–94; new agrarianism, 21, 30; organic farming movement, 110, 119–20; Southern Agrarian rhetoric, 93–95; urban community gardens, 149

Google, 224

Gordon, Constance, 13

Gore, Al, 116, 130

Govan, Thomas P., 197

Graham, Sylvester, 112, 114

Grammar, John, 96

Grange movement, 205

grape varietals, 118–19

grasshoppers, 228

grassland prairies, 231, 232

grassroots activism, 7, 110, 114, 128, 137, 142, 144, 147

Great Britain: state-sponsored propaganda campaigns, 52, 53, 54; wartime strategies, 41–42

Great Depression, 232

great plow-up, 231–32

Greek goddesses, 52

green-manure crops. *See* manure

Griffin, Charles, 91

grocery cooperatives, 3

grocery stores, 233

Gross, Daniel, 111

H

Habermas, Jürgen, 68, 69, 70

Hadsell and Stormer law firm, 146, 151

Hamilton, Alexander, 18

"hands-on" agriculture, 32–33, 214

Hanson, Russell L., 19

Hanson, Victor Davis, 17

Harbor Department (Los Angeles), 144

harmony/harmonic relations, 204

Hartshorne, Oklahoma, 190

Harvey, Paul, 189, 190, 198–99, 200, 207, 208, 213, 215

healthcare, rural, 86

health-related consequences, 120, 121, 131

heirloom plants, 155

Herman, Edward S., 213

heterosexual monogamy, 184–85

heterotopias: agrarian mythmaking, 234; characteristics and functional role, 26–27; Chipotle Mexican Grill, 177; external threats, 143; organic farming movement, 109; SCFarm, 143, 151, 155, 158; social relationships, 143, 158; Southern Agrarians, 91–92, 97, 101–8

high-fat American diet, 10, 142

highly processed industrial food, 141, 142, 155, 174–79

Hofstadter, Richard, 17, 20, 24, 130, 152, 226

Holden, Stephen, 99

holistic philosophy, 211, 212

Holt, Arthur E., 78

home-front support efforts: agrarian

solidarity, 205; communication networks, 47–48, 51; food production and conservation, 38–40, 44–48, 53–54; poster campaigns, 48–54, 57, 61; state-sponsored propaganda campaigns, 39–40, 47–54, 57. *See also* gardens
homespun allegories, 170
Homestead Act (1862), 22, 112
homestead movement, 228
"Honest Scarecrow" (satirical video), 183
honesty vs. corruption, 149–50, 158, 170–71, 177–78
Hoover, Herbert, 39, 44–46, 55, 56, 63, 67
Horowitz, Ralph, 144, 145–46, 149, 150
Howard Method, 126
Howard-Pitney, David, 116
Howard, Sir Albert, 113, 117, 118, 126
huerto familiar, 155
Huffington Post, 164
Hughes, Richard, 16, 17
Human Appeal, 236
humane treatment of animals, 177
human-made climate change, 33, 229–35, 236
hunger and starvation: agrarian mythmaking, 38; chemical fertilizer use, 111; community gardens, 142; food shortages, 38; postwar food assistance, 63; post-World War I era, 56; propaganda poster campaigns, 62; weaponization of food, 235–36; World War I era, 37–38, 39, 41–42, 44, 51, 54
Hunt, Kathleen, 13
Hunt, Scott A., 212–13
Hurt, R. Douglas, 67
hydrocarbon economy, 213, 214, 216
hydroponic facilities, 228
hyperreality, 32, 168, 172, 178–79

I

Ideas Have Consequences (Weaver), 92
idle land, 59
I'll Take My Stand (1930), 90, 91

immigrant communities: communication networks, 47; community gardens, 141, 144–45, 149; constitutional freedoms, 157; Liberty/Columbia imagery, 51–52, 53; poster campaigns, 51–52, 53; pre-World War I immigration levels, 42; westward migration movement, 194
improvised horseshoe image, 215
inauthentic foods, 167
India, 118
Indian wars, 21–22
individual rights, 196
Indore Process, 126
industrial agriculture: agrarian mythmaking, 212–14; animal suffering, 173, 175; autonomous vehicles, 225; business practices, 211; challenges, 33; chemical contamination, 169; chemical pesticides use, 111–13; corporate co-option, 191, 219; culture of convenience, 6; ethnic foodways, 155; farm equipment, 201–2, 216, 225; mega-farms, 225, 226, 232; negative representations, 170, 174–79; organic farming movement, 109–13, 115, 121, 129; pan-agrarian consciousness, 234–35; place-making vs. space-making tropes, 196; policymaking impacts, 10; post-war growth, 130; processed foods, 141, 142, 155, 174–79; reform movements, 112–13; rhetorical functions, 233–34; solidarity tropes, 203–9; structural characteristics, 233; traditional vs. new agrarianism, 11, 149, 166, 191, 204–5
industrial food system: agrarian mythmaking, 7, 33, 169, 171, 174; agrarian resistance, 12; "A Love Story" (animated video), 184–86; "Back to the Start" (animated video), 169–72, 173, 174, 177, 181; challenges, 33; community gardens, 141, 142; consumer guilt, 169, 177; cultural contamination, 168,

178; documentary-style satires, 179–81; environmental concerns, 3, 6; environmental sustainability, 195; *Farmed and Dangerous* (documentary-style satire), 179–81; globalization impacts, 195; organic farming movement, 131; RAM Super Bowl commercial, 32; satirical critiques, 182–83; "The Scarecrow" (animated video), 174–79, 181, 182–83; supply chain practices, 181, 182; violence and corruption, 168, 170

industrialism: as agrarian threat, 21, 22, 24, 30, 87; exploitive environments, 204–5, 212–13, 222; organic farming movement, 129; Southern Agrarian perspective, 89–96, 98–101, 105–6, 206

influenza epidemic, 63

informal state authority, 68

Inge, M. Thomas, 194

innocence and purity campaigns, 166, 168–71, 173, 174–87, 194

insects: control methods, 125; cultivation and consumption, 228–29

insecurity, food, 10, 33, 131, 137–38, 142

interconnectedness, 210, 211

interest-based politics, 73–74

international food trade systems, 40–42

International News Service, 37

intersectional oppression, 139

invention, 17

Irazábal, Clara, 161

"I Want It That Way" (Backstreet Boys), 184–85

J

Jack in the Box, 165

Jackson, Carlton, 114, 115

Jackson, Wes, 11, 204, 212

Jaguar (car brand), 224

James, Pearl, 49, 50, 51

James, Walter, 4th Baron Northbourne, 113–14, 118

Jefferson, Thomas: agrarian vision, 18–21, 23, 25, 38, 140, 192–93, 196, 197, 213, 233; democratic spirit, 43; plantation agriculture, 99

Jenkins, Eric, 168, 176

jeremiad: basic concepts and applications, 115–16; ecological jeremiad, 108, 110, 116–30, 195; redemption and second chances, 116, 195; secularized modern jeremiad, 116–17

Jim Crow laws, 97

J.I. Rodale: Apostle of Nonconformity (Jackson), 114

John Deere, 225

Johnstone, Paul H., 130, 213, 215

Johnston, Josée, 167

Jones, Doug, 182

just food and agriculture, 5

justice, food, 137–38, 142, 158

justified violence, 21

K

Kennedy, Scott Hamilton, 138, 146–47, 148, 157, 160

Kentucky Fried Chicken, 163

Kenyon Review, 91

Keynes, John Maynard, 46

Kiernan, Ben, 106

King, F. H., 118

King, Rodney, 144

Kitchener, Horatio Herbert, 48

kitchen gardens, 155

Kutztown, Pennsylvania, 113

Kyoto Treaty (1997), 229

L

labor exploitation. *See* exploitation

labor shortages, 56

Laclau, Ernesto, 9, 79, 172

land-as-sacred trope, 83, 86, 194–95, 211

land community/land ethic, 204

land ownership, 197–98, 201

land rush, 231

Landsberg, Alison, 49

landscape erosion, 125

land-society connection, 82–86, 97–98, 100–101

LA Regional Food Bank, 144, 145–46

Laswell, Harold, 53
Latinx community: community gardens, 31, 88, 140–41, 144–61; cultural connectedness, 154; farm workers, 200; kitchen gardens, 155; mythic vernacular narratives, 138–40, 147–52, 154, 155, 157–60; stereotypical portrayals, 150; undocumented workers, 150–51
Law of Return, 118, 119, 123, 126–27
Lazarus, Emma, 51–52
Leach, William, 51
League of Nations, 63
Lebanon, 41
Lefebvre, Henri, 143
legitimacy, democratic, 69–81
Leopold, Aldo, 204, 206
Lever Act (1914). *See* Smith-Lever Act (1914)
Lex Agraria, 193
Libaw-Horowitz Investment Company (LHIC), 145
Libertas (Roman goddess), 52
Liberty Bonds, 48, 52
Liberty/Columbia imagery, 50, 51–54, 57, 61
literacy, food, 8, 142
The Living Soil (Balfour), 114
local communities: citizen-farmers, 132–33; Country Life movement, 83–87; economic factors, 206, 215; human relations, 81, 83, 90; importance, 81; mythic vernacular narratives, 138–40, 147–52, 154, 155, 157–60; Southern Agrarian perspective, 91–92, 100–108. *See also* rural life
Lockie, Stewart, 179
Logsdon, Gene, 204, 205, 206, 226–27
Lord Kitchener. *See* Kitchener, Horatio Herbert
Los Angeles, California. *See* SCFarm/SCFarmers
Los Angeles City Energy Recovery Project (LANCER), 144
Louisiana Purchase, 18
Lowden, Frank O., 71, 72, 74–75

low-income communities, 142, 145
Lyson, Thomas, 142, 156
Lytle, Andrew, 92, 93, 100, 102

M

machine-based labor, 102–3
Machine in the Garden (Marx), 194
mail delivery, rural, 86
Major, William H., 6, 7, 12, 30–31
malnutrition, 41–42, 235
manufacture of consent, 213–14
manure, 123, 126, 127
Mares, Teresa M., 138
marginalized communities: farmers/farming, 73, 76–77, 79, 87, 133; mythic vernacular narratives, 140, 158–60; neoliberalism, 140; place-making tropes, 31; social resistance, 158. *See also* SCFarm/SCFarmers
marketing and advertising: agrarian practices, 192, 209–17; agrarian romanticization, 7–9, 202–3, 206–9; Chipotle Mexican Grill, 31–32, 164–87; Liberty/Columbia imagery, 52, 53, 61; McDonaldization, 172–73; moral and ethical principles, 191; naturalization rhetoric, 202–3; pickup trucks, 32, 191–92, 206–9, 213–19, 225; propaganda poster campaigns, 48–54; RAM truck Super Bowl commercial, 32, 189–92, 198–203, 206–9, 213–19, 225; trade wars, 229; virtue/virtuousness, 32, 167, 176–77, 181, 201–3, 206–9, 214; weaponization impacts, 235–36; World War I era, 47–51
Marxist perspective, 102–3
Marx, Leo, 194
mass-produced fast food. *See* fast-food industry
McCoy, Drew R., 19
McDonaldization, 172–73
McDonald's, 163, 172–73
McFadden, Bernarr, 114
McGuire, Michael, 25
McLuhan, Marshall, 172

meatless meals, 44, 45
mechanized agriculture, 32–33, 94, 112, 127–28, 131, 216, 224–25, 231–32
mechanized society, 89, 93–94, 103, 107, 109
medicinal plants, 155
mega-farms, 225, 226, 232
memories, prosthetic, 49, 51–52, 61
Mencken, H. L., 91
Mesoamerican immigrants, 145
mesoi, 17–18
Mexican Americans: community gardens, 145, 149; Country Life movement, 66
Mexican–American War, 159
Mexican immigrants, 144–45
microbes, beneficial, 119, 122, 128
microbial safety, 163–65, 173, 181–82, 184
micro-farmers, 196
Mignolo, Walter D., 157
militant pastoralism, 101–5, 108
military base gardens, 58–59
Milstein, Tema, 140, 177
mindfulness, 211
minority populations, 142, 145
"Mirror for Artists" (Davidson), 93
Mitchell, Margaret, 99–100
mobilized agriculture: Country Life movement, 235; industrial agriculture, 109; mass production, 133; organic farming movement, 121; SCFarm/SCFarmers, 138; war gardens, 29, 55–63; World War I, 29, 38–40, 42–43
modernization/modernity: agrarian ideologies, 12, 15, 218–19, 234; as agrarian threat, 13, 28, 30, 112–13, 129; authentic living, 168; community gardens, 56, 64; Country Life movement, 66; cultural contamination, 178; democratic participation, 56; ethnic agricultural practices, 155; organic farming movement, 115, 121, 129; SCFarm/SCFarmers, 139, 145, 153, 155, 157; Southern Agrarian perspective, 30, 90; World War I era, 38, 222

mono-crop cultivation, 225, 232
monopoly capitalism, 205
Monsanto, 111
Montmarquet, James, 24
Mooney, Patrick H., 212–13
moral righteousness and virtues, 18–21, 24, 25, 31, 56–57, 140, 171, 210
moral virtue: agrarian mythmaking, 10, 18–21, 23–24, 31, 133, 149, 152, 154, 158; ecological jeremiad, 110, 117, 123, 130–31; farmers/farming, 4, 26, 67, 89, 160, 171, 191, 200–202, 215–17; Jeffersonian perspective, 18–21, 210; war gardens, 56, 59, 60. *See also* Chipotle Mexican Grill; RAM truck Super Bowl commercial
Morgan, J. P., 46
Morrill Land Grant Acts, 112
Motter, Jeff, 191, 192
Mouffe, Chantal, 9, 79, 172
Muir, John, 116
munitions manufacture, 41, 57. *See also* gardens
mythical caretaking farmer, 110, 116, 125, 194, 200, 204–5, 207–8, 212–13, 214
mythic connectedness, 151, 171, 194
mythic vernacular narratives, 138–40, 147–52, 154, 155, 157–60, 170

N

Nash, Roderick Frazier, 194
national extension service, 65–66
National Farmers Organization, 205
National Grange, 75
National Urban Gardening Program, 141–42
National War Garden Commission (NWGC), 39, 50, 55–61
nation-building policies, 22, 24
naturalistic observational filming approach, 147–48
nature: beneficence, 130, 152; divine design, 116, 117, 118–23, 126–27; Law of Return, 118, 119, 123, 126–27; soil-food connection, 117–20; wildlife populations, 124–25

INDEX   301

Nelson, Willie, 169, 171
neoliberalism: Chipotle Mexican Grill, 179; consumerism, 159; democratic resistance, 153, 154, 156–57; economic factors, 137, 140, 149; food activism, 179–81; privatization/property rights, 149, 154, 158; racial ideologies, 159; SCFarm/SCFarmers, 137, 139, 140, 150, 153, 154, 160
new agrarianism: agrarian practices, 192, 210–17; alternative food networks, 179, 186, 233; apocalyptic perspective, 221–24, 229, 231–33; authenticity campaigns, 166–87; challenges, 33; connectedness, 4–7, 14; consumer-citizens, 166–67; democratic values, 157; globalization impacts, 195–96, 205; holistic philosophy, 211, 212; ideological perspectives, 11–13, 20–21, 191, 213; interconnectedness, 211; moral and ethical principles, 90, 112, 121, 153, 191; moral virtues, 10, 18–21, 23–24, 31, 133, 149, 152, 154, 158; mythic vernacular narratives, 138–40, 147–52, 154, 155, 157–60; pan-agrarian consciousness, 234–35; place-based resistance, 155; place-making tropes, 194–98, 206; redemption and second chances, 26, 88, 121, 165; resistance ethic, 6, 12, 21, 64, 133, 222; rhetorical functions, 233–34; romanticized imagery, 166–68, 171–72; socially resistive delinking, 157; solidarity tropes, 203–9; Southern Agrarian perspective, 30, 89, 101–8; spirituality-nature connection, 91–95, 97, 100–106, 108, 152–54, 199; sustainable agriculture, 149, 153; urban agrarianism, 31, 137–61, 210–17, 226–27; weaponization of food, 37–38, 40–42, 53–54, 235–36. *See also* Chipotle Mexican Grill; SCFarm/SCFarmers
New Deal politics, 93
new food movements: societal impacts, 5–6, 7; Southern Agrarians, 30

New York (New York) food riots, 37
New York Stock Exchange, 164
New York Times: climate change impacts, 229–30; on food riots, 37
"Next Year's Crop" (mini-documentary), 217–18
Niccol, Brian, 184
Nichols, Bill, 147
No Man's Land (Oklahoma), 231
non-combatant populations, 235–36
non-European ethnic agricultural practices, 149, 154
nonprofit food-distribution organizations, 144
norovirus outbreaks, 163, 164, 181
Northbourne, Lord. *See* James, Walter, 4th Baron Northbourne
nostalgic appeal: agrarian mythmaking, 166, 169, 202–3, 206–9; American conservatism, 189; connectedness, 23; creation stories, 198; land-as-sacred trope, 83, 86, 194–95, 211; RAM truck Super Bowl commercial, 187, 189–92, 198–203, 213–19; rural life, 19–20, 30–32, 171, 174–79, 189–92, 196–203, 206–9, 214–19. *See also* Chipotle Mexican Grill
"Notes on Liberty and Property" (Tate), 101

O

Obach, Brian K., 112
"Ode to the Confederate Dead" (Tate), 99
"off-the-grid" living, 228
Ohio Agricultural Experiment Station, 126
Olbrechts-Tyteca, Lucie, 77
Old South. *See* Southern Agrarians
oligarchies vs. democracies, 71, 72, 74
Omni magazine, 221
Ono, Kent A., 138, 139, 149, 158
Opie, John, 116, 117, 130
Oppenheimer, Robert, 118
Oravec, Christine, 160
Organic Farming and Gardening/Organic Life (Rodale), 109, 113, 114
organic farming movement: agrarian

resistance, 110, 113, 129–33; alternative food networks, 186, 233; biodiversity-affirming permaculture, 153; compost-based agricultural methods, 118, 125–30; corporate co-option, 131; democratic participation, 132, 172; ecological jeremiad, 108, 110, 116–30; *Farmed and Dangerous* (documentary-style satire), 179–81; health-related consequences, 120, 121, 131; historical perspective, 109, 111–13, 118–20; Law of Return, 118, 119, 123, 126–27; politicization, 131–32; sacred agrarian tradition, 129–30, 152–54; simplicity, 118, 122, 128, 132; soil-food connection, 117–20; tractor use, 127–28, 131; virtue/virtuousness, 30–31, 116–17, 120, 122, 125–30. *See also* Chipotle Mexican Grill; Rodale, Jerome I.
Ottoman Turks, 41
outsiders, 69, 71–73, 106. *See also* counterpublics
Owsley, Frank Lawrence, 92, 97, 99, 104

P

Pacific Railroad Acts, 112
Pack, Charles Lathrop, 39, 40, 41, 55–62, 63, 64, 223
Palen, Doug, 229–30
pan-agrarian consciousness, 234–35
Panera Bread, 186
pastiche, vernacular, 139, 148–49, 158, 170
patriotic imagery, 48–54
patriotism, 67
Patten, Simon, 42
Pay Dirt: Farming and Gardening with Composts (Rodale), 30, 110–11, 114, 115, 117–20, 123–25, 127–32
peasant class, 64, 71–80, 86–87
Peña, Devon G., 138
Perelman, Chaim, 77
Pereyni, Eleanor, 114
Perry, Jan, 145, 149–50, 156
Persia: food shortages, 41–42; organic farming movement, 118

pesticide residues, 120
Peterson, Tarla Rai, 11, 191, 212
Peters, Suzanne, 114, 131
Pfieffer, Ehrenfried, 113, 118, 126
Philadelphia (Pennsylvania) food riots, 37
Phillips, Sarah, 67
pickup trucks: agrarian mythmaking, 134, 191, 198–203; marketing and advertising, 32, 191–92, 206–9, 213–19, 225. *See also* RAM truck Super Bowl commercial
place-based resistance, 137, 148, 154, 155, 158, 194–95
place-making tropes, 23, 154, 192, 194–204, 206–7
plantation system, 92, 98–100, 101, 106
Poe, Clarence, 71
poisoning, food. *See* foodborne illness outbreaks
Poland, David, 146, 147
political resistance, 89–96, 104
political solidarity, 203, 205–6, 208–9
political virtues, 4, 197
Pollan, Michael, 7–8
Popular Science, 224–25
Populist movement, 13, 22, 27, 112, 140, 205
poster propaganda: hunger and starvation, 62; war gardens, 48–54, 57, 61
Post Farms, 226
post-modern perspectives, 32
postwar food policies, 54, 55–57, 63
poultry farming, 127
power-generating waste incinerator project, 144
practices, agrarian, 192, 209–17. *See also* place-making tropes; solidarity tropes
Prevention (Rodale), 113
price supports, 56, 63, 67
private property, 159, 197–98, 201
privatization, 149, 154, 158
processed foods, 33, 141, 142, 155, 174–79
processing and packing companies, 233
Progressivism, 22
proletariat class, 78

propaganda campaigns: food shortages, 39; World War I, 39, 48–54
property rights, 158, 159, 197–98, 201
prophecy vs. apocalypse, 32, 223–24, 233, 236
prosthetic memories, 49, 51–52, 61
prototypical farmer, 200
public authority, 68–74, 76–80
public good, 25
publics/public spheres, 26, 68–74, 76–80, 172
public wellness, 3
Punja, Anita, 161
"Pure Imagination" (soundtrack song), 175–76, 183
purity symbols, 8

Q

questionable agricultural practices, 123–25, 129, 130. *See also* chemical agriculture

R

racial ideologies, 21, 98, 99, 139, 158–59
racial stereotypes, 21, 150
racial violence, 106
Ragas, Matthew W., 178
RAM truck Super Bowl commercial: agrarian mythmaking, 32, 134, 187, 189–92, 198–203, 206–9, 213–19; agrarian practices, 192, 209–17; ideological perspectives, 207–9, 218–19; marketing and advertising, 32, 189–92, 198–203, 206–9, 213–19, 225; naturalization rhetoric, 202–3; nostalgic appeal, 187, 189–92, 198–203; place-making tropes, 192, 194–203, 206–7, 209; solidarity tropes, 206–9; virtuous farmers, 201–2, 206–8, 214
Ransom, John Crowe, 91, 92, 93, 94, 98, 100, 102, 104–5
rationing programs, 41, 54
real-estate marketing swindles, 231
real food: "Back to the Start" (animated video), 169–72, 173, 174, 177, 181; citizen-farmers, 167; *Farmed and Dangerous* (documentary-style satire), 179–81; "The Scarecrow" (animated video), 174–79, 181, 182–83
religion, 84, 94, 100
republican virtues, 19–20
resilience: agrarian mythmaking, 7, 9, 17, 25, 39, 216; community gardens, 143, 154; democratic values, 39; farmers/farming, 217; food networks, 161; free-market economies, 25; Jeffersonian perspective, 18–19; propaganda campaigns, 62; RAM truck Super Bowl commercial, 15, 190, 199–200, 203, 209, 216, 217–18; Southern Agrarian perspective, 106; virtue/virtuousness, 203
resistant agriculture: agrarian mythmaking, 137; American agrarian myth, 27; Country Life movement, 64, 68, 77, 79, 86–87, 133; decolonial resistance, 139–40; exploitive environments, 222; heterotopias, 26, 234; industrial food system, 12; land-as-sacred trope, 194–95; moral and ethical principles, 6, 12, 21, 64, 133, 222; organic farming movement, 110, 113, 129–33; place-based resistance, 137, 148, 154, 155, 158, 194–95; post-World War I era, 64; RAM truck Super Bowl commercial, 209; SCFarm/SCFarmers, 137–61; socially resistive delinking, 157; Southern Agrarian perspective, 89–96, 101–8, 133; weaponization impacts, 235. *See also* new agrarianism
restaurant chains, 233
restorative farming methods, 230
Retzinger, Jean, 201, 209
rhetoric, 9, 17, 233–34
Riordan, Richard, 145
riots, 37, 39, 41, 144
Ritzer, George, 172
Roberts, Marilyn S., 178
Rodale Institute, 113
Rodale, Jerome I.: background, 30, 87, 108, 109–11; ecological jeremiad,

108, 110, 115–30, 195; farming experimentation, 113, 114, 117, 127; food and health advocacy, 113–15; as organic prophet and leader, 12, 111, 113–15, 117–34; personal conversion experience, 117–18; philosophical perspective, 109–10, 172; publishing business, 113, 114, 131
Rodale Press, 113
Roman agrarian law, 193
Roman goddesses, 52
romanticized imagery: agrarian mythmaking, 7–9, 19, 166–68, 171–72, 194–96, 206–9, 215; agrarian solidarity, 206; American frontier, 21–23, 25, 80, 129, 194, 212; Chipotle Mexican Grill, 166–68, 172; land-as-sacred trope, 83, 86, 194–95, 211; place-making tropes, 194–203; RAM truck Super Bowl commercial, 32, 191, 195, 198–203, 206; social media campaigns, 166–68; Southern Agrarian perspective, 92, 97, 99–100; westward migration movement, 21–24, 129. *See also* nostalgic appeal; rural life
Roosevelt, Franklin D., 232
Roosevelt, Theodore, 22, 65
"routine of living" philosophy, 100, 101
Rural America (magazine), 66, 71, 74
rural life: agrarian practices, 192, 209–17; beneficial legislation, 86–88; cooperative attitudes, 83–86, 154; Country Life movement, 65–66, 67, 71–81, 86–88, 133; creation stories, 198; decline and depopulation, 12–13, 24, 66–68, 74–75, 80, 86–87, 133, 210; democratic participation, 65, 71–81, 84–85, 87, 156, 171, 236; demographic changes, 196–97; economic factors, 21, 71–80, 192–93, 215; Jeffersonian perspective, 192–93, 196, 197; land-as-sacred trope, 83, 86, 194–95, 211; marginalized communities, 73, 76–77, 79, 87, 133; marketing and advertising, 202–3, 206–9; nostalgic appeal, 19–20, 30–32, 171, 174–79, 189–92, 196–203, 206–9, 214–19; political perspectives, 197; RAM truck Super Bowl commercial, 187, 189–92, 198–203; redemption and second chances, 26, 121; sense of community, 81–85, 140, 153, 194–98, 201, 206–7; simplicity of institutions, 82–84; sociological studies, 66; solidarity tropes, 203–9; Southern Agrarian perspective, 100–108
rural nostalgia and imagery. *See* nostalgic appeal; romanticized imagery; rural life
rural sociology studies, 66
Rushing, Janice Hocker, 11, 139
Russia: Bolshevik Revolution, 78; food shortages, 41, 55
Rutherford, G. W., 83

S

sacred agrarian traditions: citizen-farmers, 121; democratic values, 83; ecological jeremiad, 116, 117–18, 129–30; gardens, 26, 236; land-as-sacred trope, 83, 86, 194–95, 211; organic farming movement, 109, 123, 127, 129–30, 132, 222; place-making tropes, 194–95, 196; SCFarm, 152–54; soil-food connection, 120; Southern Agrarian perspective, 91, 94
safety, food, 163–65, 173, 181–82, 184
salmonella outbreaks, 163, 164
Sanderson, Dwight, 65
Sandoval, Jennifer, 140
satellite publics, 26, 68, 69, 70, 77, 86
"The Scarecrow" (animated video), 174–79, 181, 182–83
SCFarmers Feeding Families, 145
SCFarm/SCFarmers: agrarian mythmaking, 133, 137–61; agrarian solidarity, 206; contested land, 143–46, 149–51; cultural connectedness, 154; democratic land-use procedures, 155–56; demographic composition, 144–45; economic

factors, 137, 145; environmental connectedness, 152, 153–54; ethnic agricultural practices, 149, 154, 155; eviction, 146, 150, 151, 157, 158; familial connectedness, 151–52, 154, 156; heterotopian space, 143, 151, 155, 158; historical perspective, 143–46; impact and legacy, 160–61; internal governance structure, 155–56; marginalized communities, 31, 88; mythic connectedness, 151, 206; mythic vernacular narratives, 139–40, 154, 155; place-based resistance, 137, 148, 154, 155, 158; political dynamics, 143; sacred agrarian tradition, 152–54; socially resistive delinking, 157; sociopolitical context and significance, 137–38; urban agrarianism, 140–41. *See also The Garden* (documentary)

Schaffner, William, 181–82
Schnapp, Jeffery T., 51, 52
schoolhouse meetings, 85
science-technical progress connection, 93–96, 113–15, 132
scientific agrarianism, 117–20, 132
"The Scientist" (Coldplay), 169
seafood harvests, 228
secularized modern jeremiad, 116–17
security, food, 10, 39, 131, 137–38, 142
self-driving cars, 224
self-sufficiency, 13, 19–20, 24, 112, 116, 154, 214, 217
sense of community, 81–86, 140, 153, 197, 201, 204, 206–7
Shabani, O. A. Payrow, 70
shellfish harvests, 228
Short, Brant, 198
short-term farm profits, 123
sick farm animals, 124
Silent Spring (Carson), 110
simplicity: agrarian mythmaking, 173, 174, 191; authentic foods, 167; brandscapes, 171; civic virtues, 19; democratic values, 82–85, 149; innocence and purity campaigns, 171, 173; local communities, 82–83;

moral and ethical principles, 84, 171, 177; nostalgic appeal, 23, 31, 191; organic farming movement, 118, 122, 128, 132; RAM truck Super Bowl commercial, 191; virtue/virtuousness, 149. *See also* Chipotle Mexican Grill; Country Life movement

Singer, Ross, 89, 191, 192
Skees, Jerry R., 24
slacker land, 59
slavery, 97, 99–100, 101, 106
Sloop, John M., 138, 139, 149, 158
Slotkin, Richard, 17, 21, 22, 23
small-scale farms, 177, 225–27
Smith, Andrew F., 114
Smith, Henry Nash, 17, 22, 194
Smith, Kimberly, 12
Smith-Lever Act (1914), 46
Smith, Tex, 190
social change movement, 13–14
social contracts, 196
social identity, 4, 7, 10, 16
social justice, 3, 28, 147, 161
socially and ecologically responsible agricultural cultivation and husbandry, 210, 212
socially resistive delinking, 157
social media campaigns. *See* Chipotle Mexican Grill
social norms: democratic legitimacy, 68, 70, 83; rural life, 21; simplicity of institutions, 83; virtue/virtuousness, 19
social resistance: SCFarm/SCFarmers, 137–61; Southern Agrarian perspective, 89–96, 101–5, 108
social stereotypes, 21
sociopolitical solidarity, 203, 205, 208–9
"So God Made a Farmer" (television commercial), 32, 189–92, 198–203, 206–9, 213–19, 225
Soil and Health Foundation, 113
soil conservation, 232
soil-food connection, 117–20
solidarity tropes, 19, 149, 191, 192, 203–9

Sontag, Susan, 49
South Central Farmers (SCFarmers). *See* SCFarm/SCFarmers
South Central Los Angeles, 143–46
Southern Agrarians: agrarian resistance, 89–96, 101–8, 133; agrarian solidarity, 206; anti-consumerism viewpoint, 101–8; anti-industrialism viewpoint, 89–96, 98–101, 195, 206; background, 17, 30, 90–91; economic factors, 91, 93–95, 98–102, 105; heterotopian space, 91–92, 101–8; historical reassembly, 97, 100; militant pastoralism, 101–5, 108; new agrarianism, 30, 89, 101–8; philosophical perspective, 89–108, 171; political manifesto, 91, 93, 94; redemption and second chances, 99, 170, 195; rural decline and depopulation, 87; sociocultural impacts, 90, 93–97; Southern Eden paradise, 96, 97, 98
Southern Plains states, 225, 231–32
space-making tropes, 196
spatial influence, 26–27
special interest politics, 73–74
The Spirit of '76 (painting), 49
spirituality-nature connection: branding strategies, 171; connectedness, 151–54, 169, 176, 177, 199, 222; historical perspective, 194–95; RAM truck Super Bowl commercial, 199; SCFarm/SCFarmers, 152–54; Southern Agrarian perspective, 91–95, 97, 100–106, 108
Squires, Catherine, 69
starvation and hunger: agrarian mythmaking, 38; chemical fertilizer use, 111; community gardens, 142; Europe, 54; food shortages, 38; postwar food assistance, 63; post-World War I era, 55, 56; propaganda poster campaigns, 62; weaponization of food, 235–36; World War I era, 37–38, 39, 41–42, 44, 51, 54
state authority, 68

state legitimacy and power, 67, 69–81, 86–88
"Statement of Principles" (Southern Agrarians), 91, 93, 94
state-sponsored propaganda campaigns: demonization portrayals, 54; Europe, 52, 53–54; food production and conservation, 47–54, 57, 62; poster campaigns, 48–54, 57, 61; war gardens, 39–40
Statue of Liberty, 51–52, 53
Steiner, Rudolf, 111–12, 113
Stereopticon, 92, 104
stereotypical portrayals, 21, 51, 150
stewardship, 67, 159, 160, 215, 217
St. John de Crèvecoeur, J. Hector, 21, 23
Stormer, Dan, 151
suitcase farmers, 231–32
Super Bowl (2013), 32, 134, 187, 189–90
surplus crops, 232
sustainable agriculture: agrarian practices, 212; biodiversity, 153, 154–55, 195, 212; ecologically responsible agricultural practices, 210; emerging trends, 33; new agrarianism, 149, 153; soil-food connection, 119; Southern Agrarian perspective, 89, 105–6; structural characteristics, 233. *See also* organic farming movement
Swanson, Louis E., 24
symbolic space, 26–27
synthetic chemicals, 109–11, 112, 127
Syria, 235

T

Taber, L. J., 75
Taco Bell, 184
Tate, Allen, 92, 98, 99, 100, 104, 211
Tate, Juanita, 150–51, 156
Taylor, John, 21
technology-based agriculture, 32–33, 112, 127–28, 131, 132, 224–25, 228, 231–32
technology-based industrialization, 93–96
television commercials. *See* Chipotle

Mexican Grill; RAM truck Super Bowl commercial
tenant-operated farms, 123
Terpenning, Walter A., 81–82
"The Hind Tit" (Lytle), 93
"The Irrepressible Conflict" (Owsley), 97
Theiss, Lewis Edwin, 77
"The Small Farm Secures the State" (Lytle), 101
Thompson, Paul B., 18, 24, 193, 195, 196, 198, 211, 233
Thoreau, Henry David, 90
Tigert, John J., 75
Times (Gadsden, Alabama), 190
topoi. *See* agrarian topoi; place-making tropes; practices, agrarian; solidarity tropes
town meetings, 85
tractors, 127–28, 131, 216, 224–25, 231–32, 235
trade wars, 229
traditional vs. new agrarianism, 11–12, 14–15, 149, 166, 191, 204–5
Treaty of Versailles (1918), 41, 63
Trentmann, Frank, 41
Trump, Donald, 163, 165
Trust for Public Land, 146
Turner, Frederick Jackson, 22, 23, 25, 194
Turnip Winter (1916-1917), 41

U

Uncle Sam, 48–49, 50, 51
underprivileged communities, 137–38
undocumented workers, 150–51
United States: chemical fertilizer use, 118; democratic values, 67–81; Liberty/Columbia imagery, 50, 51–54, 57, 61; mobilized agriculture, 42–43; postwar food policies, 54, 55–57, 63; World War I era, 42–44. *See also* state-sponsored propaganda campaigns
United States Department of Agriculture (USDA), 112, 225, 226
United States Food Administration, 39, 44–45, 67
Unnevehr, Laurian J., 24
The Unsettling of America (Berry), 89
urban agrarianism, 31, 33, 137–61, 210–17, 226–27
urban citizen participation, 40, 85–86
urban community gardens: benefits, 141–43; cultural diversity, 141, 144–61; democratic land-use procedures, 155–56; emerging trends, 227; ethnic agricultural practices, 149, 154, 155; food production and conservation, 55–61; historical perspective, 141–42; internal governance structure, 155–56; place-based resistance, 137, 148, 154, 155, 158; political dynamics, 143; South Central Farm (SCFarm), 31, 137–61; vacant lots, 59, 141, 143–46
urban farming, 88, 227
urbanization as agrarian threat, 13, 21, 22, 24, 30. *See also* community gardens
urban proletariat, 78
U.S. Department of Agriculture (USDA), 112, 225, 226
utopian countermovements, 90, 92, 97–98, 107

V

vacant land, 59, 141, 143–46
Vanderbilt University Medical Center, 181
Veit, Helen Zoe, 38, 44, 62
vernacular discourse, 138–40, 147–52, 154, 155, 157–60, 170
Victoria (Roman goddess), 52
victory gardens, 30, 40, 57, 114
Villaraigosa, Antonio, 146, 155
violent colonization, 21–22, 194
Virgin Land (Smith), 194
virtue/virtuousness: agrarian myth-making, 4, 6, 14, 25, 67, 125, 129, 210, 216–18; agrarian solidarity, 149, 181, 203; branding strategies, 191; citizen-farmers, 17, 125, 129, 140,

149, 151, 171, 194, 213; civic virtues, 18–21, 25, 26, 87, 154, 179, 199; core virtues, 236; democratic values, 81, 132, 171, 172, 174, 214, 222; ecological jeremiad, 116–17, 120, 122, 124; environmental sustainability, 149; farmers/farming, 18–21, 24, 25, 81, 124, 148, 201–2, 206–9, 213–14, 217; German propaganda, 52; home-front support efforts, 29, 38, 48; Jeffersonian perspective, 18–21, 210; Latinx community, 154–55; marketing and advertising, 32, 167, 176–77, 181, 201–3, 206–9, 214; mechanized agriculture, 216; organic farming movement, 30–31, 116–17, 120, 122, 125–30; patience, 126; place-making tropes, 23, 154, 194–204, 230; political virtues, 4, 197; private property, 159; prosthetic memory, 49, 51–52, 61; simplicity, 149; war gardens, 29; westward migration movement, 21–24, 133, 194. *See also* Country Life movement; democratic values; moral virtue; RAM truck Super Bowl commercial
voluntary meal schedules, 48
volunteerism: coercive voluntarism, 44; community gardens, 59–60, 227; farming assistance, 56; state-sponsored propaganda campaigns, 47
vulnerable populations, 235–36

W
Walsh, Lynda, 118
Wanzer-Serrano, Darrel, 138
war bonds, 48, 52
Ward, Gordon H., 79
War Garden Commission, 171
war gardens, 30, 39–40, 50–51, 55–61, 114
war zones, 235–36
Washington Post, 164
waste incinerator project, 144
waste products, 178
Waymo, 224

weaponization of food, 37–38, 40–42, 53–54, 235–36
Weaver, Richard, 92, 104
Weber, Pete, 182
Weiss, Elaine F., 44
wellness, public, 3
westward migration movement, 21–24, 129, 133, 194
"What Does the South Want?" Ransom, 101
wheat farming, 232
wheatless meals, 45
Whiskey Rebellion (1794), 205
White-Indian conflict, 21–22, 133
White population: female farmers, 56, 199–200; frontier conflicts, 21–22, 133; male farmers, 148, 149, 158, 160, 199, 200, 207–8; privilege and affluence, 5, 140, 159; racial neoliberalism, 159; yeoman farmers, 140
White, Richard, 22, 23
Who Owns America (1936), 90, 101
Wickstrom, Maurya, 168
Wiebe, Robert H., 42
Wilder, Gene, 175, 176
Wilderness and the American Mind (Nash), 194
wildlife populations, 124–25
Wild West romanticization, 22, 23–24
Willard, Archibald, 49
Williams, Raymond, 167
Willy Wonka and the Chocolate Factory (film), 175
Wilson, M. L., 80, 85, 86
Wilson, Woodrow, 39, 42, 43–44, 47, 63
windbreaks, 125
Wirzba, Norman, 9, 12
Wizard of Oz (Baum), 175
Wolfe, Dylan, 116
Woman's Land Army, 56
women: farm workers, 56, 199–200; female emancipation, 103
Wonka, Willy, 175, 176
work ethic, 199, 200–202, 207, 208
working myths, 25
World War I: agrarian solidarity, 205; communication networks, 47–48,

51; demonization portrayals, 54; food production and conservation, 38–40, 44–48, 53–54, 62; food shortages, 37, 38, 39, 41–42, 44, 62; mobilized agriculture, 38–40, 42–43; propaganda campaigns, 39, 48–54, 61; social impacts, 40–42; social reorganization impact, 62–64; technological advances, 37–38

World War II: post-war farm production, 196; victory gardens, 30, 40, 114

worn hands motif, 200, 214

Y

Yemen, 235

yeoman farmers: civic virtues, 18–21, 25, 26; democratic participation, 71–81; devolution to peasant class, 64, 71–80, 86–87; displaced farmers, 225–26; economic factors, 71–80; importance, 72; Jeffersonian perspective, 18–19, 38, 90, 156, 196, 197; moral virtues, 20; as mythic heroes, 20, 23, 206–8, 214; as noble victims, 140; plantation system, 30, 98, 106; Southern Agrarian perspective, 30, 98–100, 101, 106; urban agrarianism, 226, 227; war gardens, 61. *See also* Country Life movement

You Can't Eat That, Health Digest (Rodale), 113

Z

Zagacki, Kenneth, 106, 108

Zeiger, Robert H., 46